VOLUME TWO HUNDRED AND ONE

ADVANCES IN
IMAGING AND
ELECTRON PHYSICS

EDITOR-IN-CHIEF

Peter W. Hawkes
CEMES-CNRS
Toulouse, France

VOLUME TWO HUNDRED AND ONE

ADVANCES IN
IMAGING AND
ELECTRON PHYSICS

Edited by

PETER W. HAWKES
CEMES-CNRS
Toulouse, France

ACADEMIC PRESS

An imprint of Elsevier

Cover photo credit:
The Cover picture is taken from Fig. 10 of the chapter by Ashkan Ashrafi (p. 30).

Academic Press is an imprint of Elsevier
125 London Wall, London EC2Y 5AS, United Kingdom
525 B Street, Suite 1800, San Diego, CA 92101-4495, United States
50 Hampshire Street, 5th Floor, Cambridge, MA 02139, United States
The Boulevard, Langford Lane, Kidlington, Oxford OX5 1GB, United Kingdom

Notices

Knowledge and best practice in this field are constantly changing. As new research and experience
broaden our understanding, changes in research methods, professional practices, or medical
treatment may become necessary.

Practitioners and researchers must always rely on their own experience and knowledge in evaluating
and using any information, methods, compounds, or experiments described herein. In using such
information or methods they should be mindful of their own safety and the safety of others,
including parties for whom they have a professional responsibility.

To the fullest extent of the law, neither the Publisher nor the authors, contributors, or editors,
assume any liability for any injury and/or damage to persons or property as a matter of products
liability, negligence or otherwise, or from any use or operation of any methods, products,
instructions, or ideas contained in the material herein.

ISBN: 978-0-12-812089-7
ISSN: 1076-5670

For information on all Academic Press publications
visit our website at https://www.elsevier.com/books-and-journals

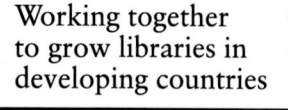 **Working together
to grow libraries in
developing countries**

www.elsevier.com • www.bookaid.org

Publisher: Zoe Kruze
Acquisition Editor: Jason Mitchell
Editorial Project Manager: Shellie Bryant
Production Project Manager: Magesh Kumar Mahalingam
Designer: Mark Rogers

Typeset by VTeX

CONTENTS

CONTRIBUTORS

Ashkan Ashrafi
Department of Electrical and Computer Engineering, San Diego State University, San Diego, CA, USA

Gaston Dupouy (1900–1985)
CNRS Laboratory of Electron Optics, Toulouse, France

Sameen Ahmed Khan
Department of Mathematics and Sciences, College of Arts and Applied Sciences, Dhofar University, Salalah, Oman

Inder Jeet Taneja
Formerly, Professor of Mathematics, Departamento de Matemática, Universidade Federal de Santa Catarina, Florianópolis, SC, Brazil

PREFACE

This volume opens with an account of the family of Walsh–Hadamard transforms. Although these have been covered here in the past, the introduction of complex forms of the transforms certainly justifies a new contribution on the subject. Ashkan Ashrafi first explains the relations between the Walsh and Haar functions and the related transforms. He then presents in detail recent developments and the reasons why these are so valuable.

This is followed by a description of quantum approaches to Maxwell optics by Sameen Ahmed Khan, who is no stranger to these *Advances*. He sets out from a matrix expression for Maxwell's equations and leads us methodically through the optics of light beams and the theory of polarization. This unified formalism is a nice extension of the Hamiltonian analogy.

A third very different chapter is concerned with information measures. There are many definitions of divergence measures in the literature, which Inder Jeet Taneja examines and at the same time shows how they can be related. This complicated material is described in detail and this article should be very helpful for anyone wishing to go deeply into the subject.

Finally, we reprint a historic article, which first appeared in *Advances in Optical and Electron Microscopy*. This is a beautifully illustrated account by the late Gaston Dupouy, who built the first high-voltage electron microscope with an accelerating voltage in the megavolt range. This gives a vivid picture of the first such instruments, their advantages and their limitations. It is an essential element in the history of the subject.

I am most grateful to all the authors for their efforts to make difficult subjects accessible.

Peter W. Hawkes

FUTURE CONTRIBUTIONS

S. Ando
Gradient operators and edge and corner detection

J. Angulo, S. Velaso-Forero
Convolution in (max, min)-algebra and its role in mathematical morphology

J. Angulo, S. Velaso-Forero
Non-negative sparse mathematical morphology

D. Batchelor
Soft x-ray microscopy

E. Bayro Corrochano
Quaternion wavelet transforms

C. Beeli
Structure and microscopy of quasicrystals

C. Bobisch, R. Möller
Ballistic electron microscopy

F. Bociort
Saddle-point methods in lens design

K. Bredies
Diffusion tensor imaging

A. Broers
A retrospective

A. Cornejo Rodriguez, F. Granados Agustin
Ronchigram quantification

J. Elorza
Fuzzy operators

R.G. Forbes
Liquid metal ion sources

P.L. Gai, E.D. Boyes
Aberration-corrected environmental microscopy

R. Herring, B. McMorran
Electron vortex beams

F. Houdellier, A. Arbouet
Ultrafast electron microscopy

M.S. Isaacson
Early STEM development

K. Ishizuka
Contrast transfer and crystal images

K. Jensen, D. Shiffler, J. Luginsland
Physics of field emission cold cathodes

U. Kaiser
The sub-Ångström low-voltage electron microscope project (SALVE)

K. Kimoto
Monochromators for the electron microscope

O.L. Krivanek
Aberration-corrected STEM

M. Kroupa
The Timepix detector and its applications

C. Krowne
Disorder modifications of the critical temperature for superconductors

C. Krowne
Critical magnetic field and its slope, specific heat and gap for superconductivity as modified by nanoscopic disorder

B. Lencová
Modern developments in electron optical calculations

H. Lichte
Developments in electron holography

M. Matsuya
Calculation of aberration coefficients using Lie algebra

J.A. Monsoriu
Fractal zone plates

L. Muray
Miniature electron optics and applications

M.A. O'Keefe
Electron image simulation

V. Ortalan
Ultrafast electron microscopy

D. Paganin, T. Gureyev, K. Pavlov
Intensity-linear methods in inverse imaging

N. Papamarkos, A. Kesidis
The inverse Hough transform

H. Qin
Swarm optimization and lens design

Q. Ramasse, R. Brydson
The SuperSTEM laboratory

B. Rieger, A.J. Koster
Image formation in cryo-electron microscopy

P. Rocca, M. Donelli
Imaging of dielectric objects

J. Rodenburg
Lensless imaging

J. Rouse, H.-n. Liu, E. Munro
The role of differential algebra in electron optics

J. Sánchez
Fisher vector encoding for the classification of natural images

P. Santi
Light sheet fluorescence microscopy

R. Shimizu, T. Ikuta, Y. Takai
Defocus image modulation processing in real time

T. Soma
Focus-deflection systems and their applications

J. Valdés
Recent developments concerning the Système International (SI)

J. van de Gronde, J.B.T.M. Roerdink
Modern non-scalar morphology

CHAPTER ONE

Walsh–Hadamard Transforms: A Review

Ashkan Ashrafi
Department of Electrical and Computer Engineering, San Diego State University, San Diego, CA, USA
e-mail address: ashrafi@mail.sdsu.edu

Contents

Advances in Imaging and Electron Physics, Volume 201
ISSN 1076-5670
http://dx.doi.org/10.1016/bs.aiep.2017.05.002

1. INTRODUCTION

Walsh functions were introduced in 1923 by Joseph L. Walsh (Walsh, 1923), where they were also used for orthogonal decomposition of continuous functions. In other words, Walsh functions construct an orthogonal basis set for the Hilbert space of functions defined in the interval $0 \leq x \leq 1$. Walsh also showed that the expansion of arbitrary continuous functions in terms of the Walsh functions is similar to their expansion in terms of Haar's orthogonal function set, which was introduced by Alfred Haar (Haar, 1910). This was the birth of the Walsh Transform. The Walsh functions can also be defined based on the Rademacher functions, which can provide a closed-form expression for the Walsh functions (Beauchamp, 1975).

A need for discrete-time implementation of the Walsh Transform to be suitable for digital computers has led to the introduction of the Walsh–Hadamard Transform, which is a discrete transform on discrete functions. The Walsh–Hadamard Transform is defined based on the Hadamard matrix that was introduced by Jacques Hadamard (Hadamard, 1893).

The Walsh–Hadamard Transform is a discrete orthogonal transform on the N-dimensional Hilbert space that is spanned by the columns of the Hadamard matrix. Since the entries of the Hadamard matrix are either $+1$ or -1, the calculation of the Walsh–Hadamard Transform requires only additions and subtractions. Therefore, the computation of the Walsh–Hadamard Transform is significantly less intense than that of the Discrete Fourier Transform. However, the Walsh–Hadamard Transform does not possesses the shift-invariant and conjugate symmetry properties of the Discrete Fourier Transform.

Another important feature of the Walsh–Hadamard Transform is the fact that it can mimic the concept of frequency as the number of zero-crossings of the function. By defining the sequency as the half of the number of zero-crossings in a function, we can generate the sequency spectra of signals using the Walsh–Hadamard Transform and study the cycling behavior of different functions.

The Walsh–Hadamard Transform has several deficiencies compared to the Discrete Fourier Transform. One deficiency is its lack of time-invariance property. In the Discrete Fourier Transform (and Fourier transform in general), the power spectrum of a signal and its delayed version is the same. A delay in the function will just change the phase of its Fourier transform. This property does not exist in the Walsh–Hadamard Transform. The Walsh–Hadamard Transform is also a real-valued transform and it dose not provide symmetric power spectra of the signals. These problems have been overcome by the introduction of Conjugate Walsh–Hadamard Transforms such as Unified Complex Hadamard Transform (Rahardja & Falkowski, 1999), Sequency-Ordered Complex Hadamard Transform (SCHT) (Aung, Ng, & Rahardja, 2008), and Conjugate Symmetric Sequency-Ordered Hadamard Transform (CS-SCHT) (Aung, Ng, & Rahardja, 2009).

The Frequency Domain Walsh functions and sequences are introduced by Ashrafi (2014). These functions construct infinite number of shift-invariant sampling Hilbert spaces and they have shown to have interesting properties suitable for defining new sampling schemes. The Frequency Domain Walsh Sequences also construct an orthogonal discrete Hilbert space, which can be used for signal decomposition.

In this article, we review old and new Walsh–Hadamard Transforms and briefly provide their applications in different areas of science and engineering. The organization of this article is as follows. In Section 2, the definitions of the Walsh functions and their representation in terms of the Rademacher functions are given. The definition of the discrete Walsh functions and the Walsh–Hadamard matrix are also presented in this section. In the last part of this section, the properties of the continuous and discrete Walsh functions are discussed. In Section 3, the continuous and discrete Walsh transforms are presented. It is shown in this section that the Discrete Walsh Transform can be presented as two separate discrete transforms namely SAL and CAL transforms. The properties of the Discrete Walsh Transform as well as the fast Walsh Transform algorithm are also discussed in this section. In Section 4, the sequency spectrum in terms of sequency periodogram and dyadic autocorrelation are discussed. Sequency filtering is another subject that is discussed in Section 4. In Section 5, the Frequency Domain Walsh Functions and Sequences are studied were it is shown how these functions can generate infinite shift-invariant sampling Hilbert spaces. In Section 6, new advances in Walsh–Hadamard Transforms, namely complex Walsh Hadamard Transforms, are reviewed. The fast implementations of these transforms are also reviewed and it is shown that they can be faster alternatives to the Fast Fourier Transform.

2. DEFINITION OF THE WALSH FUNCTIONS

The Walsh functions can be defined in different ways. We briefly discuss these definitions in this section.

2.1 Original Definition

The Walsh functions $\varphi_n^{(k)}(\theta)$ are originally defined by Walsh (1923) as follows:

$$\varphi_0(\theta) = 1, 0 \leq \theta \leq 1$$

$$\varphi_1^{(1)}(\theta) = \begin{cases} 1 & 0 \leq \theta < \frac{1}{2} \\ -1 & \frac{1}{2} < \theta \leq 1 \end{cases}$$

$$\varphi_{n+1}^{(2k-1)}(\theta) = \begin{cases} \varphi_n^{(k)}(2\theta) & 0 \leq \theta < \frac{1}{2} \\ (-1)^{k+1}\varphi_n^{(k)}(2\theta - 1) & \frac{1}{2} < \theta \leq 1 \end{cases} \tag{1}$$

$$\varphi_{n+1}^{(2k)}(\theta) = \begin{cases} \varphi_n^{(k)}(2\theta) & 0 \leq \theta < \frac{1}{2} \\ (-1)^k\varphi_n^{(k)}(2\theta - 1) & \frac{1}{2} < \theta \leq 1 \end{cases}$$

This is a recursive definition of the Walsh functions, i.e., Walsh functions at each order n are produced from the functions of the previous order. It is worth noting that the Walsh functions are ordered in ascending values of the number of zero crossings (Beauchamp, 1975). This ordering is called *sequency order*. The Walsh functions in sequency order are sorted based on their zero-crossings, which is a representation of the frequency of the functions. Harmuth was the first who defined sequency (Harmuth, 1968). We can define the sequency of a Walsh function as $\frac{1}{2}$ times of its number of zero-crossing (Lackey & Meltzer, 1971).

Another important feature of the sequency order is the fact that every other function is either even or odd (with respect to $\theta = \frac{1}{2}$). The sequency ordering can also simplify the definition of the Walsh functions so that the indices "n" and "k" in (1) can be reduce to just one index "n". Thus, the Walsh functions can be represented as $w_n(\theta)$ and the even and odd Walsh

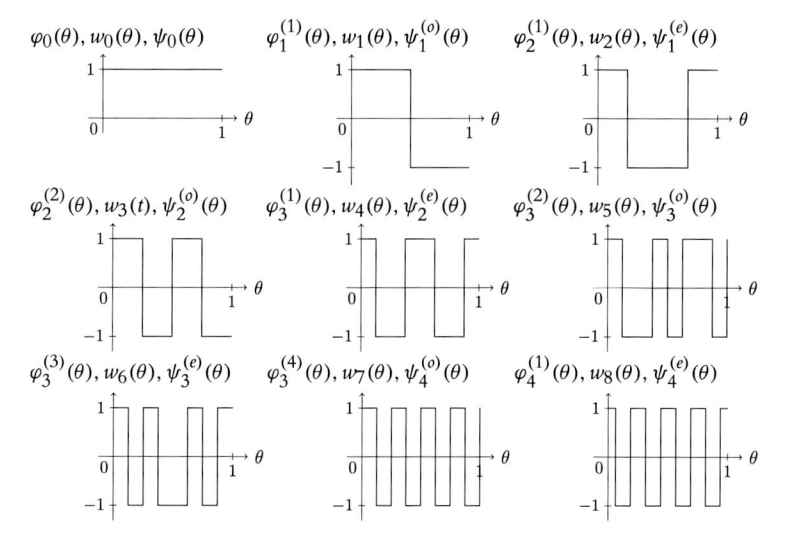

Figure 1 The first nine Walsh functions in sequence order (top to bottom, left to right).

functions can be represented as $\psi_n^{(e)}(\theta)$ and $\psi_n^{(o)}(\theta)$, respectively, such that

$$
\begin{aligned}
\psi_n^{(e)}(\theta) &= w_{2n}(\theta), \\
\psi_n^{(o)}(\theta) &= w_{2n-1}(\theta).
\end{aligned}
\tag{2}
$$

In the literature $\psi_n^{(e)}(\theta)$ and $\psi_n^{(o)}(\theta)$ are called CAL and SAL functions, respectively. Fig. 1 shows the first nine Walsh functions in the sequence order. The functions are denoted according to the original definition of the Walsh functions given in (1), the even and odd presentation of the functions given in (2) and in the sequence order represented by $w_n(\theta)$. It should be noted that all Walsh functions shown in Fig. 1 hold a positive value ($+1$) at $\theta = 0$. These functions are said to have *positive phasing*. The other phasing of the Walsh function is called *Harmuth phasing* in which $w_{4k+1}(\theta)$ and $w_{4k+2}(\theta)$ have negative values (-1) at $\theta = 0$.

It is obvious from (1) and Fig. 1 that the Walsh functions are symmetric (even or odd) with respect to $\theta = \frac{1}{2}$. Sometimes it would be beneficial to change the variable $\theta = t + \frac{1}{2}$ to make the functions symmetric with respect to 0. The resultant Walsh functions will be called $w_k(t)$. It should be noted that the usage of the variables θ and t in this article is the opposite to their usage in the literature, e.g., Chien (1975), Beauchamp (1975), and Lackey and Meltzer (1971). In this article, wherever the variable t is used, we assume that the Walsh function is defined in the interval $t \in [-1/2, 1/2]$.

2.2 Definition with Respect to Rademacher Functions

Rademacher functions $R_n(t)$ construct an orthogonal set defined as:

$$R_n(\theta) = \text{sign}[\sin(2^n \pi \theta)], \quad 0 \le \theta \le 1, \tag{3}$$

where "sign" is the sign function. Rademacher functions set is not a complete orthogonal set (a complete orthogonal set is said to be an orthogonal set for which Parseval's theorem is held) due to its lack of even functions. Nevertheless, a complete set of orthogonal functions can be created from Rademacher functions and consequently, they can be used to define the Walsh functions.

Let an integer number n be represented as an m-bit binary number $(\overline{n_{m-1}, n_{m-2}, \ldots, n_1, n_0})_2$

$$n = \sum_{i=0}^{m-1} n_i 2^i, \quad n_i \in \{0, 1\}. \tag{4}$$

Now define the product series of Rademacher functions as

$$\zeta_n(\theta) = \prod_{i=0}^{m-1} [R_i(\theta)]^{n_i}. \tag{5}$$

For example, $\zeta_{11}(\theta) = R_3(\theta) R_1(\theta) R_0(\theta)$, because the number 11 is $(\overline{1011})_2$ in binary and the locations of the 1's are at 0, 1, and 3. The product series of Rademacher functions $\zeta_n(\theta)$ is in fact the Walsh functions in the *dyadic order* with positive phasing. The dyadic-ordered Walsh functions set was first introduced by Paley (1932) so it is also called Paley-ordered Walsh functions.

The relationship between the Walsh functions in sequence order and dyadic order is established by the Gray code and Modulo-2 addition (Lackey & Meltzer, 1971). Let $g(n)$ be the Gray code of the integer number n. If n is presented in m-bit binary form defined by (4), then $g(n)$ can be defined as

$$g_i(n) = n_i \oplus n_{i+1}, \quad i = 0, 1, \ldots, m-1 \quad \text{and} \quad n_m = 0, \tag{6}$$

where $\oplus$ is Modulo-2 addition (also known as XOR in Boolean algebra) (Beauchamp, 1975; Henderson, 1964). Therefore, the relationship between the sequency and dyadic ordered Walsh functions is given as

$$\zeta_{g(n)}(\theta) = w_n(\theta). \tag{7}$$

Since Rademacher functions can be defined in terms of sine functions (3), it is also possible to define the Walsh functions in terms of sine and cosine functions as

$$w_k(\theta) = \text{sign}\left[[\sin(2\pi\theta)]^{n_0} \prod_{k=1}^{m-1} [\cos(2^k\pi\theta)]^{n_k} \right], \qquad (8)$$

where b_k and m are the same parameters defined in (4) (Ross & Kelly, 1972). Another way of defining the Rademacher functions is given as follows

$$R_n(\theta) = (-1)^{\lfloor 2^n\theta \rfloor}, \qquad 0 \leq \theta < 1, \qquad (9)$$

where $R_n(\theta)$ is the Rademacher function of order n and $\lfloor . \rfloor$ is the floor function. With this definition of the Rademacher functions, one can easily define the Walsh functions (in dyadic order) in terms of powers of -1:

$$\zeta_n(\theta) = (-1)^{(\lfloor n_{m-1}2^m\theta \rfloor + \lfloor n_{m-2}2^{m-1}\theta \rfloor + \cdots + \lfloor n_0 2\theta \rfloor)}, \qquad 0 \leq \theta < 1, \qquad (10)$$

where n_i's are the bits of the integer n in binary form (4). To show the Walsh functions in the sequency order, we can use the Gray code of n as defined by (6)

$$w_n(\theta) = (-1)^{(\lfloor g_{m-1}2^m\theta \rfloor + \lfloor g_{m-2}2^{m-1}\theta \rfloor + \cdots + \lfloor g_0 2\theta \rfloor)}, \quad 0 \leq \theta < 1. \qquad (11)$$

It is obvious that (10) and (11) satisfy (7). By changing the variable $\theta = t + \frac{1}{2}$ in (10) and (11), the resultant Walsh functions $\zeta_n(t)$ and $w_n(t)$ will have Harmuth phasing. Table 1 shows the first 16 Walsh, CAL, and SAL functions in both sequency and dyadic orders.

2.3 Recursive Representation of the Walsh Functions

Henning F. Harmuth introduced a difference equation whose solutions are the Walsh functions (Harmuth, 1968). This difference equation is given as follows:

$$w_0(t) = \begin{cases} 1 & -\frac{1}{2} \leq t < \frac{1}{2} \\ -1 & \text{elsewhere} \end{cases}$$

$$w_{2k+q}(t) = (-1)^{(\lfloor \frac{k}{2} \rfloor + q)} w_k\left(2t + \frac{1}{2}\right) + (-1)^{(k+q)} w_k\left(2t - \frac{1}{2}\right) \qquad (12)$$

$$q \in \{0, 1\}, \quad k = 0, 1, 2, \ldots$$

Table 1 Conversion between sequency and dyadic order for the first 16 Walsh functions

Sequency order		Dyadic order
$w_0(\theta)$	$\psi_0^e(\theta)$	$\zeta_0(\theta)$
$w_1(\theta)$	$\psi_1^o(\theta)$	$\zeta_1(\theta)$
$w_2(\theta)$	$\psi_1^e(\theta)$	$\zeta_3(\theta)$
$w_3(\theta)$	$\psi_2^o(\theta)$	$\zeta_2(\theta)$
$w_4(\theta)$	$\psi_2^e(\theta)$	$\zeta_6(\theta)$
$w_5(\theta)$	$\psi_3^o(\theta)$	$\zeta_7(\theta)$
$w_6(\theta)$	$\psi_3^e(\theta)$	$\zeta_5(\theta)$
$w_7(\theta)$	$\psi_4^o(\theta)$	$\zeta_4(\theta)$
$w_8(\theta)$	$\psi_4^e(\theta)$	$\zeta_{12}(\theta)$
$w_9(\theta)$	$\psi_5^o(\theta)$	$\zeta_{13}(\theta)$
$w_{10}(\theta)$	$\psi_5^e(\theta)$	$\zeta_{15}(\theta)$
$w_{11}(\theta)$	$\psi_6^o(\theta)$	$\zeta_{14}(\theta)$
$w_{12}(\theta)$	$\psi_6^e(\theta)$	$\zeta_{10}(\theta)$
$w_{13}(\theta)$	$\psi_7^o(\theta)$	$\zeta_{11}(\theta)$
$w_{14}(\theta)$	$\psi_7^e(\theta)$	$\zeta_9(\theta)$
$w_{15}(\theta)$	$\psi_8^o(\theta)$	$\zeta_8(\theta)$

Chien extended the Walsh function periodically over the entire real axis and introduced another difference equation whose solutions are the periodic Walsh functions (Chien, 1975). The Chien's difference equation is given as

$$w_0(t) = 1, \quad -\infty < t < +\infty$$

$$w_{4k+p}(t) = (-1)^{\left(\lfloor \frac{(p+1)}{2} \rfloor \lfloor 2t \rfloor \right)} w_{2k+\lfloor \frac{p}{2} \rfloor}(t) \tag{13}$$

$$p \in 0, 1, 2, 3, \quad k = 0, 1, 2, \ldots,$$

from which Chien derived the explicit expression for the Walsh functions given in (11).

There are other definitions for the Walsh functions that are appeared in the literature such as the ones given by Butin (1972) and Irshid (1986).

2.4 Discrete Walsh Sequences and the Walsh Matrix

The discrete Walsh functions (also known as Walsh sequences) can be found by sampling the Walsh functions such that there is only one sample exists between two adjacent discontinuity. To be consistent with the popular Discrete Fourier Transform notation, we consider n as the sample index in the time domain and k as the Walsh order index (corresponds to the frequency index in the Discrete Fourier Transform). Without loss of generality, we can assume that the Walsh functions have $N = 2^p$ samples, where $p = 0, 1, 2, \ldots$. Suppose the binary representations of n and k are respectively $(n)_2 = (\overline{n_{p-1}, n_{p-2}, \ldots, n_0})_2$ and $(k)_2 = (\overline{k_{p-1}, k_{p-2}, \ldots, k_0})_2$, where $n_i, k_i \in \{0, 1\}$. Pratt et al. defined the Walsh sequences with $N = 2^p$ terms (Pratt, Kane, & Andrews, 1969) such that

$$w_k[n] = \prod_{r=0}^{p-1} (-1)^{k_{p-1-r}(n_r + n_{r+1})},$$

$$\tag{14}$$

$$n, k = 0, 1, 2, \ldots, N - 1.$$

The Walsh sequences are orthogonal sequences (see Beauchamp, 1975 for the proof). They can also be defined from the discrete Rademacher functions (Henderson, 1964).

One may have noticed that the number of bits of the values n and k (p) is assumed to be the same in (14). This indicates that the number of discrete Walsh functions that can be generated is the same as the number of samples of the sequences, which is an integer power of two. Therefore, we can put the discrete Walsh functions for a given number of samples $N = 2^p$ in a square matrix called the Walsh matrix ($\mathbf{W}_N$).

The Walsh sequences obtained from (14) are in sequency order. In other words, the number of sign changes (zero crossings) of the sequences increases by increasing the order of the sequences. The Walsh sequences can be rearranged to dyadic order by using the Gray codes of the order numbers according to (6). The 8×8 Walsh matrices in sequency ($\mathbf{W}_8$) and dyadic ($\mathbf{W}_8'$) orders are shown in (15) and (16), respectively.

$$\mathbf{W}_8 = \begin{bmatrix} 1 & 1 & 1 & 1 & 1 & 1 & 1 & 1 \\ 1 & 1 & 1 & 1 & -1 & -1 & -1 & -1 \\ 1 & 1 & -1 & -1 & -1 & -1 & 1 & 1 \\ 1 & 1 & -1 & -1 & 1 & 1 & -1 & -1 \\ 1 & -1 & -1 & 1 & 1 & -1 & -1 & 1 \\ 1 & -1 & -1 & 1 & -1 & 1 & 1 & -1 \\ 1 & -1 & 1 & -1 & -1 & 1 & -1 & 1 \\ 1 & -1 & 1 & -1 & 1 & -1 & 1 & -1 \end{bmatrix}, \tag{15}$$

$$\mathbf{W}'_8 = \begin{bmatrix} 1 & 1 & 1 & 1 & 1 & 1 & 1 & 1 \\ 1 & 1 & 1 & 1 & -1 & -1 & -1 & -1 \\ 1 & 1 & -1 & -1 & 1 & 1 & -1 & -1 \\ 1 & 1 & -1 & -1 & -1 & -1 & 1 & 1 \\ 1 & -1 & 1 & -1 & 1 & -1 & 1 & -1 \\ 1 & -1 & 1 & -1 & -1 & 1 & -1 & 1 \\ 1 & -1 & -1 & 1 & 1 & -1 & -1 & 1 \\ 1 & -1 & -1 & 1 & -1 & 1 & 1 & -1 \end{bmatrix}. \tag{16}$$

Another way to find the Walsh matrix is making use of the Hadamard matrix (Beauchamp, 1975). Hadamard matrices are square matrices whose entries are either $+1$ or -1 and the size of them is $2^p \times 2^p$. They are symmetric and orthogonal matrices. The lowest order Hadamard matrix is a 2×2 matrix given as

$$\mathbf{H}_2 = \begin{bmatrix} 1 & 1 \\ 1 & -1 \end{bmatrix}. \tag{17}$$

The Hadamard matrix of order $N = 2^p$ can be found using the following recursive equation

$$\mathbf{H}_N = \mathbf{H}_2 \otimes \mathbf{H}_{N/2}. \tag{18}$$

For example, the Hadamard matrix of order 8 is given in (19).

$$\mathbf{H}_8 = \begin{bmatrix} 1 & 1 & 1 & 1 & 1 & 1 & 1 & 1 \\ 1 & -1 & -1 & 1 & -1 & 1 & 1 & -1 \\ 1 & 1 & -1 & -1 & 1 & 1 & -1 & -1 \\ 1 & -1 & -1 & 1 & 1 & -1 & -1 & 1 \\ 1 & 1 & 1 & 1 & -1 & -1 & -1 & -1 \\ 1 & -1 & 1 & -1 & -1 & 1 & -1 & 1 \\ 1 & 1 & -1 & -1 & -1 & -1 & 1 & 1 \\ 1 & -1 & 1 & -1 & 1 & -1 & 1 & -1 \end{bmatrix}. \tag{19}$$

The Hadamard matrix is arranged in bit-reversed dyadic order of the Walsh matrix and it is sometimes called the Walsh matrix in *natural order*. To find the Walsh matrix in sequency order, we should rearrange the Hadamard matrix according to the gray code of its bit-reversed row numbers. For example, the row number of the 7th row of the 8th order Hadamard matrix in binary form is $\overline{(110)}_2$. By bit reversing it, we obtain

$\overline{(011)_2}$ and the Gray code of this number is $\overline{(010)_2}$. Therefore, the 7th row of the Hadamard matrix is the 3rd row of the Walsh matrix in sequency order (note that the first row is actually row number zero). We will discuss the properties of these matrices in the next section.

2.5 Properties of the Walsh Functions

In this section we will briefly talk about some of the properties of the Walsh functions, Walsh sequences, and Walsh matrices. The proofs of the properties are not given here. An interested reader is encouraged to find the proofs in the provided references.

Here we discuss the properties of the continuous Walsh functions in sequency order $(w_k(t))$, in dyadic order $(\zeta_k(t))$, and the SAL $(\psi_k^o(t))$ and CAL $(\psi_k^e(t))$ functions.

Property 1. *The set of Walsh functions with orders less than 2^l, $l = 0, 1, 2, \ldots$, is closed under multiplications. In other words:*

$$w_r(t) = w_p(t)w_q(t), \quad p, q, r < 2^l \quad and \quad r = p \oplus q. \tag{20}$$

The operation $r = p \oplus q$ is the bit-wise Modulo-2 addition of the integers p and q. The proof is given by Chien (1975).

Property 2. *The sets of CAL $(\psi_n^e(\theta))$ and SAL $(\psi_n^o(\theta))$ functions with orders less than 2^l, $l = 0, 1, 2, \ldots$, are closed under multiplications such that*

$$\begin{aligned}
\psi_p^e(\theta)\psi_q^e(\theta) &= \psi_{p \oplus q}^e(\theta), \\
\psi_p^o(\theta)\psi_q^e(\theta) &= \psi_{q \oplus (p-1)+1}^o(\theta), \\
\psi_p^e(\theta)\psi_q^o(\theta) &= \psi_{p \oplus (q-1)+1}^o(\theta), \\
\psi_p^o(\theta)\psi_q^o(\theta) &= \psi_{(p-1) \oplus (q-1)}^e(\theta), \\
\psi_0^e(\theta) &= w_0(\theta).
\end{aligned} \tag{21}$$

The proof is given by Beauchamp (1975).

Property 3. *The Walsh functions are symmetric (odd or even) except at the discontinuities. In other words,*

$$w_k(-t) = (-1)^k w_k(t), \quad t \neq \frac{n}{2^m}, \quad m = 0, \pm 1, \pm 2, \ldots. \tag{22}$$

The proof is given by Chien (1975).

Property 4. *Let s_k be the sequency (half of the number of zero crossings) of the Walsh function of order k. Then we have*

$$s_k = \left\lfloor \frac{k+1}{2} \right\rfloor. \tag{23}$$

Moreover, the sequency s_k satisfies the following equations:

$$\begin{aligned}
s_{4k} &= 2s_{2k}, \\
s_{4k+1} &= 2s_{2k+1} + 1, \\
s_{4k+2} &= 2s_{2k+1} - 1, \\
s_{4k+3} &= 2s_{2k}, \\
s_{2k-1} &= s_{2k}.
\end{aligned} \tag{24}$$

The proofs are given by Chien (1975).

Property 5. *The Walsh functions are orthonormal.*

$$\int_0^1 w_k(t)w_m(\theta)d\theta = \delta_{k,m}, \tag{25}$$

where $\delta_{k,m}$ is the Kronecker's delta.

Two different proofs are given by Chien (1975).

Property 6. *The Walsh functions set establishes a complete orthonormal basis for the Hilbert space of functions in the interval of $\theta \in [0, 1]$ or $t \in [-1/2, 1/2]$. Therefore, the Parseval's theorem is satisfied for the orthogonal decomposition of an arbitrary signal in the space and one can write*

$$\lim_{N \to \infty} \int_0^1 \left[f(\theta) - \sum_{n=0}^{N-1} \alpha_n w_n(\theta) \right]^2 = 0, \tag{26}$$

where, $f(\theta)$ is an arbitrary function defined in $\theta \in [0, 1]$ and α_n is the coefficient obtained from inner product of $f(\theta)$ and the Walsh function of order n ($w_n(\theta)$) or

$$\alpha_n = \int_0^1 f(\theta)w_n(\theta)d\theta. \tag{27}$$

This property is the foundation of the continuous Walsh transform, which will be discussed in the next section.

Property 7. *The Walsh sequences are symmetric, i.e.,*

$$w_m[n] = w_n[m]. \tag{28}$$

Property 8. *The Walsh sequences satisfy the following identity*

$$w_m[n] = w_{\lfloor m/2 \rfloor}[2n] \cdot w_{m-2\lfloor m/2 \rfloor}[n]. \tag{29}$$

This property is useful in defining the fast discrete Walsh transform.

Property 9. *The Walsh matrix of order N ($\mathbf{W}_N$) is an orthogonal and symmetric matrix, i.e.,*

$$\mathbf{W}_N^2 = N\mathbf{I}_N, \tag{30}$$

where $\mathbf{I}_N$ is the Identity matrix of order N and $N = 2^n$.

The poof of this property is self-evident. The rows and columns of $\mathbf{W}_N$ are Walsh sequences, thus they are orthogonal to each other. Moreover, the inner product of the Walsh sequences by themselves produces a sum of 2^n ones, hence the identity matrix with the diagonals of 2^n.

3. WALSH TRANSFORMS

In general, the Walsh transform is the decomposition of an arbitrary function in the interval of interest in terms of the orthonormal Walsh functions. If the arbitrary function is a continuous function, the Walsh transform is called Continuous Walsh Transform (CWT) and if the arbitrary function is discrete, the Walsh transform is called Discrete Walsh Transform (DWT), which also known as Walsh–Hadamard Transform (WHT). We will see that these transforms are analogous to the Continuous Fourier Transform (CFT) and Discrete Fourier Transform (DFT). To make connections between Walsh and Fourier transforms we will redefine the concept of frequency and period in the Walsh transforms and discuss the concept of sequency. Finally we will discuss Walsh spectral analysis and sequency filtering.

3.1 Continuous Walsh Transform

The idea of decomposing an arbitrary continuous function into a linear combination of the Walsh functions was fist given by Walsh himself in his seminal paper (Walsh, 1923).

$$X(k) = \int_0^1 x(\theta)w_k(\theta)d\theta,$$

$$x(\theta) = \sum_{k=0}^{\infty} X(k)w_k(\theta), \quad 0 \leq \theta < 1. \tag{31}$$

In the Fourier transform, the arbitrary function is decomposed into a linear combination of sine and cosine functions. The Fourier coefficients correspond to the frequencies of the sine and cosine functions; therefore, they are called the function in the frequency domain. The definition of the frequency is valid for periodic functions and it is the number of cycles per second (cps). Another way of defining the frequency is half of the number of zero crossings in sine or cosine functions.

The concept of frequency is not valid for the Walsh functions because they are not periodic. Nevertheless, we can easily find the number of zero crossings of the Walsh functions. In fact, the sequency order of the Walsh functions (see Table 1) is based on the number of their zero crossings. Sequency of a Walsh function is defined as half of the number of its zero crossing in the interval of interest ([0, 1] or [−1/2, 1/2]). The unit of the sequency for time domain functions is "zps" (zero crossings per second), which is analogous to "cps" (cycle per second) in periodic signals. We call the CWT of a function its representation in the sequency domain.

Parseval's theorem is valid for the CWT given in (31) because the Walsh functions construct an orthonormal basis set in the Hilbert space of the functions defined in the interval [0, 1]. In other words,

$$\int_0^1 |x(\theta)|^2 d\theta = \sum_{k=0}^{\infty} |X(k)|^2. \tag{32}$$

Due to the existence of discontinuity in the Walsh functions, the CWT has limited applications and it is rarely used in science and engineering. In the next section we will discuss the Discrete Walsh Transform (DWT), which has numerous applications in different areas including optics and signal processing.

3.2 Discrete Walsh Transform

Let x[n] be a discrete signal with length N. Without loss of generality, we can assume that $N = 2^p$ (if the length is not a power of two, we can pad zeros and make it a power of two). Then we define the Discrete Walsh Transform (DWT) of $x[n]$ as

$$X_w[m] = \frac{1}{N} \sum_{n=0}^{N-1} x[n]w_m[n], \quad m = 0, 1, 2, \ldots, N-1, \tag{33}$$

where $X_w[m]$ is the DWT of $x[n]$ and $w_m[n]$ is the Walsh sequence of order m. The inverse DWT (IDWT) is defined as

$$x[n] = \sum_{m=0}^{N-1} X_w[m]w_m[n], \quad n = 0, 1, 2, \ldots, N-1. \tag{34}$$

The DWT and IDWT defined in (33) and (34) can be represented in matrix form. Let $\mathbf{W}_N$ be the Walsh matrix of order $N = 2^p$ (see for example (15)), $\mathbf{X}_w$ be the vector whose entries are the DWT of $x[n]$ and $\mathbf{x}$ be the vector whose entries are the samples of the discrete sequence $x[n]$. Then the DWT can be given as the following matrix equation

$$\mathbf{X}_w = \frac{1}{N} \mathbf{W}_N \mathbf{x}, \tag{35}$$

and the IDWT can be given as

$$\mathbf{x} = \mathbf{W}_N \mathbf{X}_w. \tag{36}$$

Since the Walsh matrix $\mathbf{W}_N$ is an orthogonal and symmetric matrix, the DWT and IDWT are essentially the same process except the factor $1/N$ in the DWT. Moreover, the matrix $\mathbf{W}_N$ contains only ± 1 thus, the process of the DWT and IDWT comprise only additions and subtractions. Therefore, the DWT is much more efficient than the Discrete Fourier Transform (DFT) and even the Fast Fourier Transform (FFT), which comprise of complex multiplications.

3.3 Discrete CAL and SAL Transforms

Like the DWT, one can define Discrete CAL transform (DCALT) and Discrete SAL transform (DSALT). The definitions of DCALT and DSALT are given here:

$$X_c[m] = \frac{1}{N} \sum_{m=0}^{N-1} x[n]\psi_m^e[n],$$

$$X_s[m] = \frac{1}{N} \sum_{m=0}^{N-1} x[n]\psi_m^o[n],$$

(37)

where

$$\psi_m^e[n] = w_{2m}[n],$$

$$\psi_m^o[n] = w_{2m-1}[n].$$

(38)

3.4 The DWT Properties

In this section, we will talk about the properties of the DWT. Many of these properties are analogous to those of the Discrete Fourier Transform but in general, they are different.

Property 10. *Linearity: If the DWT of x[n] and y[n] are X[m] and Y[m], respectively, the DWT of a linear combination of x[n] and y[n] is the same linear combination of the DWTs. In other words,*

$$a_1 x[n] + a_2 y[n] \xleftrightarrow{DWT} a_1 X_w[m] + a_2 Y_w[m].$$

(39)

Property 11. *Parseval's theorem: Since the DWT is an orthogonal decomposition on a Hilbert space (the basis functions are complete), Parseval's theorem holds, i.e.,*

$$\frac{1}{N} \sum_{n=0}^{N-1} |x[n]|^2 = \sum_{m=0}^{N-1} |X[m]|^2.$$

(40)

3.5 Dyadic Shift, Correlation, and Convolution

Unlike Discrete Fourier Transform, the DWT is not a shift invariant transform, i.e., the Parseval's theorem does not hold for shifted sequences. In other words, the energy of the shifted signal in sequency domain (the DWT coefficients) will not be the same as the energy of the original signal.

Let $y[n] = x[n - n_0]$ be a shifted signal. Then in the frequency domain we have $|Y_f[k]|^2 = |X_f[k]|^2$, where $X_f[k]$ and $Y_f[k]$ are the DFTs of the signals $x[n]$ and $y[n]$, respectively. However, in the sequency domain the signal will not have the same energy, i.e., $|Y_w[k]|^2 \neq |X_w[k]|^2$. As a result, convolution and correlation, as defined in the Fourier analysis, do not hold in the

Walsh analysis (Beauchamp, 1975). To show this, consider the convolution of two discrete functions $x[n]$ and $y[n]$ as follows

$$z[n] = x[n] * y[n] = \frac{1}{N} \sum_{q=0}^{N-1} x[q]y[n-q]. \tag{41}$$

By replacing $x[n]$ and $y[n]$ with their DWTs in (41), we can write

$$\begin{aligned} z[n] &= \frac{1}{N} \sum_{q=0}^{N-1} \left[\sum_{m=0}^{N-1} X_w[m]w_m[q] \right] \left[\sum_{l=0}^{N-1} Y_w[l]w_l[n-q] \right] \\ &= \sum_{m=0}^{N-1} \sum_{l=0}^{N-1} X_w[m]Y_w[l] \left[\frac{1}{N} \sum_{q=0}^{N-1} w_m[q]w_l[n-q] \right]. \end{aligned} \tag{42}$$

The summation inside the bracket in (42) is the convolution of the Walsh sequences, which results in no meaningful function. Therefore, the convolution based on the time shift is meaningless in the DWT analysis.

By using the Walsh sequence multiplication property (Property 1), we can define dyadic shift and dyadic convolution. Define the dyadic convolution of $x[n]$ and $y[n]$ as

$$z_d[n] = x[n] \odot y[n] = \frac{1}{N} \sum_{q=0}^{N-1} x[q]y[n \oplus q]. \tag{43}$$

Now by substituting the IDWT of $y[n \oplus q]$ into (43), we can write

$$z_d[n] = \frac{1}{N} \sum_{q=0}^{N-1} x[q] \left[\sum_{l=0}^{N-1} Y_w[l]w_l[n \oplus q] \right]. \tag{44}$$

By using the Walsh sequence multiplication property (Property 1) and rearranging the summations in (44) we can obtain

$$z_d[n] = \frac{1}{N} \sum_{l=0}^{N-1} Y_w[l] \sum_{q=0}^{N-1} x[q]w_l[q]w_l[n] = \frac{1}{N} \sum_{l=0}^{N-1} Y_w[l]X_w[l]w_l[n], \tag{45}$$

which is the IDWT of $Y_w[l]X_w[l]$. Therefore, we can summarize the above calculations in

$$x[n] \odot y[n] \xrightarrow{DWT} X_w[m]Y_w[m]. \tag{46}$$

Since the Modulo-2 addition and subtraction is the same, the dyadic correlation and convolution are also the same (Beauchamp, 1975). The dyadic autocorrelation can be defined using the dyadic convolution concept as

$$R_w[n] = \frac{1}{N} \sum_{q=0}^{N-1} x[q]x[q \oplus n]. \tag{47}$$

3.6 Fast Discrete Walsh Transform

The fast Fourier transform (FFT) is perhaps the most used algorithm in the world. Good was the first one who introduced the FFT algorithm (Good, 1958) and later Cooley and Tukey developed it to its current form (Cooley & Tukey, 1965). In this algorithm, the N^2 number of complex multiplications required in the DFT matrix operation is reduced to $N\log_2(N)$. This is a huge reduction in the computation, which can be further improved by a factor of two using the butterfly structure redundancies of the FFT algorithm (Manolakis & Ingle, 2011).

The DWT defined in (33) is inherently much more efficient than the DFT and FFT because it requires N^2 number of additions or subtractions, which are much less computationally intensive than complex multiplications. Nevertheless, the symmetry in the Walsh sequences, can enable us to further reduce the number of additions/subtractions to $N\log_2(N)$. Shanks introduced this fast transform (Shanks, 1969) where he used the multiplicative iteration equation of the Walsh sequences (14). The problem with Shanks' method is that it produces the DWT in natural order. In most practical applications, we need the spectrum of signals in sequency domain, i.e., the DWT must produce the result in sequency order.

Manz introduced another fast DWT algorithm (Manz, 1972) that produces the sequency spectrum but it requires bit-reversed input data. Carl and Swartwood developed a more efficient algorithm (Carl & Swartwood, 1973) using the matrix factorization of the Walsh matrix given in Andrews and Caspari (1970). The matrix factors are sparse and one can use the sparsity of these matrices to simplify the computation of the DWT. Although the introduced fast DWT is very efficient, it does not generate sequency-ordered DWT. Brown solved this problem (Brown, 1977) by introducing a method that will be discussed in the following paragraphs.

If we define the DWT based on the Hadamard matrix (Walsh matrix in natural ordering), we can write

$$\mathbf{X}_H = \frac{1}{N}\mathbf{H}_N\mathbf{x} = (\mathbf{L}_{N,1})^p\mathbf{x}, \tag{48}$$

where $p = \log_2(N)$ and $\mathbf{L}_{N,1}$ is a sparse matrix defined as

$$
\begin{aligned}
(\mathbf{L}_{N,1})_{i,2i-1} &= 1, & i &= 1, 2, \ldots, N/2, \\
(\mathbf{L}_{N,1})_{i,2i} &= 1, & i &= 1, 2, \ldots, N/2, \\
(\mathbf{L}_{N,1})_{i,2i-N/2-1} &= 1, & i &= N/2+1, N/2+2, \ldots, N, \\
(\mathbf{L}_{N,1})_{i,2i-N/2} &= -1, & i &= N/2+1, N/2+2, \ldots, N.
\end{aligned}
\tag{49}
$$

For example, $\mathbf{L}_{8,1}$ is

$$
\mathbf{L}_{8,1} = \begin{bmatrix}
1 & 1 & & & & & & \\
& & 1 & 1 & & & & \\
& & & & 1 & 1 & & \\
& & & & & & 1 & 1 \\
1 & -1 & & & & & & \\
& & 1 & -1 & & & & \\
& & & & 1 & -1 & & \\
& & & & & & 1 & -1
\end{bmatrix}, \tag{50}
$$

and $\mathbf{H}_8 = (\mathbf{L}_{8,1})^3$. The fast DWT produced by (48), (49), and (50) was introduced by Carl and Swartwood (1973) and its flow diagram for $N = 8$ is shown in Fig. 2.

The problem with Carl–Swartwood algorithm is that the output sequence is in natural order. Brown introduced a sparse factorization for the Walsh matrix in sequency order that can be used to develop a fast algorithm (Brown, 1977). Unfortunately, he just discussed the 8-point DWT in his paper and he did not provide the recursive formula to find the general case for $N = 2^p$. The 8-point sparse factorization introduced by Brown (1977) is given here:

$$\mathbf{X}_w = \frac{1}{8}\mathbf{W}_8\mathbf{x} = \frac{1}{8}\left(\prod_{k=3}^{1}\mathbf{L}_{8,k}\right)\mathbf{x}, \tag{51}$$

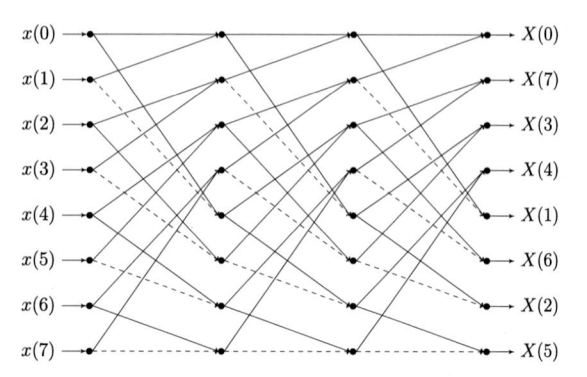

Figure 2 Flow diagram of Carl–Swartwood Fast Discrete Walsh Transform (Carl & Swartwood, 1973). The dashed lines represent subtraction.

where $\mathbf{L}_{8,k}$ are sparse matrices and $\mathbf{L}_{8,1}$ it is defined by (49). Properties 9 and 7 indicate that we can use either of the following orders

$$\mathbf{W}_8 = \prod_{k=3}^{1} \mathbf{L}_{8,k} = \prod_{k=1}^{3} \mathbf{L}_{8,k}^T. \tag{52}$$

The matrices $\mathbf{L}_{8,3}$ and $\mathbf{L}_{8,2}$ are given as

$$\mathbf{L}_{8,3} = \begin{bmatrix} 1 & 1 & & & & & & \\ 1 & -1 & & & & & & \\ & & 1 & -1 & & & & \\ & & 1 & 1 & & & & \\ & & & & 1 & 1 & & \\ & & & & 1 & -1 & & \\ & & & & & & 1 & -1 \\ & & & & & & 1 & 1 \end{bmatrix}, \tag{53}$$

$$\mathbf{L}_{8,2} = \begin{bmatrix} 1 & 1 & & & & & & \\ & & 1 & 1 & & & & \\ 1 & -1 & & & & & & \\ & & 1 & -1 & & & & \\ & & & & 1 & -1 & & \\ & & & & & & 1 & -1 \\ & & & & 1 & 1 & & \\ & & & & & & 1 & 1 \end{bmatrix}, \tag{54}$$

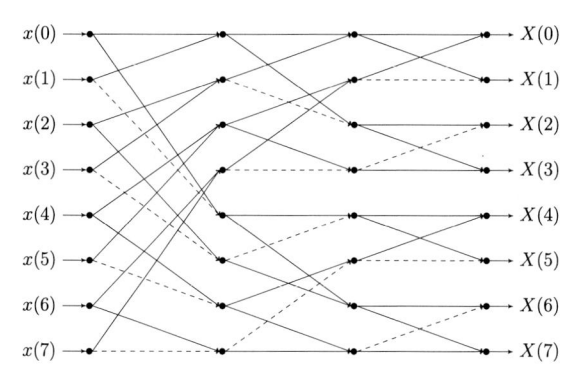

Figure 3 Flow diagram of Brown's Fast Discrete Walsh Transform according to $\mathbf{W}_N = \mathbf{L}_{8,3}\mathbf{L}_{8,2}\mathbf{L}_{8,1}$ factorization. The dashed lines represent subtraction.

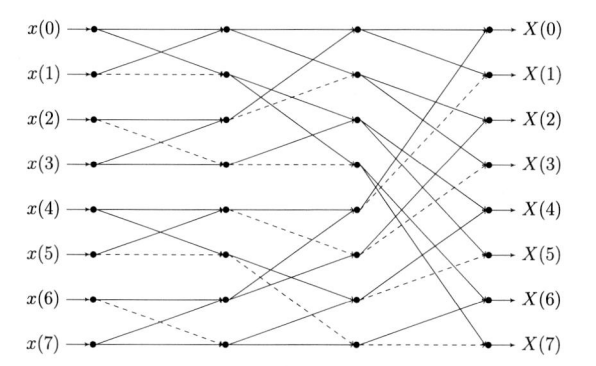

Figure 4 Flow diagram of Brown's Fast Discrete Walsh Transform according to $\mathbf{W}_N = \mathbf{L}_{8,1}^T\mathbf{L}_{8,2}^T\mathbf{L}_{8,3}^T$ factorization. The dashed lines represent subtraction.

and $\mathbf{L}_{8,1}$ is shown in (50). Therefore, one can design two flow diagrams for the fast DWT in sequency order. Fig. 3 and Fig. 4 illustrate these flow diagrams for $\mathbf{W}_N = \mathbf{L}_{8,3}\mathbf{L}_{8,2}\mathbf{L}_{8,1}$ and $\mathbf{W}_N = \mathbf{L}_{8,1}^T\mathbf{L}_{8,2}^T\mathbf{L}_{8,3}^T$, respectively.

Lee and Kaveh introduced a matrix factorization for the Hadamard matrix of order $N = 2^p$ (Lee & Kaveh, 1986). This matrix factorization is similar to the one given by Brown (1977) but the output of the algorithm will be in natural order. Nevertheless, the method has a closed-form formula for $N = 2^p$ as follows

$$\mathbf{H}_N = \prod_{n=1}^{p} \mathbf{I}_{2^{n-1}} \otimes \mathbf{H}_2 \otimes \mathbf{I}_{2^{p-n}} = \prod_{n=1}^{p} \mathbf{I}_{2^{p-n}} \otimes \mathbf{H}_2 \otimes \mathbf{I}_{2^{n-1}}, \qquad (55)$$

where $\mathbf{I}_i$ is the identity matrix of order i. It can be seen from (55) that there are two different implementations of the algorithm similar to the ones illustrated in Fig. 3 and Fig. 4. The output of the algorithm can be easily rearranged to the sequence order with no additional cost to the computations.

The *Radix-R* algorithm to factorize the Hadamard matrix can be directly derived from (55) as

$$\mathbf{H}_{R^p} = \prod_{n=1}^{p} \mathbf{I}_{R^{p-n}} \otimes \mathbf{H}_R \otimes \mathbf{I}_{R^{n-1}}, \tag{56}$$

where $R = 2^m$. The major problem with the fast algorithms obtained by the Hadamard matrix factorizations (55) and (56), is that the stages of the fast algorithm (stages are enumerated by the n) are not identical (similar to Fig. 3 and Fig. 4).

Marti-Puig introduced two *Radix-R* Hadamard matrix factorization formulas that provide identical stages in the signal flow graphs (Marti-Puig, 2006). The first factorization is given by

$$\mathbf{H}_{R^p} = \prod_{n=1}^{p} \mathbf{P}_{R^p}^{\log_2(R)} (\mathbf{I}_{R^{p-1}} \otimes \mathbf{H}_R), \tag{57}$$

where $\mathbf{P}$ is a permutation matrix defined as

$$\mathbf{P}_N = [\mathbf{e}_1 \mathbf{e}_3 \dots \mathbf{e}_{N-1} \mathbf{e}_2 \mathbf{e}_4 \dots \mathbf{e}_N]^T, \tag{58}$$

and $\mathbf{e}_i$ is the ith column of the identity matrix of order N. The second factorization that Marti-Puig proposed is

$$\mathbf{H}_{R^p} = \prod_{n=1}^{p} (\mathbf{I}_{R^{p-1}} \otimes \mathbf{H}_R) \mathbf{P}_{R^p}^{-\log_2(R)}. \tag{59}$$

It is obvious from (57) and (59), the stages of the fast algorithms (the factored matrices) do not depend on the index n thus, the stages are identical and their implementations will also be identical.

Other methods of implementing the DWT have been introduced. An enthusiastic reader is referred to many existing publications such as (Rao, Devarajan, Vlasenko, & Narasimhan, 1978), (Geadah & Corinthios, 1977), and (Zohar, 1973).

3.7 Two-Dimensional Discrete Walsh Transform

The two-dimensional Discrete Walsh Transform (2DDWT) is defined as the application of the DWT on the rows of the 2D signal with N^2 samples and then the application of the DWT on the columns of the results of the first step

$$X_w(k, m) = \frac{1}{N^2} \sum_{i=0}^{N-1} \sum_{j=0}^{N-1} x[i,j] w_k[i] w_m[j]. \tag{60}$$

The inverse 2DDWT is also defined as

$$x[i,j] = \sum_{k=0}^{N-1} \sum_{m=0}^{N-1} X_w[k, m] w_m[i] w_k[j]. \tag{61}$$

The I2DDWT has different application in image processing.

4. SEQUENCY SPECTRUM

Polyak and Schneider introduced the concept of sequency spectrum as oppose to frequency spectrum in Fourier analysis (Polyak & Shreider, 1962). Then Gibbs and Millard (Gibbs, 1967; Gibbs & Millard, 1969) developed the theory of sequency spectral analysis similar to Wiener–Khintchine theorem (Wiener, 1930; Khintchine, 1934). Later, Kennett explained some of the properties of the Walsh transform related to sequency spectrum including the dyadic autocorrelation (Kennett, 1970).

Perhaps the most important feature of the sequency spectrum (also known as Walsh spectrum) is the fact that a time-limited signal can also have a sequency-limited spectrum. The Fourier spectrum does not have this property according to the Fourier transform uncertainty principle (Oppenheim, Willsky, & Nawab, 1983). This feature would be very useful for analyzing nonstationary signals using the sliding window (like the short time Fourier transform (STFT); Quatieri, 2006) as well as filtering in the Walsh domain (sequency filtering).

In this section, we will discuss sequency spectrum and its different definitions, the appropriateness of sequency spectrum and sequency filtering.

4.1 Suitability of the Sequency Spectrum

Sequency spectrum has advantages and disadvantages like the frequency spectrum. As explained before, it has both time-limited and sequency-

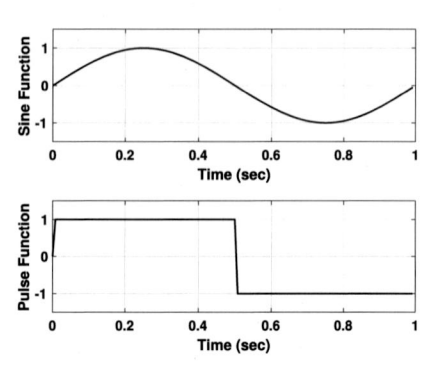

Figure 5 One period of a sinusoid and a pulse function.

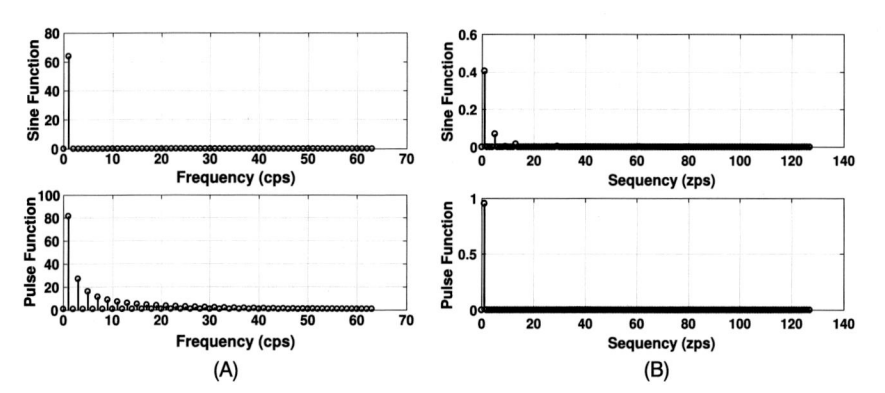

Figure 6 (A) The frequency spectra and (B) the sequency spectra of the signals shown in Fig. 5.

limited feature, which makes it suitable for small window analysis of any signal. On the other hand, due to the existence of discontinuities in the Walsh functions, the DWT is more suitable for signals with discontinuities. This is a feature that frequency spectrum (Fourier analysis) does not posses, as it is well-known that the Fourier transform of a sharp change in a signal produces high frequency components in the frequency spectrum even if the signal is slow-varying almost everywhere.

To show this phenomenon, consider one period of a sinusoid and one period of a pulse (Fig. 5). In Fig. 6A the frequency spectra of these signals are shown. It can be seen from Fig. 6A that the frequency spectrum of the sine function has only one harmonic while the frequency spectrum of the pulse contains odd harmonics. It can be easily deduced that the

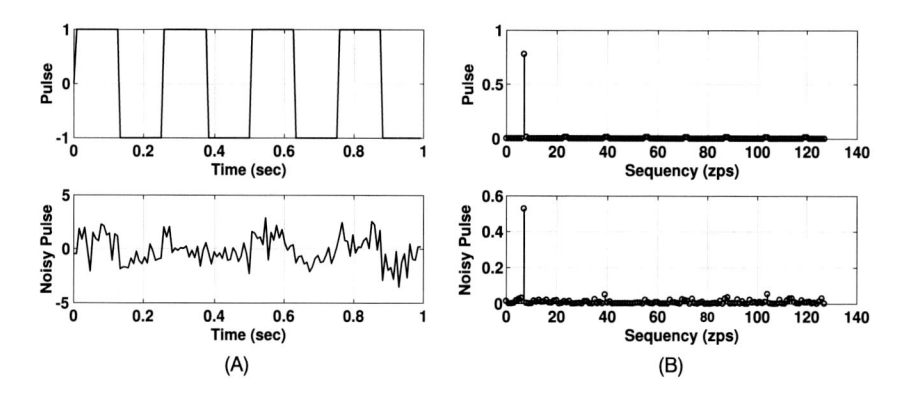

Figure 7 (A) The clean and noisy pulse signal. (B) The sequency spectrum of the clean and noisy signal.

Fourier analysis is more appropriate for smooth signals, than for signals with discontinuities.

Fig. 6B shows the sequency spectra of the sine and pulse signals. It can be seen from the figure that the sequency spectrum of the pulse signal has only one component (sequency harmonic) while the sequency spectrum of the sine signal contains several sequency harmonics. This example clearly shows that the DWT is more appropriate to analyze signals with discontinuities.

To further show the suitability of using the DWT on signals with discontinuity, we can study the DWT of a pulse corrupted by noise. Fig. 7 shows a square wave signal buried in a white Gaussian noise with the variance of 0.5. As it can be seen from the figure, the shape of the pulse is not quite visible but its sequency spectrum has a dominant sequency harmonic indicating that it comprises only one Walsh sequence. The sequency spectrum of the noisy signal is very close to the sequency spectrum of the noiseless signal. Therefore, noise reduction will be much easier done using the DWT than the DFT, for signals with discontinuities.

4.2 Determining Sequency Spectrum by Periodogram

The easiest way to find the sequency spectrum is to calculate the DWT of a signal and then find the sequency periodogram. This is similar to the definition of the periodogram in frequency analysis (Hayes, 1996) as defined by

$$\hat{P}_f[k] = |X[k]|^2 = \Re^2(X[k]) + \Im^2(X[k]), \quad k = 0, 1, 2, \ldots, N-1. \quad (62)$$

Analogous to (62), we can define the sequency periodogram as

$$\check{P}_w[0] = X_c^2[0],$$
$$\check{P}_w[m] = X_c^2[m] + X_s^2[m], \quad m = 1, 2, \ldots, \frac{N}{2} - 1, \tag{63}$$
$$\check{P}_w[N/2] = X_s^2[N/2],$$

where $X_c[m]$ and $X_s[m]$ are the Discrete CAL and SAL Transforms (DCALT and DSALT) given in (37) (Beauchamp, 1975). To avoid using the DCALT and DSALT, we can represent the sequency periodogram in terms of the DWT,

$$\check{P}_w[0] = X_w^2[0],$$
$$\check{P}_w[m] = X_w^2[2m - 1] + X_w^2[2m], \quad m = 1, 2, \ldots, \frac{N}{2} - 1, \tag{64}$$
$$\check{P}_w[N/2] = X_w^2[N - 1],$$

where $X_w[m]$ and $\check{P}_w[m]$ are the DWT and the sequency periodogram of the signal, respectively. The philosophy of arranging the sequency periodogram as a square sum of CAL and SAL functions as it was given in (64) will be further discussed in the following section.

4.3 Determining Sequency Spectrum by Dyadic Autocorrelation

To find the sequency spectrum, one can use Wiener–Khintchine theory on the DWT, i.e., calculate the sequency spectrum from the DWT of the dyadic autocorrelation given in (47) or

$$R_w[n] \xleftrightarrow{\text{DWT}} \hat{P}_w[m], \tag{65}$$

where $\hat{P}_w[m]$ is the sequency power spectrum of the signal. If the signal is deterministic, $\hat{P}_w[m] = X_w^2[m]$ and the sequency periodogram (64) can be directly found. If the signal is wide sense stationary, one can find an estimate of the arithmetic autocorrelation from the known methods (Hayes, 1996). The arithmetic autocorrelation is defined as

$$R_x[n] = \frac{1}{N} \sum_{q=0}^{N-1} x[q] x[\langle q + n \rangle_n], \tag{66}$$

where $\langle . \rangle_n$ is Modulo-n operator. Yuen and Robinson showed how one can find the estimate of the dyadic autocorrelation from the arithmetic autocorrelation (Yuen, 1973; Robinson, 1972). This method is explained in the following paragraphs.

Suppose vectors $\mathbf{r}_d$ and $\mathbf{r}_a$ are $N \times 1$ vectors representing the dyadic and arithmetic autocorrelations of the signal $\mathbf{x}$, respectively. Then the following relationship can be defined between $\mathbf{r}_d$ and $\mathbf{r}_a$

$$\mathbf{r}_d = \mathbf{D}_N \mathbf{T}_N \mathbf{r}_a, \tag{67}$$

where $\mathbf{D}_N$ and $\mathbf{T}_N$ are $N \times N$ matrices. The matrix $\mathbf{D}_N$ is a diagonal matrix and $\mathbf{T}_N$ can be found using an iterative formula:

$$(\mathbf{D})_{i,i} = 2^{-V_i + 1 - \delta(i,0)}, \qquad i = 0, 1, 2, \ldots, N-1,$$

$$\mathbf{T}_N = \begin{bmatrix} \mathbf{T}_{N/2} & \mathbf{0}_{N/2} \\ \mathbf{T}_{N/2}\mathbf{\Delta}_{N/2} & \mathbf{T}_{N/2} \end{bmatrix}, \quad \mathbf{T}_1 = 1, \tag{68}$$

where V_i is the number of bits in the binary representation of i, $\delta(i, 0)$ is the Kronecker delta, and $\mathbf{\Delta}_N$ is an $N \times N$ matrix whose lower anti-diagonal values are all equal to 1. For example, $\mathbf{\Delta}_4$ is

$$\mathbf{\Delta}_4 = \begin{bmatrix} 0 & 0 & 0 & 0 \\ 0 & 0 & 0 & 1 \\ 0 & 0 & 1 & 0 \\ 0 & 1 & 0 & 0 \end{bmatrix}. \tag{69}$$

Ahmed and Natarajan introduced a new method to reduce the number of arithmetic operations (Ahmed & Natarajan, 1974) where instead of N^2 number of additions required in (67), only $N/2 \times (\log_2(N) - 1)$ number of additions is required. It is worth noting that the matrices $\mathbf{\Delta}_N$ and $\mathbf{T}_N$ have significant sparsities and the new techniques in handling sparse matrices can render the aforementioned fast algorithm irreverent.

The major problem with the sequency spectrum is that it is not invariant to the time shift, unlike frequency spectrum. To show this phenomenon, consider Fig. 8 that shows the delayed sine and pulse signals given in Fig. 5. Fig. 9A illustrates the frequency spectra of the shifted (delayed) signals. It can be seen from the figure that the frequency spectra of the shifted signals are exactly the same as the frequency spectra of the original signals shown in Fig. 6A. However, Fig. 9B clearly shows that the sequency spectra of the signals are different than those of the original signals shown in Fig. 6B.

Fig. 9B reveals a very important aspect of the sequency spectrum. As it can be seen in Fig. 9B, the coefficients corresponding to CAL and SAL

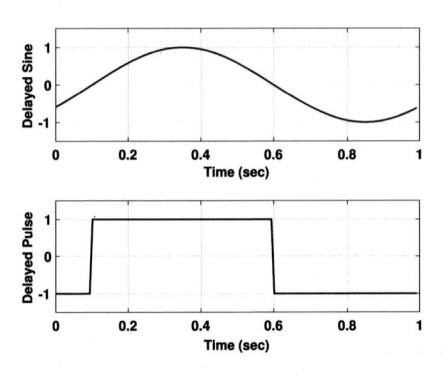

Figure 8 The delayed sine and pulse signals.

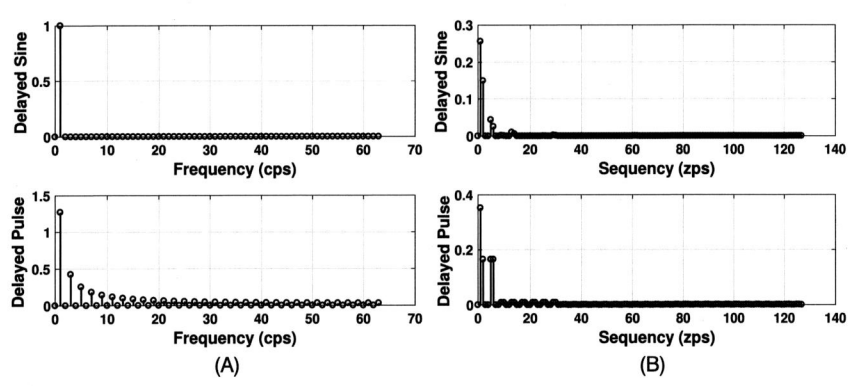

(A)   (B)

Figure 9 (A) The frequency spectra and (B) the sequency spectra of the signals shown in Fig. 8.

functions of the same order in the sequency spectrum have almost similar values. This phenomenon indicates that the components of the sequency spectrum corresponding to the same CAL or SAL orders are redundant. Therefore, one can combine these coefficients and reduce the size of the sequency spectrum by half. This is the justification for defining the sequency periodogram defined in (64).

4.4 Sequency Filtering

As it was mentioned earlier in this section, a function can be both time-limited and sequency-limited. This is in contrast with the Fourier analysis where a function cannot be both time-limited and band-limited. The justification of this behavior is not difficult. The Walsh Transform is defined on a finite interval, i.e., both the functions and their Walsh transforms are finite-length. Therefore, if a function is a linear combination of Walsh functions, its Walsh Transform contains finite functions confined in the $[0, 1]$ (or $[-1/2, 1/2]$) interval. Therefore, ideal sequency-domain filtering (brick-wall filtering) is possible and there is no need to smoothen the edges by tapering, like windowing in the Fourier analysis.

Sequency filtering can be defined in both continuous and discrete domains. However, we will only focus on discrete sequency filtering as the continuous sequency filtering is no longer attractive, due to the dominance of digital computers.

Suppose $\mathbf{x}$ is a discrete signal with the length of N (generally $N = 2^p$) and $\mathbf{X}$ is its transform (Fourier or Walsh), then the filtering process can be defined as

$$\mathbf{Y} = \mathbf{HX}, \tag{70}$$

where $\mathbf{H}$ is an $N \times N$ matrix that represents the filter and $\mathbf{Y}$ is the transform of the output of the filter. Suppose we use the Fourier transform of $\mathbf{x}$, i.e., $\mathbf{X}_f = \mathbf{F}_N\mathbf{x}$, where matrix $\mathbf{F}_N$ is the N-point Discrete Fourier Transform (DFT) matrix, then the output of the filter can be derived as

$$\mathbf{y} = \frac{1}{N}\mathbf{F}_N^*\mathbf{H}_f\mathbf{F}_N\mathbf{x}, \tag{71}$$

because $\mathbf{F}^{-1} = \frac{1}{N}\mathbf{F}^*$, where "*" is the complex conjugate operator. On the other hand, if we use Walsh transform, we can define the filtering process as

$$\mathbf{y} = \frac{1}{N}\mathbf{W}_N\mathbf{H}_w\mathbf{W}_N\mathbf{x}, \tag{72}$$

knowing the fact that $\mathbf{W}_N^{-1} = \frac{1}{N}\mathbf{W}_N$.

The filter matrix $\mathbf{H}$ could be as simple as a diagonal matrix whose diagonal vector is the weights given at each frequency (sequency) or it could be a complete matrix that represents a complex filtering process.

There has been a lot of work on designing frequency-domain filters but a little work has been done on designing sequency-domain filters. Here we discuss several sequency filtering method and provide examples.

4.4.1 Brick-Wall Filtering

In Fourier analysis, brick-wall filtering refers to filters whose frequency responses contain discontinuities. These filters are sometime called ideal

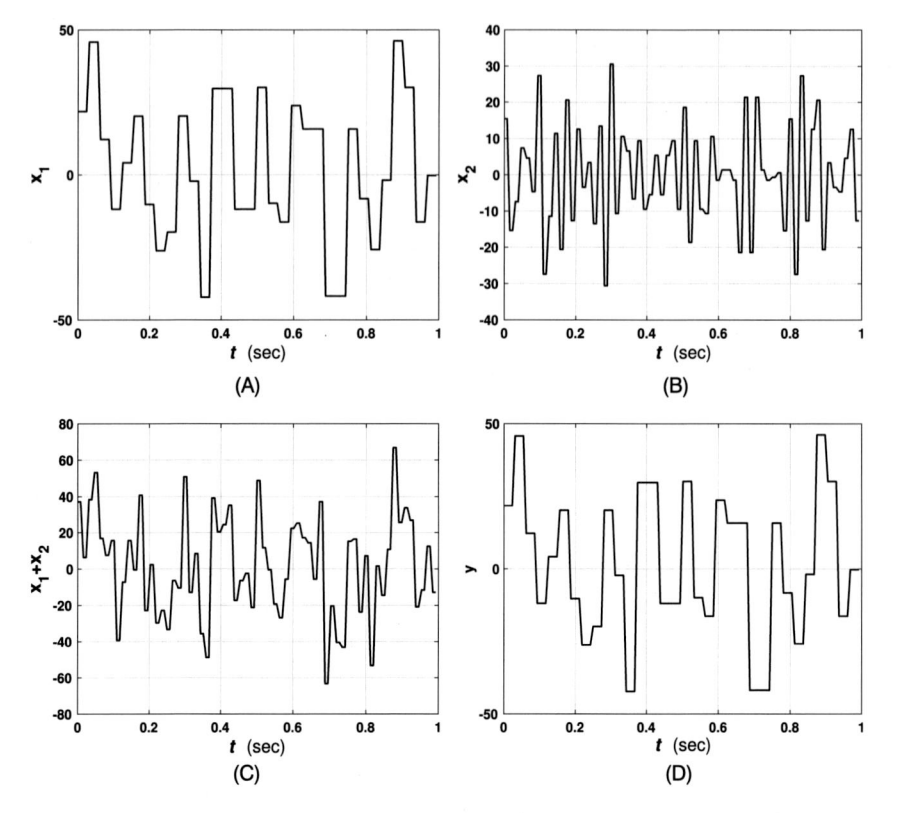

Figure 10 Sequency filtering of a low-pass signal from an additive high-pass signal.
(A) The signal $x_1(t)$. (B) The signal $x_2(t)$. (C) The signal $x_1(t) + x_2(t)$. (D) The output
of the filter $y(t)$.

filters because in the Fourier analysis they cannot be implemented unless
the filters are non-causal (Oppenheim et al., 1983). Unlike Fourier Transform, Walsh Transform is based on discontinuous functions thus, brick-wall
filtering can be easily performed and it has some interesting properties as
well as applications.

To show the effect of brick-wall filtering, we consider two signals $x_1(t)$
and $x_2(t)$ shown in Fig. 10A and Fig. 10B, respectively. The signal $x_1(t)$
contains sequency coefficients up to 32 zpc and $x_2(t)$ contains sequences
above 32 zpc. By applying a brick-wall filter that allows only sequences
below 32 zpc to pass, one can exactly extract $x_1(t)$ (Fig. 10D) from the sum
of the two signals (Fig. 10C).

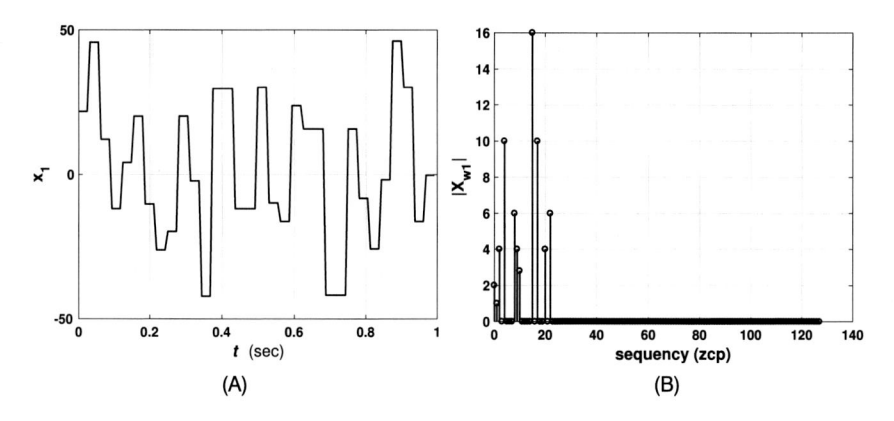

Figure 11 (A) The signal $x_1(t)$ and (B) its magnitude DWT $|X_{w1}|$.

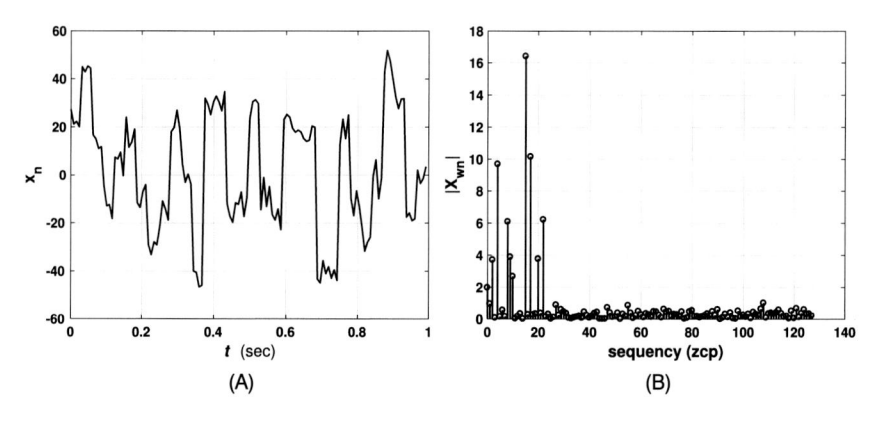

Figure 12 (A) The noisy signal $x_n(t)$ and (B) its magnitude DWT $|X_{wn}|$.

4.4.2 Sequency Masking

Another method of sequency filtering is where the sequency coefficients with magnitudes below a certain number is replaced by zeros. This method is particularly useful in noise reduction. The following example sheds more light on this method.

Suppose $x_n(t)$ is the noisy version of signal $x_1(t)$ (shown in Fig. 10A). The added noise is a white Gaussian noise with the variance of 12 and $X_{wn}[k]$ and $X_{w1}[k]$ are their DWTs, respectively. The signal-to-noise ratio (SNR) of the noisy signal is around 16dB. The signals $x_1(t)$ and $x_n(t)$ and their magnitude DWTs are shown in Fig. 11 and Fig. 12, respectively.

In Fig. 12B, the noise is manifested itself as small values across the entire sequency range and the signal is presented by larger components. If we

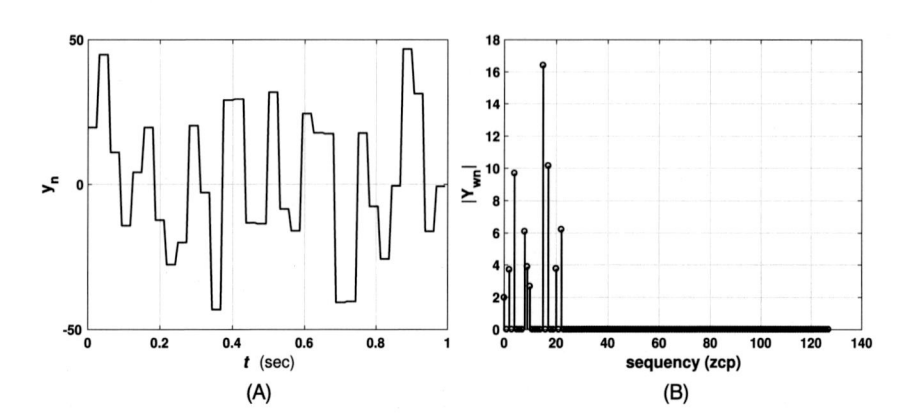

Figure 13 (A) The filtered signal $y_n(t)$ and (B) its magnitude DWT $|Y_w|$.

remove all values below a certain threshold, we can remove the noise. By removing the sequency coefficients below 10% of the maximum magnitude of the sequency spectrum, we can obtain signal $y_n(t)$. Fig. 13 illustrates the filtered signal and its sequency spectrum. It can be seen that the filtered signal is close to the original signal $x_1(t)$. It is worth mentioning that the noise removal process presented here cannot be performed using the DFT of the signal because thresholding and any other abrupt change on the DFT of the signal (such as brick-wall filtering mentioned in Section 4.4.1) will result in the appearance of many oscillatory components in the filtered signal (Gibbs effect) and will render a signal which is not even close to the original one.

4.4.3 Implementing Frequency-Domain Filters Using Sequency Filtering

It has been shown in Section 3.6 that the DWT can be calculated much faster than the DFT due to fact that the Walsh matrix only contains $+1$ and -1 and they represent only additions and subtractions in the DWT algorithm as oppose to complex multiplications in DFT. Given that fact, one may wonder whether the DFT can be calculated by using the Walsh matrix and Discrete Walsh Transform to reduce the cost of the computation and speed up the algorithm. Kahveci and Hall showed in their paper that this is possible (Kahveci & Hall, 1974). We will briefly discuss this in the following paragraphs.

Considering the Fourier and Walsh filtering of **x** as shown in (71) and (72), we want to make the output of both filters the same. Therefore, we should have

$$\mathbf{W}_N \mathbf{H}_w \mathbf{W}_N = \mathbf{F}_N^* \mathbf{H}_f \mathbf{F}_N. \tag{73}$$

Given the fact $\mathbf{F}_N^{-1} = \frac{1}{N}\mathbf{F}_N^*$, we can find $\mathbf{H}_w$ in terms of $\mathbf{H}_f$ as

$$\mathbf{H}_w = \mathbf{B}^H \mathbf{H}_f \mathbf{B}, \tag{74}$$

where $\mathbf{B}_N^H$ is conjugate transpose (Hermitian) of $\mathbf{B}$ and the matrix $\mathbf{B}$ is defined as

$$\mathbf{B} = \frac{1}{N}\mathbf{F}_N \mathbf{W}_N. \tag{75}$$

It is also not difficult to show $\mathbf{B}^{-1} = \mathbf{B}^H$. By carefully designing $\mathbf{H}_f$ using some known digital filter design techniques, we can transform the filtering to the sequency domain by (75) and then apply the fast DWT algorithm to apply the filter to the input signal $\mathbf{x}$. Since the matrix $\mathbf{H}_f$ is usually a diagonal matrix, the matrix $\mathbf{H}_w$ would be sparse matrix and the cost of calculation will be lower than that of the FFT. For example, considering $N = 4$, one can show the matrix $\mathbf{H}_f$ as

$$\mathbf{H}_f = \begin{bmatrix} h_{f0} & 0 & 0 & 0 \\ 0 & h_{f1} & 0 & 1 \\ 0 & 0 & h_{f2} & 0 \\ 0 & 0 & 0 & h_{f1}^* \end{bmatrix}, \tag{76}$$

where $h_{f1} = a + jb$. On the other hand, the matrix $\mathbf{B}$ would be

$$\mathbf{B} = \begin{bmatrix} 1 & 0 & 0 & 0 \\ 0 & \frac{1}{2}(1-j) & \frac{1}{2}(1+j) & 1 \\ 0 & 0 & 0 & 1 \\ 0 & \frac{1}{2}(1+j) & \frac{1}{2}(1-j) & 0 \end{bmatrix}. \tag{77}$$

From (76) and (77), we can obtain

$$\mathbf{H}_w = \begin{bmatrix} h_{f1} & 0 & 0 & 0 \\ 0 & a & -b & 1 \\ 0 & b & a & 0 \\ 0 & 0 & 0 & h_{f2} \end{bmatrix}, \tag{78}$$

according to Robinson (1972). We can see that the matrix $\mathbf{H}_w$ is a sparse matrix thus, the operation of sequency filtering (72) can be performed by

$2N \log_2(N)$ additions/subtractions and $M \ll N$ real multiplications. This method is particularly useful for two-dimensional functions, e.g., images. More information and details can be found in (Zarowski & Yunik, 1985), (Gerheim & Stoughton, 1987), and (Zarowski, Yunik, & Martens, 1988).

4.5 Applications of the DWT in Other Areas

The DWT has numerous applications in many different areas such as Signal and Image Processing, Image Coding (Pratt et al., 1969), Communication Systems (Yarlagadda & Hershey, 2012), Power Systems (Kish & Heydt, 1994), Statistical Analysis (Pearl, 1971), Logic Circuits Design (Ghaleb, 2016), Audio and Speech Processing (Shum, Elliott, & Brown, 1973), Optics (Harwit, 2012), Data Compression (Kekre, Sarode, Thepade, & Shroff, 2011) and Neuroscience (Adjouadi et al., 2004). An enthusiastic reader is referred to the aforementioned references for more information.

5. FREQUENCY DOMAIN WALSH FUNCTIONS

In his seminal paper, Shannon showed that the sinc function (defined as $\phi(t) = \text{sinc}(t) = \frac{\sin(\pi t)}{\pi t}$) is the generating function of the shift invariant Hilbert space of the sampling process (Shannon, 1949). This Hilbert space is defined as

$$\mathcal{H} = \text{span}\{\phi_n(t) = \phi(t - n)\}_{n \in \mathbb{N}}, \tag{79}$$

where $\mathbb{N}$ is the set of all integer numbers. The sinc function is a band-limited function and it is orthogonal to its shifted versions, thus they are the basis for the Hilbert space $\mathcal{H}$ and the projection of an arbitrary band-limited function to this space produces the samples of the function $f(t)$ shown as c_n as follows

$$f(n) = \sum_{n=-\infty}^{+\infty} c_n \phi(t - n)$$

$$c_n = \int_{-\infty}^{+\infty} f(t)\phi(t - n) dt. \tag{80}$$

Unser explained other Hilbert spaces representing other sampling schemes in his paper (Unser, 2000). In (80), the integral that generates the samples of the function $f(t)$ is nothing but filtering the signal with a band-limited (brick-wall) filter bank whose impulse responses are $\phi(t - n)$. Therefore, if the signal $f(t)$ is not a band-limited signal, the process will not be accurate (aliasing).

The Hilbert transform of the sinc function is called the *cosc* (Sukkar, LoCicero, & Picone, 1989) or *cosinc* (Abromson, 1977) function. The cosc function can be easily derived as

$$\mathrm{cosc}(t) = \frac{1 - \cos(\pi t)}{\pi t}. \tag{81}$$

A combination of the sinc and cosc functions is called the *zinc* function (Sukkar et al., 1989) and it is defined as

$$\mathrm{zinc}(t) = A \, \mathrm{sinc}(t) + B \, \mathrm{cosc}(t), \tag{82}$$

where A and B are arbitrary real numbers. It is also shown that $\mathrm{zinc}_n(t) = \mathrm{zinc}(t - n)$ functions are also a complete orthogonal basis for the Paley–Wiener space (space of band-limited functions on L_2) (Sukkar et al., 1989).

Although the sinc and cosc functions are respectively even and odd, their power spectra are the same, which is equal to "one" for the entire bandwidth. Moreover, the Fourier transforms of the sinc and cosc function are, in fact, equal to the first two Walsh functions in the frequency domain i.e., $w_0(\omega)$ and $w_1(\omega)$. Therefore, one can expect that the other Walsh functions in the frequency domain ($w_n(\omega)$) have also corresponding time-domain functions that share similar properties with the sinc, cosc, and zinc functions and can be used to define different time-invariant Paley–Wiener spaces. In this section, we define these functions, which we call them Frequency Domain Walsh Functions or FDWF (Ashrafi, 2014). Later, we will discuss the discrete version of these functions and explain how they can be derived by sampling and orthogonalizing of the FDWFs.

5.1 Definition of the FDWFs

The goal here is to find time-domain functions whose Fourier transforms are the Walsh functions. To achieve this goal, we can make use of the Fourier transform duality principle (Oppenheim et al., 1983), i.e., we can find the Fourier transform of the Walsh functions first and then use the results to find the FDWFs.

The Fourier transform of the Walsh functions was derived by a recursive formula (Schreiber, 1970). It can also be derived from the Walsh transform of sinusoids discussed (Blachman, 1971). However, these methods are complicated and counterintuitive. Siemens and Kitai introduced a non-recursive formula for the Fourier transform of the Walsh functions (Siemens & Kitai, 1973). This formula uses the Gray code of the orders of the Walsh functions to avoid the necessity for the recursive formula. This formula is given as follows

$$\mathcal{F}[w_m(t)] = W_m(\omega) = (-1)^{g_0}(-j)^{\alpha} \left[\prod_{k=0}^{M-1} \cos\left(\frac{\omega}{2^{k+2}} - g_k \frac{\pi}{2}\right) \right] \text{sinc}\left(\frac{\omega}{2^{M+1}}\right),$$

(83)

where $j = \sqrt{-1}$, M is the number of bits representing M, $G = (g_{M-1}g_{M-2}\ldots g_1 g_0)_2$ is the Gray code representation of m, g_k is the kth bit of G, and α is the number of Gray code bits of value ONE in G.

By changing the frequency range of the Walsh functions from $[-\frac{1}{2}, \frac{1}{2}]$ to $[-\omega_c, \omega_c]$ and using the duality principle of the Fourier transform given as

$$x(t) \Leftrightarrow X(\omega),$$
$$X(t) \Leftrightarrow 2\pi x(-\omega),$$

(84)

we can define the FDWFs as

$$\phi_m(t) = \frac{\omega_c(-1)^{g_0}}{\pi} \left[\prod_{k=0}^{M-1} \cos\left(\frac{\omega_c t}{2^{k+1}} - \frac{\pi g_k}{2}\right) \right] \text{sinc}\left(\frac{\omega_c t}{\pi 2^M}\right).$$

(85)

It should be noted that (83) is scaled by $(j)^{\alpha}$ to make the FDWFs real-valued. Therefore, the Fourier transform of the FDWSs are

$$\Phi_m(\omega) = (j)^{\alpha}(-1)^m W_m(\omega).$$

(86)

Fig. 14 illustrates the first four FDWFs.

5.2 Properties of the FDWFs

The FDWFs have very interesting properties. These properties are discussed with their proofs by Ashrafi (2014). We will just list these properties here with some explanations.

Property 12. *The functions $\phi_{2k}(t)$ and $\phi_{2k+1}(t)$ are, respectively, even and odd functions ($k = 0, 1, 2, \ldots$). The function $\phi_0(t)$ is the sinc function and $\phi_1(t)$ is the cosc function. In other words, the zinc function is a member of the FDWF set.*

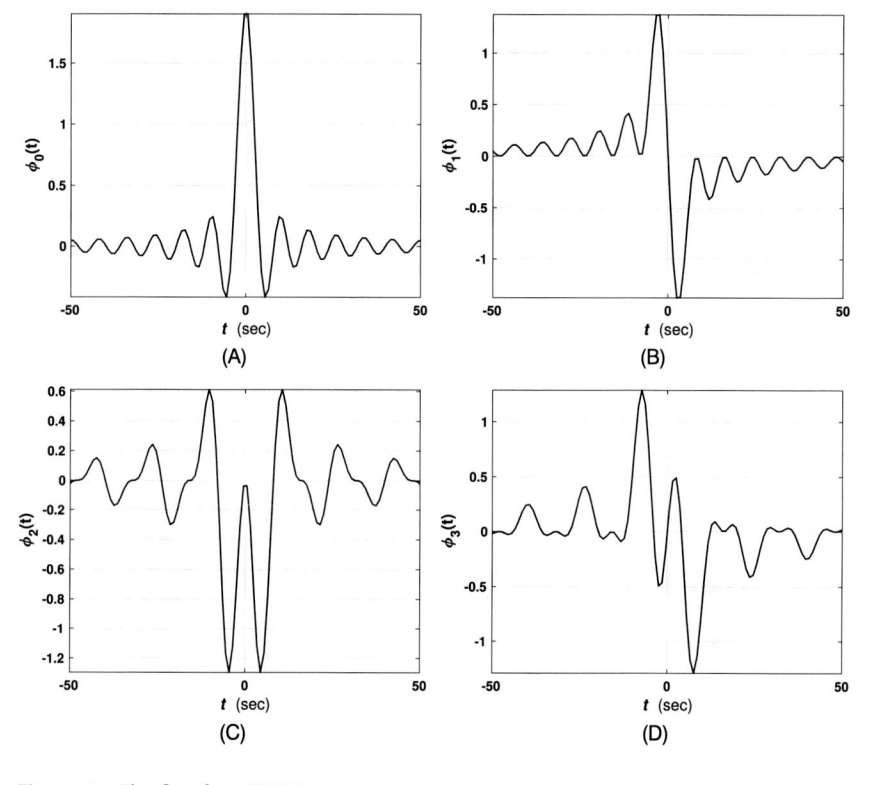

Figure 14 The first four FDWFs.

Property 13. *Only $\phi_0(t)$ is nonzero at the origin. In other words, $\phi_m(0) = 0$, if $m \neq 0$.*

Property 14. *The first derivative of the $\phi_m(t)$ is nonzero only for $m = 2^{M-1}$, where M is the number of bits representing the order m.*

Property 15. *The zeros of $\phi_m(t)$, i.e., the solutions to $\phi_m(t) = 0$ are*

$$t_z = \frac{\pi \ell 2^M}{\omega_c}, \quad \ell \in \mathbb{N}, \tag{87}$$

and

$$t = \frac{(2\ell + 1 + g_k)\pi 2^k}{\omega_c}, \quad \ell \in \mathbb{N}, \quad k = 0, 1, 2, \ldots, M - 1. \tag{88}$$

It should be noted that if $g_{M-1} = 1$, the zeros given by (88) coincide with the zeros given by (87).

Property 16. *The FDWFs are mutually orthogonal, i.e.*

$$\int_{-\infty}^{+\infty} \phi_m(t)\phi_k^*(t)\,dt = \int_{-\omega_c}^{+\omega_c} \Phi_m(\omega)\Phi_k^*(\omega)\,d\omega = 2\omega_c\delta_{mk}. \tag{89}$$

Property 17. *The functions $\phi_k(t)$ and $\phi_m(t-n\tau)$ are orthogonal if $n\tau$ coincides with the zeros of $\phi_r(t)$ where r is the bitwise Modulo-2 of k and m, i.e., $r = k \oplus m$.*

This property indicates that the FDWFs are not only mutually orthogonal, their shifted versions are also mutually orthogonal under certain conditions. This may have some applications in defining unusual shift invariant Hilbert spaces for special sampling schemes.

Property 18. *Even-ordered FDWFs or $\phi_{2m}(t)$, where $m = 0, 1, 2, \ldots$, are the generating functions of the shift-invariant Hilbert spaces defined by $\mathrm{span}\{\phi_n(t) = \phi(t-nT)\}_{n\in\mathbb{N}}$, where $T = \pi/\omega_c$.*

This property indicates that the functions $\phi_{2m}(t)$ can be the generating functions of infinite number of shift-invariant Hilbert spaces and each of these spaces can be a separate sampling space. A special case of this property is the Hilbert space for the generating function $\phi_0(t) = \mathrm{sinc}(t)$, which is the Shannon's sampling space.

Property 19. *The global extrema of the FDWFs occur at*

$$t_e = \pm\frac{0.50475\pi k}{\omega_c}, \tag{90}$$

where k is the order of the FDWF.

The global extrema of the FDWFs are the points where the absolute value of the function is maximum. This property indicates that the extrema of the FDWFs are moving away from the origin as the order of the function increases. This phenomenon can be observed in Fig. 14. It should be noted that the proof of this property has not been found, yet.

5.3 Frequency Domain Walsh Sequences

Sampling and truncation of the FDWFs can produce the discrete version of the function. However, the obtained discrete signals do not posses the same properties as the continuous versions do, because of the effect of the truncation. One of these lost properties is the mutual orthogonality of the sequences. Different orthogonalization methods such as Gram–Schmidt orthogonalization algorithm (Meyer, 2000) can be used to restore the orthonormal property of the discrete functions. Nevertheless, the Fourier transform of the restored sequences will not posses the Walsh like behavior. To restore the Walsh like behavior of the sequences' spectra, we need to rotate the subspace spanned by the orthogonalized sequences toward the original sampled sequences. We will explain the process in the following paragraphs.

Suppose the columns of the $N \times N$ matrix $\mathbf{P}$ are the samples of the FDWFs shown in (85). We can make the columns of $\mathbf{P}$ orthonormal using the following method

$$\mathbf{Z} = \mathbf{P}(\mathbf{P}^T\mathbf{P})^{-\frac{1}{2}}. \tag{91}$$

In other words, the matrix $\mathbf{Z}$ is a unitary matrix. To restore the behavior of the columns of $\mathbf{Z}$ to those of the columns of $\mathbf{P}$, we need to rotate the subspace spanned by the columns of $\mathbf{Z}$ toward the subspace spanned by the column of $\mathbf{P}$ such that the resultant matrix is also unitary. This can be achieved by orthogonal Procrustes algorithm (Golub & Van Loan, 1996). The rotating $N \times N$ matrix $\mathbf{X}$ is the solution to the following optimization problem

$$\begin{aligned} \text{minimize} \quad & ||\mathbf{P} - \mathbf{Z}\mathbf{X}||_F, \\ \text{subject to} \quad & \mathbf{X}^T\mathbf{X} = \mathbf{I}, \end{aligned} \tag{92}$$

where $||.||_F$ is the Frobenius norm and $\mathbf{I}$ is the $N \times N$ identity matrix. The rotating matrix $\mathbf{X}$ is also a unitary matrix according to the constraint of (92). The solution of (92) can be found by using the method of Lagrange multipliers method as

$$\mathbf{X} = \mathbf{U}\mathbf{V}^T, \tag{93}$$

where $\mathbf{U}$ and $\mathbf{V}$ are found by taking the singular value decomposition of $\mathbf{Z}^T\mathbf{P}$ such that $\mathbf{Z}^T\mathbf{P} = \mathbf{U}\Sigma\mathbf{V}$ and Σ is a diagonal matrix. The FDWSs are the columns of the matrix $\mathbf{E} = \mathbf{Z}\mathbf{X}$, which are the bases of the rotated subspace. Since both $\mathbf{X}$ and $\mathbf{Z}$ are unitary matrices, their product is also a unitary matrix, hence the columns are orthonormal sequences. Since the matrix $\mathbf{E}$ is unitary and the closest matrix to $\mathbf{P}$ in the Frobenius norm

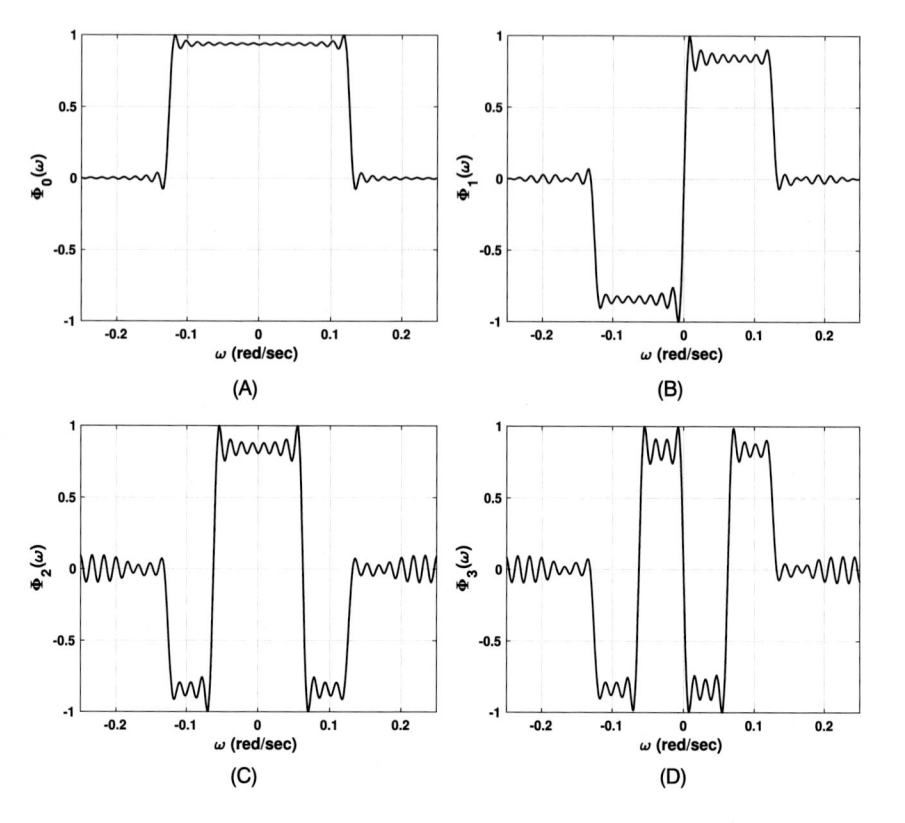

Figure 15 The Fourier transform of the first four FDWSs.

sense; therefore, the columns of **E** are closest orthonormal sequences to the truncated versions of the FDWFs.

It has been shown that the dimension of a vector space is the same as the length of the vectors when the normalized bandwidth ($\Omega = \omega_c/\omega_s$) of the basis vectors is $\frac{1}{2}$ (Slepian, 1978). Obviously this will never happen (ideal case). The dimension of a vector space with basis vectors having a normalized bandwidth of Ω is $r = 2N\Omega$. Therefore, only the first r columns of **C** are valid solutions of the problem and they are the sequences we are looking for. These sequences are called Frequency Domain Walsh Sequences (FDWS). In Fig. 15, the Fourier transform of the first four FDWSs are shown. The resultant sequences do not posses the Fourier transforms exactly the same as the Walsh functions due to the truncation of the sampled signals and the Fourier transform uncertainty principle.

6. COMPLEX WALSH–HADAMARD TRANSFORMS

The DWT (or Walsh–Hadamard Transform) is a real-valued transform, which is suitable for real-valued signals. If the signal being processed is complex-valued, we need to use a complex transform. In this section, we will review the unitary transform directly or indirectly derived or inspired by the DWT.

6.1 Complex Hadamard Transforms

Ahmed and Rao were the first who introduced the Complex Hadamard Transform (CHT) (or as they called it Complex BIFORE Transform (CBT)) (Rao & Ahmed, 1971). The CHT and its inverse are defined as

$$\mathbf{X}_c = \frac{1}{N}\mathbf{M}_{2^n}\mathbf{x}, \tag{94}$$

$$\mathbf{x} = \mathbf{M}_{2^n}^H\mathbf{X}_c, \tag{95}$$

where $\mathbf{M}_{2^n}$ is the $2^n \times 2^n$ CHT matrix, which can be obtained using the following recursive formula

$$\mathbf{M}_{2^n} = \begin{bmatrix} \mathbf{M}_{2^{n-1}} & \mathbf{L} \otimes \mathbf{H}_{2^{n-2}} \\ \mathbf{M}_{2^{n-1}} & -\mathbf{L} \otimes \mathbf{H}_{2^{n-2}} \end{bmatrix}, \tag{96}$$

where $\mathbf{H}$ is the Hadamard matrix (18) and $\mathbf{L}$ is

$$\mathbf{L} = \begin{bmatrix} 1 & 1 \\ -j & j \end{bmatrix}. \tag{97}$$

Ahmed et al. introduced a matrix factorization for the matrix $\mathbf{M}$ and developed a fast algorithm for the CHT (Ahmed, Rao, & Schultz, 1971). Rao and Ahmed introduced the Modified CHT (MCHT) where the matrix is sparsest than the CDWT matrix, which leads to a faster algorithm (Rao & Ahmed, 1972). The matrix of the MCHT ($\mathbf{M}'$) is defined by the following recursive formula

$$\mathbf{M}'_{2^n} = \begin{bmatrix} \mathbf{M}'_{2^{n-1}} & \mathbf{M}'_{2^{n-1}} \\ 2^{\frac{n-2}{2}}\mathbf{L}' \otimes \mathbf{H}_{2^{n-2}} & -2^{\frac{n-2}{2}}\mathbf{L}' \otimes \mathbf{H}_{2^{n-2}} \end{bmatrix}, \tag{98}$$

where $\mathbf{L}'$ is defined as

$$\mathbf{L}' = \begin{bmatrix} 1 & -j \\ 1 & j \end{bmatrix}. \tag{99}$$

The fast algorithm and the signal flow graph of this algorithm are also given in Rao and Ahmed (1972).

6.2 Complex Haar Transform

Rao et al. introduced the Complex Haar Transform (CHAT) (Rao, Revuluri, Narasimhan, & Ahmed, 1976). The matrix of the CHAT ($\mathbf{H}_c(n)$) is a $2^n \times 2^n$ matrix and it was defined as

$$\mathbf{H}_c(n) = \begin{bmatrix} \mathbf{H}_c(n-1) \otimes \begin{bmatrix} 1 & 1 \end{bmatrix} \\ 2^{(n-2)/2} \begin{bmatrix} 1 & -j \\ 1 & j \end{bmatrix} \otimes \mathbf{I}_{2^{(n-2)}} \otimes \begin{bmatrix} 1 & -1 \end{bmatrix} \end{bmatrix}. \tag{100}$$

For example, $\mathbf{H}_c(2)$ is

$$\mathbf{H}_c(2) = \begin{bmatrix} 1 & 1 & 1 & 1 \\ 1 & 1 & -1 & -1 \\ 1 & -1 & -j & j \\ 1 & -1 & j & -j \end{bmatrix}. \tag{101}$$

The matrix factorization of the Complex Haar matrix and its fast transform signal flow graph are also given in Rao et al. (1976) only for up to 8×8 matrices. The CHT, MCHT, and the CHAT were the motivation to develop a unified DWT, which will be discussed in the next section.

6.3 Unified Complex Hadamard Transforms

Rahardja and Kalkowski introduced a unified complex Hadamard transform (UCHT) (Rahardja & Falkowski, 1999). These transforms encompass all possible complex Walsh–Hadamard transforms by considering different 2×2 seed matrices to generate any $2^n \times 2^n$ transform matrix.

Suppose $\mathbf{C}_0 = 1$ and the seed matrix $\mathbf{C}_1$ is given as

$$\mathbf{C}_1 = \begin{bmatrix} c_{00} & c_{01} \\ c_{10} & c_{11} \end{bmatrix}, \tag{102}$$

where $\mathbf{c}_{i,j} \in \{+1, -1, +j, -j\}$. Then the UCHT matrix of order n can be obtained from the following recursive equation

$$\mathbf{C}_n = \mathbf{C}_1 \otimes \mathbf{C}_{n-1}. \tag{103}$$

Obviously, the entries of the matrix $\mathbf{C}_n$ are also from the set $\{+1, -1, +j, -j\}$. A special case of UCHT is the Hadamard transform (18) when $\mathbf{C}_1 = \mathbf{H}_2$. Rahardja and Kalkowski discussed the properties of the UCHT and the associated matrices (Rahardja & Falkowski, 1999, 2000). We will briefly talk about some of these properties without their proofs. For detail analysis and proofs of these properties please refer to Rahardja and Falkowski (1999, 2000).

Property 20. *The matrix $\mathbf{C}_n$ is orthogonal matrix. In other words,*

$$\mathbf{C}_n \mathbf{C}_n^H = \mathbf{C}_n^H \mathbf{C}_n = N\mathbf{I}_{2^n}, \tag{104}$$

where $\mathbf{C}_n^H$ is the Hermitian (conjugate transposed) of $\mathbf{C}_n$.

Property 21. *The absolute value of the determinant of the UCHT matrix of order n is $|\det \mathbf{C}_n| = N^{(1/2)N}$, where $N = 2^n$.*

Property 22. *There are $4^3 = 64$ different UCHTs that can be defined from* (102).

Property 23. *The seed matrix $\mathbf{C}_1$ can be defined as*

$$\mathbf{C}_1 = \mathbf{A}_L \mathbf{H}_2, \tag{105}$$

where

$$\mathbf{A}_L = \frac{1}{2} \begin{bmatrix} \tau_1(1 + \tau_3) & \tau_1(1 - \tau_3) \\ \tau_2(1 - \tau_3) & \tau_2(1 + \tau_3) \end{bmatrix}, \quad \tau_1, \tau_2, \tau_3 \in \{+1, -1, +j, -j\}, \tag{106}$$

and $\mathbf{H}_2$ is the lowest order Hadamard matrix (17).

Property 24. *If τ_3 is either $+j$ or $-j$, the UCHT has the half-spectrum property (HSP), i.e., half of the spectrum of every signal transformed by the UCHT can be calculated by the other half. In this case the (k, m) entry of the matrix $\mathbf{C}_n$, i.e., $c_n(k, m)$ satisfies the following formula*

$$c_n(N - 1 - k, m) = (\tau_1 \tau_2)^{(n \bmod 4)} c_n^*(k, m). \tag{107}$$

Therefore the UCHT of the signal $\mathbf{x}$, *i.e.,* $\mathbf{X}_u = \mathbf{C}_n\mathbf{x}$ *satisfies the following property*

$$\mathbf{X}_u(N - 1 - k) = (\tau_1\tau_2)^{(n \bmod 4)}\mathbf{X}_u^*(k). \tag{108}$$

This property is similar to the conjugate symmetric property of the DFT.

6.4 Sequency-Ordered Complex Hadamard Transform

The CHAT, CHT, MCHT, and UCHT matrices are not arranged by sequency order. In fact, sequency ordered matrix never discussed in those transforms. Since sequency-ordered transforms are very important in cases where we want to compare the transformed signals with their frequency spectra obtained by the DFT, we need to define complex Walsh–Hadamard transforms where the row (columns) of the transform matrices are arranged in ascending sequency order. Aung et al. introduced this transform and called it sequency-ordered complex Hadamard transform (SCHT) (Aung et al., 2008).

To define the SCHT, we need to define complex Rademacher functions and matrices. This can be achieved by extending the definition of the real-valued Rademacher functions (9) to define the complex Rademacher functions as

$$\mathrm{CRAD}(0, t) = \begin{cases} 1, & t \in [0, \frac{1}{4}) \\ j, & t \in [0\frac{1}{4}, \frac{1}{2}) \\ -1, & t \in [0\frac{1}{2}, \frac{3}{4}) \\ -j, & t \in [0\frac{3}{4}, 1) \end{cases} \tag{109}$$

$$\mathrm{CRAD}(0, t + 1) = \mathrm{CRAD}(0, t).$$

Moreover, the complex Rademacher function of order r can be represented as the scaled version of the complex Rademacher function of order zero or

$$\mathrm{CRAD}(r, t) = \mathrm{CRAD}(0, 2^r t). \tag{110}$$

The complex Rademacher matrices can be defined by sampling the complex Rademacher functions. The complex Rademacher matrix of order n can be defined as the following $n \times 2^n$ matrix

$$
\mathbf{R}_n(r, k) = \mathrm{CRAD}\left(r, \frac{4k+1}{2^{n+2}}\right),
$$
$$
r = 0, 1, 2, \ldots, n - 1,
$$
$$
k = 0, 1, 2, \ldots, 2^n - 1,
$$
(111)

where $\mathbf{R}_n(r, k)$ is the (r, k) entry of $\mathbf{R}_n$ (Aung et al., 2008). The rows of the complex Rademacher matrix are mutually orthogonal, i.e.,

$$
\mathbf{R}_n \mathbf{R}_n^H = N\mathbf{I}.
$$
(112)

Now we can define the SCHT matrix $\mathbf{H}_N$ from the complex Rademacher matrix.

$$
\mathbf{S}_N(m, k) = \prod_{r=0}^{n-1} \mathbf{R}_n^{m_r}(r, k), \quad m = 0, 1, 2, \ldots, N - 1,
$$
$$
k = 0, 1, 2, \ldots, N - 1,
$$
(113)

where $N = 2^n$, $(m)_2 = (\overline{m_{n-1}, m_{n-2}, \ldots, m_0})_2$, and m_r is the rth bit of the binary representation of m. Using the SCHT matrix, we can define the SCHT and inverse SCHT (ISCHT) of a signal $\mathbf{x}$ as

$$
\mathbf{X}_s = \frac{1}{N}\mathbf{S}_N^H \mathbf{x},
$$
$$
\mathbf{x} = \mathbf{S}_N \mathbf{X}_s.
$$
(114)

The SCHT is not a shift–invariant transform. In other words, the SCHT power spectrum of a signal and its circularly shifted version are not identical (unlike the DFT). The SCHT has inherited this feature from its older brother the DWT.

A fast implementation algorithm of the SCHT is also presented by Aung et al. (2008). The cost of the computation for SCHT is $N\log_2(N)$ number of additions/subtractions and $N/4\log_2(N/2)$ complex multiplications when $N = 2^n$. We will briefly discuss a few properties of the SCHT matrix in the following paragraphs. The detailed proofs of theses properties are given in (Aung et al., 2008).

Property 25. *The SCHT matrix is a symmetric and orthogonal matrix i.e.,*

$$
\mathbf{S}_N = \mathbf{S}_N^T,
$$
$$
\mathbf{S}_N \mathbf{S}_N^H = \mathbf{S}_N^H \mathbf{S}_N = N\mathbf{I}_N.
$$
(115)

Property 26. *The SCHT matrix can also be found using the UCHT matrix as*

$$\mathbf{S}_N = \mathbf{H}_2 \otimes \mathbf{I}_{2^n-1} \left[\prod_{r=0}^{n-2} \left(\mathbf{I}_{2^r} \otimes \begin{bmatrix} \mathbf{H}_2 & \mathbf{0} \\ \mathbf{0} & \mathbf{C}_1 \end{bmatrix} \otimes \mathbf{I}_{2^{n-r-2}} \right) \right] \mathbf{P}_b, \qquad (116)$$

where $\mathbf{P}_b$, $\mathbf{H}_2$, and $\mathbf{C}_1$ are, respectively, the bit-reversal permutation matrix, the Hadamard matrix of second order (17), and the UCHT seed matrix (102) defined as

$$\mathbf{C}_1 = \begin{bmatrix} 1 & 1 \\ j & -j \end{bmatrix}. \qquad (117)$$

Property 27. *The rows of the SCHT matrix are arranged in an ascending order of the sequency. The sequency is defined as the transition from a real number (+1 or −1) to an imaginary number (+j or −j). In other words, zero crossing means crossing the imaginary axis when rotating around the unit circle.*

6.5 Conjugate-Symmetric Sequency-Ordered Complex Hadamard Transform

The SCHT matrix discussed in the previous section is arranged to have an ascending sequency order. Nevertheless, the rows do not possess the conjugate symmetry property like that of the DFT matrix and as a result, the obtained sequency spectrum of any transformed signal will not have the half spectrum property (HSP). To remedy this deficiency, Aung et al. introduced a new SCHT transform that has this property (Aung et al., 2009). The name of this new transform is Conjugate Symmetric Sequency-Ordered Complex Hadamard Transform (CS-SCHT).

To define the CS-SCHT, we need to first define Conjugate Symmetric Natural-Ordered Complex Hadamard Transform (CS-NCHT). The CS-SCHT will be the bit reversal of CS-NCHT. The CS-NCHT matrix can be found by recursive formulas explained here. Assume $N = 2^n$ and $\mathbf{\Psi}_N$ is the CS-NCHT matrix that can be calculated by this recursive formula

$$\mathbf{\Psi}_N = \begin{bmatrix} \mathbf{\Psi}_{N/2} & \mathbf{\Psi}_{N/2} \\ \mathbf{\Psi}'_{N/2}\mathbf{J}_{N/2} & -\mathbf{\Psi}'_{N/2}\mathbf{J}_{N/2} \end{bmatrix}, \qquad (118)$$

where $\mathbf{J}_{N/2} = \mathbf{J}_{2^n-1}$ is defined as

$$\mathbf{J}_{2^n-1} = \begin{bmatrix} \mathbf{I}_{2^{n-2}} & \mathbf{0} \\ \mathbf{0} & j\mathbf{I}_{2^{n-2}} \end{bmatrix}, \qquad (119)$$

and $\boldsymbol{\Psi}'_{N/2}$ is a Hadamard matrix with a specific row arrangement defined as

$$\boldsymbol{\Psi}'_{N/2} = \begin{bmatrix} \boldsymbol{\Psi}'_{N/4} & \boldsymbol{\Psi}'_{N/4} \\ \boldsymbol{\Psi}'_{N/4}\mathbf{I}'_{N/4} & -\boldsymbol{\Psi}'_{N/4}\mathbf{I}'_{N/4} \end{bmatrix}, \qquad (120)$$

where $\mathbf{I}'_{N/4}$ is also defined as

$$\mathbf{I}'_{N/4} = \begin{bmatrix} \mathbf{I}_{N/8} & \mathbf{0} \\ \mathbf{0} & -\mathbf{I}_{N/8} \end{bmatrix}. \qquad (121)$$

In the above formulas, we have $\boldsymbol{\Psi}_2 = \boldsymbol{\Psi}'_2 = \mathbf{H}_2$, where $\mathbf{H}_2$ is the Hadamard matrix defined in (17). According to the above formulas, the lowest order CS-NCHT matrix is a 4×4 matrix and it is

$$\boldsymbol{\Psi}_4 = \begin{bmatrix} 1 & 1 & 1 & 1 \\ 1 & -1 & 1 & -1 \\ 1 & j & -1 & -j \\ 1 & -j & -1 & j \end{bmatrix}. \qquad (122)$$

The CS-SCHT matrix can be easily found by rearranging the rows according to the bit-reversed row numbers of the CS-NCHT matrix, i.e.,

$$\boldsymbol{\Gamma}_N(p, k) = \boldsymbol{\Psi}_N(b(p), k), \qquad (123)$$

where $b(p)$ is the bit-reversed of p. For example, the bit reversed $6 = \overline{(110)}_2$ is $3 = \overline{(011)}_2$.

The CS-SCHT and its inverse (ICS-SCHT) can be defined as

$$\mathbf{X}_c = \frac{1}{N}\boldsymbol{\Gamma}_N^*\mathbf{x}, \qquad (124)$$
$$\mathbf{x} = \boldsymbol{\Gamma}_N^T\mathbf{X}_c.$$

We will briefly discuss some of the properties of the CS-SCHT in the following paragraphs. The details and proofs of these properties are given by Aung et al. (2009).

Property 28. *The determinant of the CS-SCHT matrix* $\boldsymbol{\Gamma}_N$ *is*

$$|\det(\boldsymbol{\Gamma}_N)| = N^{N/2}. \qquad (125)$$

Property 29. *The CS-SCHT is an orthogonal transform and the CS-SCHT matrix is orthogonal, i.e.,*

$$\boldsymbol{\Gamma}_N\boldsymbol{\Gamma}_N^H = \boldsymbol{\Gamma}_N^H\boldsymbol{\Gamma}_N = N\mathbf{I}_N. \qquad (126)$$

Property 30. *The Parseval's theorem is valid for the CS-SCHT,*

$$\frac{1}{N} \sum_{n=0}^{N-1} |x[n]|^2 = \sum_{k=0}^{N-1} |X_c[n]|^2, \tag{127}$$

where $x[n]$ is the signal and $X_c[k]$ is its CS-SCHT transform.

Property 31. *The CS-SCHT has the conjugate symmetric property, i.e.,*

$$X_c[k] = X_c[N - k], \tag{128}$$

where $X_c[k]$ is the CS-SCHT of $x[n]$.

Property 32. *The CS-SCHT power spectrum is dyadic shift-invariant. In other words,*

$$\boldsymbol{\Gamma}_N(p, k \oplus m) = \begin{cases} \boldsymbol{\Gamma}_N(p, m)\boldsymbol{\Gamma}_N^*(p, m), & \text{if both entries are imaginary} \\ \boldsymbol{\Gamma}_N(p, m)\boldsymbol{\Gamma}_N(p, m), & \text{otherwise} \end{cases} \tag{129}$$

The CS-NCHT matrix can be factorized into several sparse matrices and then the factorized matrices can be used to derive fast CS-NCHT algorithm (Aung et al., 2009). The fast CS-SCHT can be easily derived by bit-reversing the CS-NSCHT output indices.

Consider 8-point CS-NCHT matrix defined in (118),

$$\boldsymbol{\Psi}_8 = \begin{bmatrix} \boldsymbol{\Psi}_4 & \boldsymbol{\Psi}_4 \\ \boldsymbol{\Psi}_4'\mathbf{J}_4 & -\boldsymbol{\Psi}_4'\mathbf{J}_4 \end{bmatrix}, \tag{130}$$

where

$$\mathbf{J}_4 = \begin{bmatrix} \mathbf{I}_2 & \mathbf{0} \\ \mathbf{0} & j\mathbf{I}_2 \end{bmatrix} \quad \text{and} \quad \boldsymbol{\Psi}_4' = \begin{bmatrix} \boldsymbol{\Psi}_2 & \boldsymbol{\Psi}_2 \\ \boldsymbol{\Psi}_2\mathbf{I}_2' & -\boldsymbol{\Psi}_2\mathbf{I}_2' \end{bmatrix}. \tag{131}$$

By factoring (130) to

$$\boldsymbol{\Psi}_8 = \begin{bmatrix} \boldsymbol{\Psi}_4 & \mathbf{0} \\ \mathbf{0} & \boldsymbol{\Psi}_4 \end{bmatrix} \begin{bmatrix} \mathbf{I}_4 & \mathbf{0} \\ \mathbf{0} & \mathbf{J}_4 \end{bmatrix} \begin{bmatrix} \mathbf{I}_4 & \mathbf{I}_4 \\ \mathbf{I}_4 & -\mathbf{I}_4 \end{bmatrix}, \tag{132}$$

and continuing the factorization until reaching 2×2 sub-matrices, we can obtain

$$
\boldsymbol{\Psi}_8 =
\begin{bmatrix}
\boldsymbol{\Psi}_2 & & & \\
& \boldsymbol{\Psi}_2 & & \\
& & \boldsymbol{\Psi}_2 & \\
& & & \boldsymbol{\Psi}_2
\end{bmatrix}
\begin{bmatrix}
\mathbf{I}_2 & & & \\
& \mathbf{J}_2 & & \\
& & \mathbf{I}_2 & \\
& & & \mathbf{I}'_2
\end{bmatrix}
\cdot
$$

$$
\begin{bmatrix}
\mathbf{I}_2 & \mathbf{I}_2 & & \\
\mathbf{I}_2 & -\mathbf{I}_2 & & \\
& & \mathbf{I}_2 & \mathbf{I}_2 \\
& & \mathbf{I}_2 & -\mathbf{I}_2
\end{bmatrix}
\begin{bmatrix}
\mathbf{I}_4 & \\
& \mathbf{J}_4
\end{bmatrix}
\begin{bmatrix}
\mathbf{I}_4 & \mathbf{I}_4 \\
\mathbf{I}_4 & -\mathbf{I}_4
\end{bmatrix}
\tag{133}
$$

Bouguezel et al. introduced the closed-form formula for the factorization of the CS-SCHT matrix (Bouguezel, Ahmad, & Swamy, 2011). This formula is given as

$$
\boldsymbol{\Gamma}_N = \left(\prod_{i=0}^{N-2} (\mathbf{I}_{N-i} \otimes \mathbf{H}_2 \otimes \mathbf{I}_{i-1}) \times \left(\begin{bmatrix} \begin{bmatrix} \mathbf{I}_1 & \mathbf{0} \\ \mathbf{0} & \mathbf{J}_1 \end{bmatrix} & \mathbf{0} \\ \mathbf{0} & \hat{\mathbf{I}}_i \otimes \begin{bmatrix} \mathbf{I}_1 & \mathbf{0} \\ \mathbf{0} & \mathbf{I}'_1 \end{bmatrix} \end{bmatrix} \otimes \mathbf{I}_{i-1} \right) \right)
$$

$$
\times (\mathbf{I}_1 \otimes \mathbf{H}_2 \otimes \mathbf{I}_{N-2}) \left(\begin{bmatrix} \mathbf{I}_1 & \mathbf{0} \\ \mathbf{0} & \mathbf{J}_1 \end{bmatrix} \otimes \mathbf{I}_{N-2} \right) \times (\mathbf{H}_2 \otimes \mathbf{I}_{N-1}),
\tag{134}
$$

where $\hat{\mathbf{I}}_i$ denotes the identity matrix of order $2^{N-i-1} - 1$. To obtain the CS-SCHT, we should replace j with $-j$ in (134) and bit-reverse the order of indices of the output $X[k]$. The computation cost of the CS-SCHT is $N \log_2(N)$ complex additions/subtractions and $N/2 - 1$ number of complex multiplications, which is smaller that the number of complex multiplications of the SCHT ($N/4 \log_2(N/2)$) and much smaller than the complex multiplications of the DFT ($N \log_2(N)$). Fig. 16 illustrates the signal flow graph of the 8-point CS-SCHT. A general factorization scheme is also introduced by Kyochi and Tanaka (2014).

The CS-SCHT spectrum is the closest one among Walsh–Hadamard transforms to the DFT. To show that, the CS-SCHT and DFT spectra of pulse and sinusoidal signals as well as delayed version of them with frequency of 14 Hz are shown in Fig. 17.

Since the CS-SCHT and DFT are both conjugate symmetric, only half of the spectra is shown in Fig. 17. It can be seen from the figure that

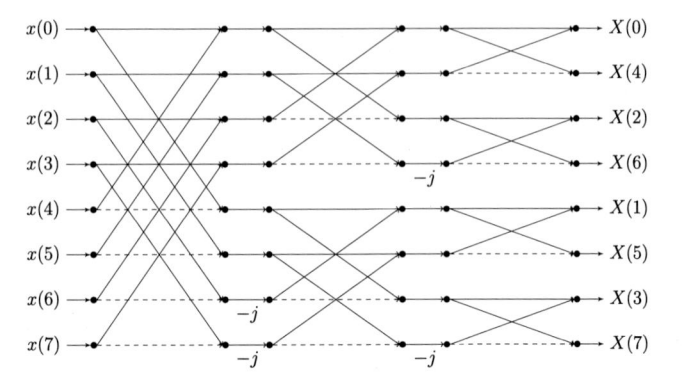

Figure 16 Signal flow graph of the 8-point CS-SCHT (the dashed lines represent multiplication by −1).

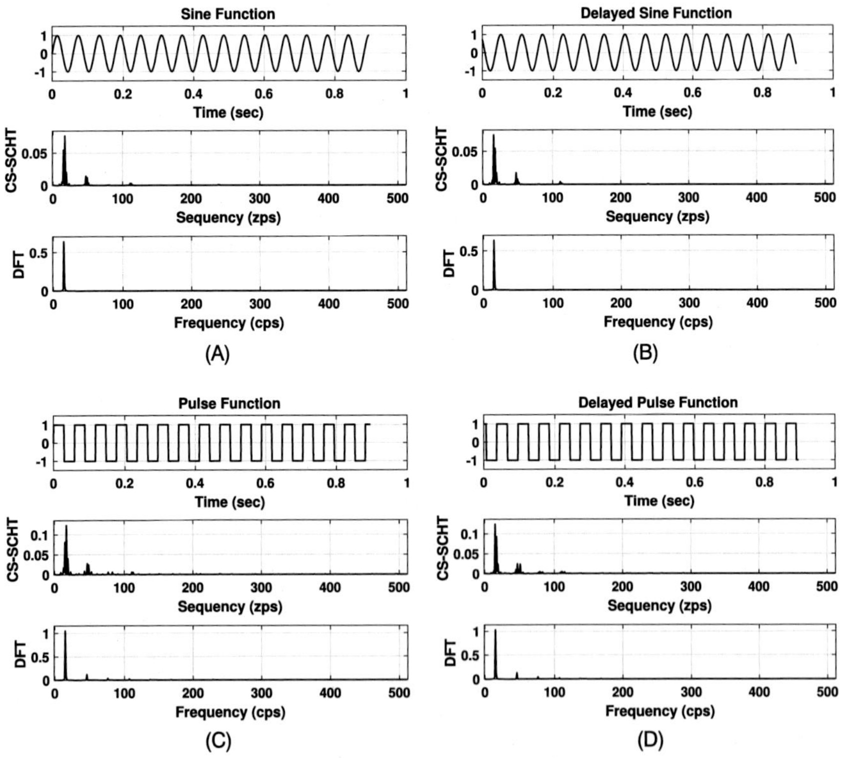

Figure 17 The CS-SCHT and DFT of sinusoid and pulse signals and their delayed versions.

the spectra of the signals are all concentrated around the frequency and sequency of the signals (14 Hz and 14 zps). Although the CS-SCHT of the signals and their delayed versions are not exactly the same, their spectra cover the same range and one can say the envelopes of the spectra are very similar. This is a very important feature and it shows that CS-SCHT can replace the DFT in applications where the speed of calculations is more important than the accuracy of the obtained spectra such as neural signal processing. A sliding CS-SCHT method is introduced in Wu et al. (2012), which is suitable for nonstationary signals.

6.6 Some of the Applications of the Complex Walsh–Hadamard Transforms

The complex nature of the Walsh–Hadamard Transoms make them serious contenders to the DFT. They have been used in different applications. The UCHTs are used to study and analyze digital logic (Boolean) functions (Rahardja & Falkowski, 2000). The SCHT is used for image processing, particularly in Watermarking (Aung et al., 2008). The CS-SCHT has shown promising results in image compression (Aung et al., 2009). We should expect to see more applications of these transforms in future.

7. CONCLUSIONS

Walsh–Hadamard transforms have been applied in different areas including Signal and Image Processing, Pattern Recognition, Data Compression, Information Theory, Filtering, Communication Systems, Audio and Speech Processing, Digital Logic Design, Image Watermarking, and many more. A simple Google Scholar search for "Walsh Hadamard Transform Applications" produces more than 15,000 results. This is a testament of the importance of Walsh–Hadamard Transforms.

The importance of the Walsh–Hadamard Transforms lie mainly on their computational efficiency as they involve only fourth roots of unity ($+1$, -1, $+j$, $-j$), which all have magnitude of one. This structure makes the calculations much less complicated compared to the Discrete Fourier Transform and as a result the Walsh–Hadamard Transforms are much suitable for high-volume calculations for the future challenges in Big Data and parallel computing (Lu, 2016). In the years to come, the importance of Walsh–Hadamard Transforms becomes more apparent, which will result in more attention to these efficient tools.

REFERENCES

Abromson, R. (1977). The sinc and cosinc transform. *EMC*, *19*(2), 88–94.

Adjouadi, M., Sanchez, D., Cabrerizo, M., Ayala, M., Jayakar, P., Yaylali, I., et al. (2004). Interictal spike detection using the Walsh transform. *IEEE Trans. Biomed. Eng.*, *51*(5), 868–872.

Ahmed, N., & Natarajan, T. (1974). On logical and arithmetic autocorrelation functions. *IEEE Trans. Electromagn. Compat.*, *3*, 177–183.

Ahmed, N., Rao, K. R., & Schultz, R. (1971). Fast complex BIFORE transform by matrix partitioning. *IEEE Trans. Comput.*, *20*(6), 707–710.

Andrews, H. C., & Caspari, K. L. (1970). A generalized technique for spectral analysis. *IEEE Trans. Comput.*, *100*(1), 16–25.

Ashrafi, A. (2014). Frequency domain Walsh functions and sequences: An introduction. *Appl. Comput. Harmon. Anal.*, *36*(2), 348–353.

Aung, A., Ng, B. P., & Rahardja, S. (2008). Sequency-ordered complex Hadamard transform: Properties, computational complexity and applications. *IEEE Trans. Signal Process.*, *56*(8), 3562–3571.

Aung, A., Ng, B. P., & Rahardja, S. (2009). Conjugate symmetric sequency-ordered complex Hadamard transform. *IEEE Trans. Signal Process.*, *57*(7), 2582–2593.

Beauchamp, K. G. (1975). *Walsh functions and their applications*. New York: Academic Press.

Blachman, N. M. (1971). Spectral analysis with sinusoids and Walsh functions. *IEEE Trans. Aerosp. Electron. Syst.*, *5*, 900–905.

Bouguezel, S., Ahmad, M. O., & Swamy, M. (2011). An efficient algorithm for the conjugate symmetric sequency-ordered complex Hadamard transform. In *2011 IEEE international symposium on circuits and systems (ISCAS)* (pp. 1516–1519). IEEE.

Brown, R. (1977). A recursive algorithm for sequency-ordered fast Walsh transforms. *IEEE Trans. Comput.*, *26*(8), 819–822.

Butin, H. (1972). A compact definition of Walsh functions. *IEEE Trans. Comput.*, *6*(C-21), 590–592.

Carl, J. W., & Swartwood, R. V. (1973). A hybrid Walsh transform computer. *IEEE Trans. Comput.*, *22*(7), 669–672.

Chien, T. A. (1975). On representations of Walsh functions. *EMC*, *17*(3), 170–176.

Cooley, J. W., & Tukey, J. W. (1965). An algorithm for the machine calculation of complex Fourier series. *Math. Comput.*, *19*(90), 297–301.

Geadah, Y. A., & Corinthios, M. (1977). Natural, dyadic, and sequency order algorithms and processors for the Walsh–Hadamard transform. *IEEE Trans. Comput.*, *26*(5), 435–442.

Gerheim, A., & Stoughton, J. (1987). Further results in Walsh domain filtering. *IEEE Trans. Acoust. Speech Signal Process.*, *35*(3), 394–397.

Ghaleb, F. A. M., et al. (2016). The Walsh spectrum and the real transform of a switching function: A review with a Karnaugh-map perspective. *J. Eng. Comput. Sci.*, *7*(2).

Gibbs, J. (1967). *Walsh spectrometry: A form of spectral analysis well suited to binary digital computation* (NPLDES reports). Teddington, England: National Physics Laboratory.

Gibbs, J., & Millard, M. (1969). *Walsh functions as solutions of a logical differential equation* (NPLDES reports). Teddington, England: National Physics Laboratory.

Golub, G. H., & Van Loan, C. F. (1996). *Matrix computations* (3rd edition). Baltimore, MD: The Johns Hopkins University Press.

Good, I. J. (1958). The interaction algorithm and practical Fourier analysis. *J. R. Stat. Soc. B, Methodol.*, 361–372.

Haar, A. (1910). Zur theorie der orthogonalen funktionensysteme (erste mitteilung). *Math. Ann.*, *69*, 331–371.

Hadamard, J. (1893). Résolution d'une question relative aux déterminants. *Bull. Sci. Math.*, *17*(1), 240–246.

Harmuth, H. (1968). A generalized concept of frequency and some applications. *IEEE Trans. Inf. Theory*, *14*(3), 375–382.

Harwit, M. (2012). *Hadamard transform optics*. Elsevier.

Hayes, M. (1996). *Statistical signal processing and modeling* (Chapter 6).

Henderson, K. W. (1964). Some notes on the Walsh functions. *IEEE Trans. Electron. Comput.*, *1*, 50–52.

Irshid, M. I. (1986). A simple recursive definition for Walsh functions. *IEEE Trans. Electromagn. Compat.*, *4*(28), 276–279.

Kahveci, A., & Hall, E. (1974). Sequency domain design of frequency filters. *IEEE Trans. Comput.*, *C-23*(9), 976–981.

Kekre, H., Sarode, T. K., Thepade, S. D., & Shroff, S. (2011). Instigation of orthogonal wavelet transforms using Walsh, cosine, Hartley, Kekre transforms and their use in image compression. *Int. J. Comput. Sci. Inf. Secur.*, *9*(6), 125.

Kennett, B. (1970). A note on the finite Walsh transform (corresp.). *IEEE Trans. Inf. Theory*, *16*(4), 489–491.

Khintchine, A. (1934). Korrelationstheorie der stationären stochastischen prozesse. *Math. Ann.*, *109*(1), 604–615.

Kish, D., & Heydt, G. (1994). A generalization of the concept of impedance and a novel Walsh domain immittance. *IEEE Trans. Power Deliv.*, *9*(2), 970–976.

Kyochi, S., & Tanaka, Y. (2014). General factorization of conjugate-symmetric Hadamard transforms. *IEEE Trans. Signal Process.*, *62*(13), 3379–3392.

Lackey, R., & Meltzer, D. (1971). A simplified definition of Walsh functions. *IEEE Trans. Comput.*, *20*(2), 211–213.

Lee, M., & Kaveh, M. (1986). Fast Hadamard transform based on a simple matrix factorization. *IEEE Trans. Acoust. Speech Signal Process.*, *34*(6), 1666–1667.

Lu, Y. (2016). Practical tera-scale Walsh–Hadamard transform. In *2016 future technologies conference (FTC)* (pp. 1230–1236).

Manolakis, D. G., & Ingle, V. K. (2011). *Applied digital signal processing: Theory and practice*. Cambridge University Press.

Manz, J. (1972). A sequency-ordered fast Walsh transform. *IEEE Trans. Audio Electroacoust.*, *20*(3), 204–205.

Marti-Puig, P. (2006). A family of Fast Walsh Hadamard algorithms with identical sparse matrix factorization. *IEEE Signal Process. Lett.*, *13*(11), 672–675.

Meyer, C. D. (2000). *Matrix analysis and applied linear algebra (Vol. 2)*. SIAM.

Oppenheim, A. V., Willsky, A. S., & Nawab, S. H. (1983). *Signals and systems (Vol. 2)*. Englewood Cliffs, NJ: Prentice-Hall.

Paley, R. (1932). A remarkable series of orthogonal functions (i). *Proc. Lond. Math. Soc.*, *2*(1), 241–264.

Pearl, J. (1971). Application of Walsh transform to statistical analysis. *IEEE Trans. Syst. Man Cybern.*, *1*(2), 111–119.

Polyak, B., & Shreider, Yu. A. (1962). The application of Walsh functions in approximate calculations. In Bazilevskii (Ed.), *Voprosy teorii matematicheskikh mashin 2*. Moscow: Fizmatgiz.

Pratt, W. K., Kane, J., & Andrews, H. C. (1969). Hadamard transform image coding. *Proc. IEEE, 57*(1), 58–68.

Quatieri, T. F. (2006). *Discrete-time speech signal processing: Principles and practice.* Pearson Education India.

Rahardja, S., & Falkowski, B. J. (1999). Family of unified complex Hadamard transforms. *IEEE Trans. Circuits Syst. II, Analog Digit. Signal Process., 46*, 1094–1099.

Rahardja, S., & Falkowski, B. J. (2000). Complex composite spectra of unified complex Hadamard transform for logic functions. *IEEE Trans. Circuits Syst. II, Analog Digit. Signal Process., 47*(11), 1291–1297.

Rao, K., & Ahmed, N. (1971). Complex BIFORE transform. *Int. J. Syst. Sci., 2*(2), 149–162.

Rao, K., & Ahmed, N. (1972). Modified complex BIFORE transform. *Proc. IEEE, 60*(8), 1010–1012.

Rao, K., Devarajan, V., Vlasenko, V., & Narasimhan, M. (1978). Cal–Sal Walsh–Hadamard transform. *IEEE Trans. Acoust. Speech Signal Process., 26*(6), 605–607.

Rao, K., Revuluri, K., Narasimhan, M., & Ahmed, N. (1976). Complex Haar transform. *IEEE Trans. Acoust. Speech Signal Process., 24*(1), 102–104.

Robinson, G. (1972). Logical convolution and discrete Walsh and Fourier power spectra. *IEEE Trans. Audio Electroacoust., 20*(4), 271–280.

Ross, I., & Kelly, J. J. (1972). A new method for representing Walsh functions. In *Applications of Walsh functions: 1972 proceedings* (p. 359). National Technical Information Service.

Schreiber, H. (1970). Bandwidth requirements for Walsh functions. *IEEE Trans. Inf. Theory, 16*(4), 491–493.

Shanks, J. L. (1969). Computation of the fast Walsh–Fourier transform. *IEEE Trans. Comput., 18*(5), 457–459.

Shannon, C. E. (1949). Communication in the presence of noise. *Proc. IRE, 37*(1), 10–21.

Shum, F., Elliott, A., & Brown, W. (1973). Speech processing with Walsh–Hadamard transforms. *IEEE Trans. Audio Electroacoust., 21*(3), 174–179.

Siemens, K. H., & Kitai, R. (1973). A nonrecursive equation for the Fourier transform of a Walsh function. *EMC, 15*(2), 81–83.

Slepian, D. (1978). Prolate spheroidal wave functions, Fourier analysis, and uncertainty – V: The discrete case. *Bell Syst. Tech. J., 54*(5), 1371–1429.

Sukkar, R. A., LoCicero, J. L., & Picone, J. W. (1989). Decomposition of the LPC excitation using the zinc basis functions. *IEEE Trans. Acoust. Speech Signal Process., 37*(9), 1329–1341.

Unser, M. (2000). Sampling – 50 years after Shannon. *Proc. IEEE, 88*(4), 569–587.

Walsh, J. L. (1923). A closed set of normal orthogonal functions. *Am. J. Math., 45*, 5–24.

Wiener, N. (1930). Generalized harmonic analysis. *Acta Math., 55*(1), 117–258.

Wu, J., Wang, L., Yang, G., Senhadji, L., Luo, L., & Shu, H. (2012). Sliding conjugate symmetric sequency-ordered complex Hadamard transform: Fast algorithm and applications. *IEEE Trans. Circuits Syst. I, Regul. Pap., 59*(6), 1321–1334.

Yarlagadda, R. K., & Hershey, J. E. (2012). *Hadamard matrix analysis and synthesis: With applications to communications and signal/image processing (Vol. 383).* Springer Science & Business Media.

Yuen, C. (1973). A fast algorithm for computing Walsh power spectrum. *Applications of Walsh functions, 16*(18), 279–283.

Zarowski, C., & Yunik, M. (1985). Spectral filtering using the fast Walsh transform. *IEEE Trans. Acoust. Speech Signal Process., 33*(5), 1246–1252.

Zarowski, C., Yunik, M., & Martens, G. (1988). DFT spectrum filtering. *IEEE Trans. Acoust. Speech Signal Process.*, *36*(4), 461–470.

Zohar, S. (1973). Fast hardware Fourier transformation through counting. *IEEE Trans. Comput.*, *22*(5), 433–441.

CHAPTER TWO

Quantum Methodologies in Maxwell Optics

Sameen Ahmed Khan
Department of Mathematics and Sciences, College of Arts and Applied Sciences, Dhofar University, Salalah, Oman
e-mail address: rohelakhan@yahoo.com

Contents

Advances in Imaging and Electron Physics, Volume 201
ISSN 1076-5670
http://dx.doi.org/10.1016/bs.aiep.2017.05.003

57

1. INTRODUCTION

Historically, the scalar wave theory of optics is based on the *Fermat's principle of least time* (Lakshminarayanan, Ghatak, & Thyagarajan, 2002). In this approach the beam-optical Hamiltonians (paraxial as well as the aberrating Hamiltonians to all orders) are derived using the Fermat's principle of least time. This approach is purely geometrical and works effectively in the scalar regime. Later, it was realized that the whole of optics is governed by Maxwell's equations of electromagnetism (e.g., see Born & Wolf, 1999). Then, all the laws of laws of geometrical optics were deduced from Maxwell's equations. This deduction is generally not done directly from Maxwell's equations, but through the Helmholtz equation, which is an excellent approximation to Maxwell's equations. The deduction is based on the *square-root* of the Helmholtz operator (Dragt, Forest, & Wolf, 1986; Dragt, 1988). The square-root approach transforms the original *boundary value problem* to a *first order initial value problem*. This switchover results in the powerful system or the Fourier optic approach and facilitates many applications (Nazarathy & Shamir, 1980, 1982; Goodman, 1996). But it is interesting to note, that the beam-optical Hamiltonian in the square-root approach is *identical* to the one obtained using the Fermat's principle. Mathematically, the deduction of the beam-optical Hamiltonian using the square-root of the Helmholtz operator is not rigorous enough and exact (see Khan, Jagannathan, & Simon, 2002; Khan, 2005a, 2016g and the references therein). So, there have been attempts to examine the Helmholtz equation using diverse techniques. In the context of Helmholtz optics, we note that the square-root operators have been extensively examined (Fishman, De Hoop, & Van Stralen, 2000; Gill & Zachary, 2005).

Helmholtz equation governing scalar optics and the Klein–Gordon equation for a spin-0 particle have a striking mathematical similarity. This mathematical similarity between the two systems has been successfully exploited to develop an alternative to the traditional square-root approach. This approach makes use of a scheme similar to the Feshbach–Villars procedure used for linearizing the Klein–Gordon equation (Feshbach & Villars, 1958; Bjorken & Drell, 1964). The Feshbach–Villars-*like* procedure casts the Helmholtz equation into a Dirac-*like* form. This enables us to use the standard and powerful techniques from relativistic quantum mechanics. The chief technique in this case is the Foldy–Wouthuysen expansion. The quantum methodologies lead to a *non-traditional* prescription of Helmholtz optics (Khan et al., 2002; Khan, 2005a, 2016g), providing an alternative to the traditional square-root procedure. Different ways of linearization of wave equations lead to different prescriptions (Khan, 2017b). The non-traditional prescription of Helmholtz optics naturally leads to the wavelength-dependent modifications of the traditional prescriptions. For instance, all the aberration coefficients are modified by a wavelength-dependent part (Khan, 2005a, 2016g). The algebraic machinery of the non-traditional Helmholtz optics is adopted from the one used in development of the *quantum theory of charged-particle beam optics* (Jagannathan, Simon, Sudarshan, & Mukunda, 1989; Jagannathan, 1990, 1993, 1999, 2002, 2003; Jagannathan & Khan, 1995, 1996, 1997, 2018; Khan, 1997, 1999a, 1999b, 2001, 2002a, 2002b, 2002c, 2003d, 2016d; Khan & Jagannathan, 1993, 1994, 1995).

Maxwell's equations are linear but coupled and constrained. They contain the permittivity and permeability of the medium, which may have spatial and temporal variations. These variations are responsible for the variation of the refractive index of the medium. The spatial and temporal derivatives of the permittivity and permeability are omitted in the derivation of the Helmholtz equation. So, the Helmholtz equation is approximate to start with. Moreover, the widely used procedure of the square-root of the Helmholtz operator has been under question. Any study of light polarization has to be necessarily based on Maxwell's equations. Even in Maxwell optics, the beam optics and light polarization are usually addressed separately using entirely different techniques. For instance, a group theoretical analysis leads to a systematic procedure for the passage from scalar optics to vector wave optics (Mukunda, Simon, & Sudarshan, 1983a, 1983b, 1985a, 1985b; Simon, Sudarshan, & Mukunda, 1986, 1987). This procedure is based on the Poincaré invariance of the paraxial system. With this as the motivation, we develop a formalism of beam optics based entirely on Maxwell's equations, taking into account the vectorial nature of light. The resulting formalism is sure to provide a deeper understanding of beam optics and light polarization in a unified manner.

Any attempt to deal with Maxwell's equations directly in their usual form is bound to be limited by approximations. Helmholtz equation is one of the most widely used approximations. A possible way to overcome this situation is to express Maxwell's equations in a matrix form: a single matrix equation containing all the four Maxwell's equations. So, we require an exact matrix representation of Maxwell's equations for a medium with varying refractive index. This is accomplished only when we pay due regards to the spatial and temporal variations of the permittivity $\epsilon(r, t)$ and permeability $\mu(r, t)$. The available matrix representations do not meet this crucial requirement completely! The possible reasons are perhaps that the various matrix representations were derived with very different motivations, for instance: spinor analysis or seeking symmetries and the photon wave function (Bialynicki-Birula, 1994, 1996a, 1996b; Dvoeglazov, 1993; Esposito, 1998; Giannetto, 1985; Good, 1957; Inskeep, 1988; Ivezć, 2006; Lomont, 1958; Laporte & Uhlenbeck, 1931; Majorana, 1974; Mohr, 2010; Moses, 1959; Ohmura, 1956; Oppenheimer, 1931; Sachs & Schwebel, 1962). Hence, an exact matrix representation of Maxwell's equations using eight-dimensional matrices was specially developed to meet the stringent requirements of the Maxwell optics (Khan, 2005b). This representation makes use of the Riemann–Silberstein vector (Silberstein, 1907a, 1907b), which is described in Appendix A. The specially derived matrix representation and its priority over the numerous other representations is presented in detail in Appendix B. The exact matrix representation leads to the general beam-optical Hamiltonian without any assumptions on the refractive index. The derived beam-optical Hamiltonian is exact and has an algebraic structure of the Dirac equation. This enables us to make use of the quantum methodologies such as the Foldy–Wouthuysen transformation technique.

From the onset, let us borne in mind that the six-dimensional electromagnetic field and the four-dimensional Dirac field are two completely different physical entities. Let us note, that the Foldy–Wouthuysen transformation technique was historically developed for the Dirac equation of the electron. But it can also be applied to a certain class of matrix equations. The specially developed matrix representation of Maxwell's equations falls within this special class of matrix equations (Khan, 2006b, 2008). The extreme similarities in the underlying mathematical structures of the two systems ensures that we can use the powerful machinery of relativistic quantum mechanics. The use of quantum methodologies (particularly, the Foldy–Wouthuysen transformation technique) has served as a powerful analytic tool leading to well-established results and even in predicting new affects. The matrix formulation of Maxwell optics was decisively influenced by the 'quantum theory of charged-particle beam optics' (Conte, Jagannathan, Khan, & Pusterla, 1996; Jagannathan et al., 1989; Jagannathan, 1990, 1993, 1999, 2002, 2003; Jagannathan & Khan, 1995, 1996, 1997, 2018; Khan, 1997, 1999a, 1999b, 2001, 2002a, 2002b, 2002c, 2002d, 2016d; Khan &

Jagannathan, 1993, 1994, 1995). So, we present an outline of the quantum prescriptions in Appendix E.

The idea to use the Foldy–Wouthuysen transformation technique to examine the Helmholtz equation was made as a remark (Fishman & McCoy, 1984). The same idea was independently outlined by Jagannathan and Khan (see p. 277 in Jagannathan & Khan, 1996). But it is only in the recent years, that the idea of the Foldy–Wouthuysen transformation was systematically used to examine the specific beam-optical systems in the Helmholtz case (Khan et al., 2002; Khan, 2005a, 2016g). The same idea has been extended to the matrix formulation of Maxwell optics. The Foldy–Wouthuysen transformation technique leads to new approaches, which we call as the non-traditional prescriptions of Helmholtz optics (Khan, 2005a, 2016g) and Maxwell optics (Khan, 2006c, 2010, 2014a, 2016f, 2017c) respectively. Both the formalism are algebraically similar to the Lie algebraic formalism of light beam optics. An outline of the Foldy–Wouthuysen transformation technique is presented in Appendix C. Detailed coverage to the Foldy–Wouthuysen transformation in the context of optics is to be found in Khan (2006b, 2008). In this article, we shall develop the formalism without making any assumptions on the form of the refractive index. It is seen that the Hamiltonians in the non-traditional prescription of Maxwell optics additionally contain wavelength-dependent parts. We shall consider the following applications:

1. *Medium with Constant Refractive Index.* This is an ideal example and possesses an exact solution. Moreover, it enables us to acquaint with the various techniques being used.

2. *Axially Symmetric Graded-Index Medium.* Two points are worth mentioning about this system. *Image Rotation*: our formalism gives rise to the wavelength-dependent image rotation, and we have derived an expression for it. The other pertains to the aberrations: in our formalism, we get all the nine aberrations permitted by the axial symmetry. The traditional approaches have only six aberrations. Our formalism modifies these six aberrations coefficients by wavelength-dependent contributions. Moreover, we also obtain the remaining three aberrations permitted by the axial symmetry. The existence of the nine aberrations and image rotation are well-known in the case of the *axially symmetric magnetic lenses*. The quantum treatment of this system modifies the aberration coefficients by wavelength-dependent parts (Jagannathan & Khan, 1996). The axially symmetric graded-index medium in the non-traditional prescription of Helmholtz optics has the usual six aberrations

which are modified by the wavelength-dependent parts but does not have the image rotation (Khan et al., 2002; Khan, 2005a, 2016g). The image rotation and the three extra aberrations is an exclusive outcome of the quantum methodologies, when applied to Maxwell optics (Khan, 2006c, 2010, 2014a, 2016f, 2017c).

3. *Thick Lens*: Explicit formulae relating the incident and emergent light rays for the general case of a thick lens are presented for this system (Khan, 2016g).

4. *Polarization*: As for the light polarization, the elegant Mukunda–Simon–Sudarshan rule (MSS-rule) for transition from scalar to vector wave optics is obtained as the paraxial limit of the general formalism presented here (Khan, 2016f). The new formalism is a suitable candidate to extend traditional theory of polarization beyond the paraxial approximation. The MSS-rule is applied to the Gaussian beams (Khan, 2017c).

The formalism of Maxwell optics resulting from the use quantum methodologies is consistent with other approaches using varying techniques including: the Eikonal expansion in geometrical optics (Lakshminarayanan et al., 2002); the Lie algebraic formalism (Dragt, 1982, 1988; Dragt et al., 1986; Lakshminarayanan, Sridhar, & Jagannathan, 1998); and the group-theoretical techniques in Maxwell optics (Mukunda et al., 1983a, 1983b, 1985a, 1985b; Simon et al., 1986, 1987). In the limit of low wavelength ($\lambdabar \longrightarrow 0$), our formalism leads to the Lie algebraic formalism of light beam optics (Dragt, 1982, 1988; Dragt et al., 1986; Rangarajan, Dragt, & Neri, 1990) We call this as the *traditional limit* of our formalism. This is analogous to the *classical limit* obtained by taking $\hbar \longrightarrow 0$ in the quantum prescriptions. The scheme of using the Foldy–Wouthuysen machinery in our formalism is adopted from the one used in the *quantum theory of charged-particle beam optics* (Conte et al., 1996; Jagannathan et al., 1989; Jagannathan, 1990, 1993, 1999, 2002, 2003; Jagannathan & Khan, 1995, 1996, 1997, 2018; Khan, 1997, 1999a, 1999b, 2001, 2002a, 2002b, 2002c, 2003d, 2016d; Khan & Jagannathan, 1993, 1994, 1995). There too, the classical limit leads to the Lie algebraic formalism of charged-particle beam optics (Dragt & Forest, 1986; Dragt, Neri, Rangarajan, Douglas, Healy, & Ryne, 1988; Forest, Berz, & Irwin, 1989; Forest & Hirata, 1992; Radlička, 2008; Rangarajan et al., 1990; Rangarajan and Sachidanand, 1997; Rangarajan & Sridharan, 2010; Ryne & Dragt, 1991; Turchetti, Bazzani, Giovannozzi, Servizi, & Todesco, 1989; Todesco, 1999). The classical limit is obtained by $\lambdabar_0 \longrightarrow 0$, where

$\lambdabar_0 = \hbar/p_0$ is the de Broglie wavelength and p_0 is the design momentum of the system.

As for the light polarization, our formalism readily leads to the elegant Mukunda–Simon–Sudarshan rule for passage from scalar to vector wave optics in the case of paraxial systems (Khan, 2016f, 2016h, 2017c). The unified formalism light beam-optics and light polarization further strengthens the Hamilton's optical-mechanical analogy, particularly in the wavelength-dependent regime (Khan, 2002c, 2017a). A short note on the Hamilton's optical-mechanical analogy is presented in Appendix F.

2. TRADITIONAL PRESCRIPTIONS

In this section, we shall briefly review the traditional prescription of scalar optics. The beam is assumed to be monochromatic and quasiparaxial. The optic-axis is along the z-axis. Furthermore, it is assumed that a Schrödinger-*like* equation given below describes the z-evolution of the corresponding optical wavefunction $\psi(\boldsymbol{r})$,

$$i\lambdabar \frac{\partial}{\partial z} \psi(\boldsymbol{r}) = \widehat{H} \psi(\boldsymbol{r}), \tag{1}$$

where $\lambdabar = \lambda/2\pi = c/\omega$ is the reduced wavelength and the analogue of $\hbar$, the reduced Planck's constant in quantum mechanics. The beam-optical Hamiltonian $\widehat{H}$ is

$$\widehat{H} = -\left(n^2(\boldsymbol{r}) - \widehat{\boldsymbol{p}}_\perp^2\right)^{1/2}, \tag{2}$$

where $\widehat{\boldsymbol{p}} = -i\lambdabar\boldsymbol{\nabla}$ and $\widehat{\boldsymbol{p}}_\perp = -i\lambdabar\boldsymbol{\nabla}_\perp$. It is assumed that the light rays are propagating very close to the optic-axis (z-axis). Mathematically, $|\widehat{\boldsymbol{p}}_\perp| \ll p_z \approx 1$. The only condition on the refractive index is that it varies smoothly around a constant value n_0, satisfying $|n(\boldsymbol{r}) - n_0| \ll n_0$. The beam-optical Hamiltonian in Eq. (2) is traditionally obtained by two distinct procedures. The first procedure predating Maxwell is to derive it from the *Fermat's principle of least time* (Lakshminarayanan et al., 2002). The starting point for the second procedure is the Helmholtz equation

$$\left(\boldsymbol{\nabla}^2 - \frac{n^2(\boldsymbol{r})}{\lambdabar^2}\right)\psi(\boldsymbol{r}) = 0. \tag{3}$$

The terms in the Helmholtz equation are rearranged as

$$\left(i\lambdabar\frac{\partial}{\partial z}\right)^2 \psi(\boldsymbol{r}) = \left(n^2(\boldsymbol{r}) - \widehat{\boldsymbol{p}}_\perp^2\right)\psi(\boldsymbol{r}). \tag{4}$$

Then, one takes the *square-root* (Dragt et al., 1986) as

$$i\hbar \frac{\partial}{\partial z} \psi(\boldsymbol{r}) = -\left(n^2(\boldsymbol{r}) - \widehat{\boldsymbol{p}}_\perp^2\right)^{1/2} \psi(\boldsymbol{r}), \tag{5}$$

corresponding to the requirement that the propagation be entirely in the positive z-direction; if the propagation is in the negative z-direction, the right hand side of Eq. (5) will have the opposite sign. It is to be noted, that the passage from Eq. (3) to Eq. (5) replaces the original *boundary value problem* with a *first-order initial value problem* in z. This switchover results in the powerful Fourier optic approach and enables many applications (Goodman, 1996). It is interesting to note that the purely geometric approach of the Fermat's principle of least time and the Helmholtz equation based on Maxwell's equations of electromagnetism lead to the same beam-optical Hamiltonian.

The Hamiltonian in Eq. (2) or equivalently in Eq. (5) is expanded in a power series in the transverse components, $\boldsymbol{r}_\perp$ and $\widehat{\boldsymbol{p}}_\perp$ respectively. In order to have an Hermitian $\widehat{H}$, it is required to incorporate a suitable ordering and symmetrization, of the polynomials resulting from the expansion (Dragt, 1982, 1988; Dragt et al., 1986; Gloge & Marcuse, 1969). For a homogeneous medium, the refractive index, $n(\boldsymbol{r})$ reduces to a constant value n_0. Then, the Hamiltonian has the simplified Taylor expansion

$$
\begin{aligned}
\widehat{H} &= -\left(n^2(\boldsymbol{r}) - \widehat{\boldsymbol{p}}_\perp^2\right)^{1/2} \\
&= -\left(n_0^2 - \widehat{\boldsymbol{p}}_\perp^2\right)^{1/2} \\
&= -n_0 \left\{1 - \frac{1}{n_0^2}\widehat{\boldsymbol{p}}_\perp^2\right\}^{1/2} \\
&= n_0 \left\{-1 + \frac{1}{2n_0^2}\widehat{\boldsymbol{p}}_\perp^2 + \frac{1}{8n_0^4}\widehat{\boldsymbol{p}}_\perp^4 + \frac{1}{16n_0^6}\widehat{\boldsymbol{p}}_\perp^6 \right. \\
&\qquad\qquad \left. + \frac{5}{128n_0^8}\widehat{\boldsymbol{p}}_\perp^8 + \frac{7}{256n_0^{10}}\widehat{\boldsymbol{p}}_\perp^{10} + \cdots\right\} \\
&= n_0 \sum_{m=0}^{\infty} \frac{(2m-3)!!}{(2m)!!} \frac{1}{n_0^m}\widehat{\boldsymbol{p}}_\perp^m,
\end{aligned} \tag{6}
$$

where we recall that $1!! = 1$, $0!! = 1$, $(-1)!! = 1$, and $(-3)!! = -1$. In practice, the refractive index does not have a constant value. In such situations, one makes a Taylor expansion of the varying refractive index $n(\boldsymbol{r}) = n(\widehat{\boldsymbol{r}}_\perp, z)$. Then the radical is expanded to the required degree of accuracy in powers of $\widehat{\boldsymbol{p}}_\perp^2/n_0^2$, consistently taking into account the terms arising from the expansion of $n(\mathbf{r})$. The aforementioned expansion of the Hamiltonian has quadratic terms and higher-order terms. The quadratic terms govern the paraxial or ideal behavior. The higher-order terms describe the aberrations of corresponding order, order-by-order (Dragt et al., 1986). When using the quantum methodologies, the resulting beam-optical Hamiltonians are automatically Hermitian and there is absolutely no need to incorporate any ordering and symmetrization, of the polynomials in the expansion.

3. THE BEAM-OPTICAL FORMALISM

The Matrix representations of Maxwell's equations have a long history (Bialynicki-Birula, 1994, 1996a, 1996b; Dvoeglazov, 1993; Esposito, 1998; Giannetto, 1985; Good, 1957; Inskeep, 1988; Ivezć, 2006; Kulyabov, 2016; Laporte & Uhlenbeck, 1931; Lomont, 1958; Majorana, 1974; Mohr, 2010; Moses, 1959; Ohmura, 1956; Oppenheimer, 1931; Ovsiyuk, Kisel, & Red'kov, 2013; Sachs & Schwebel, 1962; Veko, Vlasii, Ovsiyuk, Red'kov, & Sitenko, 2014). But these representations are inadequate for the Maxwell optics. Some of them are for a medium with a constant refractive index, hence they are not exact for a medium with varying refractive index. Others employ a *pair* of matrix equations (Bialynicki-Birula, 1994, 1996a, 1996b). For Maxwell optics, we require a general matrix representation containing all the four Maxwell's equations in medium with varying refractive index. Such a matrix representation was specially developed with due regard to the spatial and temporal variations of the permittivity $\epsilon(\mathbf{r}, t)$ and the permeability $\mu(\mathbf{r}, t)$ of the medium (Khan, 2005b).

Maxwell's equations (Jackson, 1998; Panofsky & Phillips, 1962) in an inhomogeneous medium with sources and currents are

$$\nabla \cdot \mathbf{D}(\mathbf{r}, t) = \rho\,,$$

$$\nabla \times \mathbf{H}(\mathbf{r}, t) - \frac{\partial}{\partial t}\mathbf{D}(\mathbf{r}, t) = \mathbf{J}\,,$$

$$\nabla \times \mathbf{E}(\mathbf{r}, t) + \frac{\partial}{\partial t}\mathbf{B}(\mathbf{r}, t) = 0\,,$$

$$\nabla \cdot \mathbf{B}(\mathbf{r}, t) = 0\,. \tag{7}$$

In the present study of optics, we assume that the media is linear implying $\mathbf{D} = \epsilon\mathbf{E}$, and $\mathbf{B} = \mu\mathbf{H}$. Here, $\epsilon = \epsilon(\mathbf{r}, t)$ is the *permittivity* and $\mu = \mu(\mathbf{r}, t)$ is the *permeability* of the medium. The speed of light in this medium is $v(r, t) = 1/\sqrt{\epsilon(r, t)\mu(r, t)}$. In optics, the chief quantity is the refractive index of the medium, $n(r, t) = c/v(r, t) = c\sqrt{\epsilon(r, t)\mu(r, t)}$. The quantity $h(r, t) = \sqrt{\mu(r, t)/\epsilon(r, t)}$ has the dimensions of resistance. In vacuum, we have, $\epsilon_0 = 8.85 \times 10^{-12} C^2/N.m^2$ and $\mu_0 = 4\pi \times 10^{-7} N/A^2$. In the present study, it is realistic to set aside the resistance function. So, we shall use the 'basic optical quantity', the refractive index $n(r, t)$ and the ignorable quantity, the resistance $h(r, t)$ in place of $\mu(r, t)$ and $\epsilon(r, t)$ in the matrix representation of Maxwell's equations. This switchover leads to the physical basis of the approximations. This will be demonstrated through our exact matrix representation of Maxwell's equations. In terms of these quantities $\epsilon = 1/hv = n/ch$ and $\mu = h/v = nh/c$.

Following the notation in Bialynicki-Birula (1996b) and Khan (2005b) we use the Riemann–Silberstein vector (Silberstein, 1907a, 1907b) given by

$$F^{\pm}(r, t) = \frac{1}{\sqrt{2}} \left(\frac{1}{\sqrt{\epsilon(r)}} D(r, t) \pm i \frac{1}{\sqrt{\mu(r)}} B(r, t) \right)$$
$$= \frac{1}{\sqrt{2}} \left(\sqrt{\epsilon(r)} E(r, t) \pm i \frac{1}{\sqrt{\mu(r)}} B(r, t) \right). \tag{8}$$

In order to derive the matrix representation, we define

$$\Psi^{\pm}(r, t) = \begin{bmatrix} -F_x^{\pm} \pm i F_y^{\pm} \\ F_z^{\pm} \\ F_z^{\pm} \\ F_x^{\pm} \pm i F_y^{\pm} \end{bmatrix}. \tag{9}$$

The sources and currents are incorporated through the following vectors

$$W^{\pm} = \left(\frac{1}{\sqrt{2\epsilon}} \right) \begin{bmatrix} -J_x \pm i J_y \\ J_z - v\rho \\ J_z + v\rho \\ J_x \pm i J_y \end{bmatrix}. \tag{10}$$

The various matrices arising in our matrix representation are

$$M_x = \begin{bmatrix} 0 & 1 \\ 1 & 0 \end{bmatrix}, \quad M_y = \begin{bmatrix} 0 & -i1 \\ i1 & 0 \end{bmatrix}, \quad M_z = \beta = \begin{bmatrix} 1 & 0 \\ 0 & -1 \end{bmatrix},$$

$$\Sigma = \begin{bmatrix} \sigma & 0 \\ 0 & \sigma \end{bmatrix}, \quad \alpha = \begin{bmatrix} 0 & \sigma \\ \sigma & 0 \end{bmatrix}, \quad I = \begin{bmatrix} 1 & 0 \\ 0 & 1 \end{bmatrix}, \tag{11}$$

where $\mathbb{1}$ is the 2×2 unit matrix and $\boldsymbol{\sigma}$ are the Pauli triplets

$$\boldsymbol{\sigma} = \left(\sigma_x = \begin{bmatrix} 0 & 1 \\ 1 & 0 \end{bmatrix}, \ \sigma_y = \begin{bmatrix} 0 & -i \\ i & 0 \end{bmatrix}, \ \sigma_z = \begin{bmatrix} 1 & 0 \\ 0 & -1 \end{bmatrix} \right). \tag{12}$$

The following two coupling functions arise very naturally in our matrix representation

$$\boldsymbol{u}(\boldsymbol{r}, t) = \frac{1}{2v(\boldsymbol{r}, t)} \boldsymbol{\nabla} v(\boldsymbol{r}, t) = \frac{1}{2} \boldsymbol{\nabla} \{\ln v(\boldsymbol{r}, t)\} = -\frac{1}{2} \boldsymbol{\nabla} \{\ln n(\boldsymbol{r}, t)\}$$

$$\boldsymbol{w}(\boldsymbol{r}, t) = \frac{1}{2h(\boldsymbol{r}, t)} \boldsymbol{\nabla} h(\boldsymbol{r}, t) = \frac{1}{2} \boldsymbol{\nabla} \{\ln h(\boldsymbol{r}, t)\}. \tag{13}$$

As we shall see shortly, the coupling functions shall facilitate the approximations. Following the notation in Khan (2005b) the exact matrix representation of Maxwell's equations is

$$\frac{\partial}{\partial t} \begin{bmatrix} I & 0 \\ 0 & I \end{bmatrix} \begin{bmatrix} \Psi^+ \\ \Psi^- \end{bmatrix} - \frac{\dot{v}(\boldsymbol{r}, t)}{2v(\boldsymbol{r}, t)} \begin{bmatrix} I & 0 \\ 0 & I \end{bmatrix} \begin{bmatrix} \Psi^+ \\ \Psi^- \end{bmatrix}$$

$$+ \frac{\dot{h}(\boldsymbol{r}, t)}{2h(\boldsymbol{r}, t)} \begin{bmatrix} 0 & i\beta\alpha_y \\ i\beta\alpha_y & 0 \end{bmatrix} \begin{bmatrix} \Psi^+ \\ \Psi^- \end{bmatrix}$$

$$= -v(\boldsymbol{r}, t) \begin{bmatrix} \{\boldsymbol{M} \cdot \boldsymbol{\nabla} + \boldsymbol{\Sigma} \cdot \boldsymbol{u}\} & -i\beta \left(\boldsymbol{\Sigma} \cdot \boldsymbol{w}\right) \alpha_y \\ -i\beta \left(\boldsymbol{\Sigma}^* \cdot \boldsymbol{w}\right) \alpha_y & \{\boldsymbol{M}^* \cdot \boldsymbol{\nabla} + \boldsymbol{\Sigma}^* \cdot \boldsymbol{u}\} \end{bmatrix} \begin{bmatrix} \Psi^+ \\ \Psi^- \end{bmatrix}$$

$$- \begin{bmatrix} I & 0 \\ 0 & I \end{bmatrix} \begin{bmatrix} W^+ \\ W^- \end{bmatrix}, \tag{14}$$

where '*' stands for complex-conjugation and the time derivatives are $\dot{v} = \partial v / \partial t$ and $\dot{h} = \partial h / \partial t$.

The matrix representation has thirteen 8×8 matrices, of which ten are Hermitian. The remaining three containing the logarithmic gradient of the resistance function $\boldsymbol{w}(\boldsymbol{r}, t)$ are antihermitian. In our matrix representation of Maxwell's equations, the refractive index is the dominant function and the resistance function occurs very weakly. This vindicates our choice of using the refractive index and resistance function over the permittivity and permeability of the inhomogeneous medium respectively. By approximating the logarithmic gradient of the resistance function to zero, we can reduce the representation from eight to four dimensions. This is precisely, the methodology, we follow in developing the matrix formalism of Maxwell optics. It is fascinating to note, that Maxwell's equations can be obtained from Fermat's principle of geometrical optics through the process of 'wavization', in analogy to the quantization of classical mechanics (Pradhan, 1987). Matrix methods have been used in other areas of optics and provide great advantages of simplifying and presenting the equations of optics (Gerrard & Burch, 1994). Same is true for other algebraic techniques (Khan & Wolf, 2002; Sunilkumar, Bambah, Jagannathan, Panigrahi, & Srinivasan, 2000).

In the context of beam-optics, there are some simplifications. The sources and currents are zero (i.e., $W^{\pm} = 0$). It is reasonable to assume that the refractive index, $n(\mathbf{r}, t) = c/v(\mathbf{r}, t)$ and the resistance function, $h(\mathbf{r}, t)$ of the medium under study are time-independent. Consequently, the general expression for the exact matrix representation of Maxwell's equations in Eq. (14) simplifies to

$$
\frac{\partial}{\partial t}
\begin{bmatrix} \Psi^+ \\ \Psi^- \end{bmatrix}
= -v(\mathbf{r})
\begin{bmatrix}
\{\mathbf{M} \cdot \nabla + \mathbf{\Sigma} \cdot \mathbf{u}\} & -\mathrm{i}\beta\,(\mathbf{\Sigma} \cdot \mathbf{w})\,\alpha_y \\
-\mathrm{i}\beta\,(\mathbf{\Sigma}^* \cdot \mathbf{w})\,\alpha_y & \{\mathbf{M}^* \cdot \nabla + \mathbf{\Sigma}^* \cdot \mathbf{u}\}
\end{bmatrix}
$$

$$
\times
\begin{bmatrix} \Psi^+ \\ \Psi^- \end{bmatrix}.
\tag{15}
$$

We shall focus on the monochromatic case,

$$
\Psi^{\pm}(\mathbf{r}, t) = \psi^{\pm}(\mathbf{r})\,e^{-\mathrm{i}\omega t}, \qquad \omega > 0.
\tag{16}
$$

Then after rearranging Eq. (15), we have

$$
\begin{bmatrix} M_z & \mathbf{0} \\ \mathbf{0} & M_z \end{bmatrix}
\frac{\partial}{\partial z}
\begin{bmatrix} \psi^+ \\ \psi^- \end{bmatrix}
$$

$$
= \mathrm{i}\frac{\omega}{v(\mathbf{r})}
\begin{bmatrix} \psi^+ \\ \psi^- \end{bmatrix}
$$

$$
- \begin{bmatrix}
\{\mathbf{M}_\perp \cdot \nabla_\perp + \mathbf{\Sigma} \cdot \mathbf{u}\} & -\mathrm{i}\beta\,(\mathbf{\Sigma} \cdot \mathbf{w})\,\alpha_y \\
-\mathrm{i}\beta\,(\mathbf{\Sigma}^* \cdot \mathbf{w})\,\alpha_y & -\{\mathbf{M}_\perp^* \cdot \nabla_\perp + \mathbf{\Sigma}^* \cdot \mathbf{u}\}
\end{bmatrix}
\begin{bmatrix} \psi^+ \\ \psi^- \end{bmatrix}.
\tag{17}
$$

Now, we introduce the process of *wavization* (Gloge & Marcuse, 1969), through the very familiar Schrödinger replacement

$$
-\mathrm{i}\lambdabar\nabla_\perp \longrightarrow \widehat{\mathbf{p}}_\perp, \qquad -\mathrm{i}\lambdabar\frac{\partial}{\partial z} \longrightarrow p_z,
\tag{18}
$$

where the reduced wavelength $\lambdabar = \lambda/2\pi = c/\omega$ is the analogue of the reduced Planck's constant, $\hbar = h/2\pi$. We note, that the spatial variable 'z'

along the optic axis (z-axis) has the 'role of time'. In quantum mechanics, the fundamental commutation relation is $[p, q] = -i\hbar$. The analogue of this relation in wave optics is the commutation relation $[p, q] = pq - qp = -i\lambdabar$. The spatially varying refractive index $n(\boldsymbol{r})$ and the transverse momenta $\widehat{\boldsymbol{p}}_\perp = -i\lambdabar \boldsymbol{\nabla}_\perp$ need not commute. It is precisely this non-commutativity, which leads to the wavelength-dependent affects.

Noting, that $M_z^{-1} = M_z = \beta$, we multiply both sides of Eq. (17) by

$$\begin{bmatrix} M_z & \boldsymbol{0} \\ \boldsymbol{0} & M_z \end{bmatrix}^{-1} = \begin{bmatrix} \beta & \boldsymbol{0} \\ \boldsymbol{0} & \beta \end{bmatrix} \tag{19}$$

and $(i\lambdabar)$, then, we obtain

$$i\lambdabar \frac{\partial}{\partial z} \begin{bmatrix} \psi^+(\boldsymbol{r}_\perp, z) \\ \psi^-(\boldsymbol{r}_\perp, z) \end{bmatrix} = \widehat{H}_g \begin{bmatrix} \psi^+(\boldsymbol{r}_\perp, z) \\ \psi^-(\boldsymbol{r}_\perp, z) \end{bmatrix}. \tag{20}$$

This is the basic beam-optical equation, where

$$\widehat{H}_g = -n_0 \begin{bmatrix} \beta & \boldsymbol{0} \\ \boldsymbol{0} & -\beta \end{bmatrix} + \widehat{\mathcal{E}}_g + \widehat{\mathcal{O}}_g$$

$$\widehat{\mathcal{E}}_g = -(n(\boldsymbol{r}) - n_0) \begin{bmatrix} \beta & \boldsymbol{0} \\ \boldsymbol{0} & \beta \end{bmatrix} \beta_g$$

$$+ \begin{bmatrix} \beta \{ \boldsymbol{M}_\perp \cdot \boldsymbol{p}_\perp - i\lambdabar \boldsymbol{\Sigma} \cdot \boldsymbol{u} \} & \boldsymbol{0} \\ \boldsymbol{0} & \beta \{ \boldsymbol{M}_\perp^* \cdot \boldsymbol{p}_\perp - i\lambdabar \boldsymbol{\Sigma}^* \cdot \boldsymbol{u} \} \end{bmatrix}$$

$$\widehat{\mathcal{O}}_g = \begin{bmatrix} \boldsymbol{0} & -\lambdabar (\boldsymbol{\Sigma} \cdot \boldsymbol{w}) \alpha_y \\ -\lambdabar (\boldsymbol{\Sigma}^* \cdot \boldsymbol{w}) \alpha_y & \boldsymbol{0} \end{bmatrix}, \tag{21}$$

where 'g' stands for *grand*, signifying the eight dimensions and

$$\beta_g = \begin{bmatrix} \boldsymbol{I} & \boldsymbol{0} \\ \boldsymbol{0} & -\boldsymbol{I} \end{bmatrix}. \tag{22}$$

The optical Hamiltonian thus derived is exact. The approximations arise only during the calculations and even these approximations are minimal due to the powerful techniques adopted from relativistic quantum mechanics. Moreover, the optical Hamiltonian is in complete algebraic correspondence with the Dirac equation accompanied with relevant physical interpretations. In the context of the Foldy–Wouthuysen transformation technique

$$\beta_g \widehat{\mathcal{E}}_g = \widehat{\mathcal{E}}_g \beta_g, \qquad \beta_g \widehat{\mathcal{O}}_g = -\widehat{\mathcal{O}}_g \beta_g. \tag{23}$$

The upper and lower components (ψ^+ and ψ^-) are coupled through the weak logarithmic divergence of the resistance function. As noted earlier, the weak coupling function, w can be approximated to be zero. This leads to the two independent and equivalent four dimensional equations. Thus the problem is reduced from eight dimensions to four. We shall develop the formalism in four dimensions and later see how to incorporate the resistance function. This will require the application of the Foldy–Wouthuysen transformation technique in *cascade* as we shall see. This decoupling with a physical meaning validates our choice of the two derived laboratory functions in place of permittivity and permeability respectively.

We drop the '$+$' throughout, then the beam-optical Hamiltonian is

$$i\hbar \frac{\partial}{\partial z} \psi\,(\boldsymbol{r}) = \widehat{H} \psi\,(\boldsymbol{r})$$

$$\widehat{H} = -n_0 \beta + \widehat{\mathcal{E}} + \widehat{\mathcal{O}}$$

$$\widehat{\mathcal{E}} = -\,(n\,(\boldsymbol{r}) - n_0)\,\beta - i\hbar\beta\boldsymbol{\Sigma} \cdot \boldsymbol{u}$$

$$\widehat{\mathcal{O}} = i\,\bigl(M_y p_x - M_x p_y\bigr)$$

$$= \beta\,\bigl(\boldsymbol{M}_\perp \cdot \widehat{\boldsymbol{p}}_\perp\bigr). \tag{24}$$

The term, $-i\hbar\beta\boldsymbol{\Sigma} \cdot \boldsymbol{u}$ is a consequence of the exact treatment. When we compare the beam-optical Hamiltonian with the Dirac equation, this extra term is analogous to the anomalous magnetic/electric moment term coupled to the magnetic/electric field respectively in the Dirac equation. The term we omitted to reduce the problem from eight dimensions to four dimensions is analogous to the anomalous magnetic/electric moment term coupled to the electric/magnetic fields respectively. But there is one major difference, the extra terms in our beam optics formalism are derived from Maxwell's equations. Where as in the relativistic quantum mechanics, the anomalous terms are added to the Dirac equation using experimental results (some even predating the Dirac equation) and on certain arguments of invariances (Conte et al., 1996). The exact treatment of Maxwell optics has only these two terms, where as in the Dirac equation the scheme of invariances permits addition of numerous terms!

One of the other similarities with the Dirac equation, worth noting, relates to the square of the optical Hamiltonian.

$$
\begin{aligned}
\widehat{H}^2 &= \left\{ n^2\left(\boldsymbol{r}\right) - \widehat{\boldsymbol{p}}_\perp^2 \right\} - \lambdabar^2 u^2 + \left[\boldsymbol{M}_\perp \cdot \widehat{\boldsymbol{p}}_\perp , n\left(\boldsymbol{r}\right) \right] \\
&\quad + 2\mathrm{i}\lambdabar n(\boldsymbol{r}) \boldsymbol{\Sigma} \cdot \boldsymbol{u} + \mathrm{i}\lambdabar \left[\boldsymbol{M}_\perp \cdot \widehat{\boldsymbol{p}}_\perp , \boldsymbol{\Sigma} \cdot \boldsymbol{u} \right] \\
&= \left\{ n\left(\boldsymbol{r}\right) + \mathrm{i}\lambdabar \boldsymbol{\Sigma} \cdot \boldsymbol{u} \right\}^2 - \widehat{\boldsymbol{p}}_\perp^2 \\
&\quad + \left[\boldsymbol{M}_\perp \cdot \widehat{\boldsymbol{p}}_\perp , \left\{ n\left(\boldsymbol{r}\right) + \mathrm{i}\lambdabar \boldsymbol{\Sigma} \cdot \boldsymbol{u} \right\} \right] , \tag{25}
\end{aligned}
$$

where, $[A, B] = (AB - BA)$ is the commutator. It is to be noted that the square of the Hamiltonian in our formalism differs from the square of the Hamiltonian in the square-root approaches (Dragt et al., 1986; Dragt, 1988) and the scalar approach in Khan et al. (2002) and Khan (2005a, 2016g). This is essentially the same type of difference, which exists in the Dirac case. There too, the square of the Dirac Hamiltonian gives rise to extra terms, such as, $-\hbar q \boldsymbol{\Sigma} \cdot \boldsymbol{B}$, the Pauli term which couples the spin to the magnetic field. These extra terms are absent in the Schrödinger and the Klein–Gordon descriptions respectively. It is this difference in the square of the Hamiltonians which give rise to the various extra wavelength-dependent contributions in our formalism. These differences persist even in the paraxial approximation.

The beam optical Hamiltonian derived in Eq. (24) has a very close algebraic correspondence with the Dirac equation, accompanied by the analogous physical interpretations. This enables us to employ the machinery of the Foldy–Wouthuysen transformation technique. The details are available in Appendix C. To the leading order, i.e., to order, $\left(\widehat{\boldsymbol{p}}_\perp^2 / n_0^2 \right)$ the beam-optical Hamiltonian in terms of $\widehat{\mathcal{E}}$ and $\widehat{\mathcal{O}}$ is formally given by

$$
\mathrm{i}\lambdabar \frac{\partial}{\partial z} |\psi\rangle = \widehat{\mathcal{H}}^{(2)} |\psi\rangle ,
$$

$$
\widehat{\mathcal{H}}^{(2)} = -n_0 \beta + \widehat{\mathcal{E}} - \frac{1}{2n_0} \beta \widehat{\mathcal{O}}^2 . \tag{26}
$$

Note that $\widehat{\mathcal{O}}^2 = -\widehat{\boldsymbol{p}}_\perp^2$ and $\widehat{\mathcal{E}} = -\left(n\left(\boldsymbol{r}\right) - n_0 \right) \beta - \mathrm{i}\lambdabar \beta \boldsymbol{\Sigma} \cdot \boldsymbol{u}$. As we are interested in a beam propagating in the forward direction, the β has to be omitted while computing the transfer maps. The formal Hamiltonian in Eq. (26), in terms of the phase-space variables is

$$
\widehat{\mathcal{H}}^{(2)} = -\left\{ n\left(\boldsymbol{r}\right) - \frac{1}{2n_0} \widehat{\boldsymbol{p}}_\perp^2 \right\} - \mathrm{i}\lambdabar \beta \boldsymbol{\Sigma} \cdot \boldsymbol{u} . \tag{27}
$$

We keep only the terms up to quadratic in the Taylor expansion of the refractive index $n(r)$, to be consistent with the order of $(\widehat{p}_\perp^2/n_0^2)$. This is the paraxial Hamiltonian, which also contains an extra matrix dependent term. Rest of the beam optical Hamiltonian is similar to the one obtained in the traditional approaches.

The second iteration in the Foldy–Wouthuysen expansion gives the leading-order aberrating Hamiltonian describing the third-order aberrations. Note that $\widehat{O}$ is the order of $\widehat{p}_\perp$. To order $(\widehat{p}_\perp^2/n_0^2)^2$, the beam-optical Hamiltonian in terms of $\widehat{\mathcal{E}}$ and $\widehat{O}$ is formally given by

$$i\hbar \frac{\partial}{\partial z}|\psi\rangle = \widehat{\mathcal{H}}^{(4)}|\psi\rangle \,,$$

$$\widehat{\mathcal{H}}^{(4)} = -n_0\beta + \widehat{\mathcal{E}} - \frac{1}{2n_0}\beta\widehat{O}^2$$
$$-\frac{1}{8n_0^2}\left[\widehat{O}, \left([\widehat{O},\widehat{\mathcal{E}}] + i\hbar\frac{\partial}{\partial z}\widehat{O}\right)\right]$$
$$+\frac{1}{8n_0^3}\beta\left\{\widehat{O}^4 + \left([\widehat{O},\widehat{\mathcal{E}}] + i\hbar\frac{\partial}{\partial z}\widehat{O}\right)^2\right\}. \tag{28}$$

Note, that $\widehat{O}^4 = \widehat{p}_\perp^4$, and $\partial\widehat{O}/\partial z = 0$. The formal Hamiltonian in Eq. (28) in terms of the phase-space variables is

$$\widehat{\mathcal{H}}^{(4)} = -\left\{n(r) - \frac{1}{2n_0}\widehat{p}_\perp^2 - \frac{1}{8n_0^3}\widehat{p}_\perp^4\right\}$$
$$-\frac{1}{8n_0^2}[\widehat{O},[\widehat{O},\widehat{\mathcal{E}}]] + \frac{1}{8n_0^3}\beta\left([\widehat{O},\widehat{\mathcal{E}}]\right)^2 + \cdots. \tag{29}$$

The first parenthesis contains the expressions obtained in the traditional prescriptions. Rest of the terms are wavelength-dependent. Any further simplification would require information about the refractive index $n(r)$.

Note that the paraxial Hamiltonian in Eq. (27) and the leading order aberration Hamiltonian in Eq. (29) differ from the ones derived in the traditional approaches. These differences arise by the presence of the wavelength-dependent contributions, which occur in two guises. One set occurs totally independent of the matrix terms in the basic Hamiltonian. This set is a multiple of the unit matrix or at most the matrix β. The other set involves the contributions coming from the matrix terms in the starting optical Hamiltonian. This gives rise to both matrix contributions and the non-matrix contributions, as the square of each of the matrices is ($\pm$)unity. These wavelength-dependent contributions are absent in the traditional prescriptions.

3.1 Incorporating the Resistance Function in Beam-Optics

In the preceding sections, we assumed that the logarithmic gradient of the resistance function, $\boldsymbol{w} = 0$, and this enabled us to build a formalism using 4×4 matrices *via* the Foldy–Wouthuysen machinery. The Foldy–Wouthuysen transformation technique enables us to eliminate the odd part in the 4×4 Hamiltonian, to any desired order of accuracy. In eight dimensions, the problem is mathematically very similar. Hence, we can use the Foldy–Wouthuysen transformation technique to reduce the strength of the odd part in eight dimensions.

We start with the grand beam-optical equation in Eq. (20) and proceed with the Foldy–Wouthuysen transformations as before, but with each quantity in double the number of dimensions. Symbolically this means

$$\widehat{H} \longrightarrow \widehat{H}_g, \qquad \psi \longrightarrow \psi_g = \begin{bmatrix} \psi^+ \\ \psi^- \end{bmatrix},$$

$$\widehat{\mathcal{E}} \longrightarrow \widehat{\mathcal{E}}_g, \qquad \widehat{\mathcal{O}} \longrightarrow \widehat{\mathcal{O}}_g,$$

$$n_0 \longrightarrow n_g = n_0 \begin{bmatrix} \beta & 0 \\ 0 & -\beta \end{bmatrix}. \tag{30}$$

The first Foldy–Wouthuysen iteration gives

$$\widehat{\mathcal{H}}_g^{(2)} = -n_0 \begin{bmatrix} \beta & 0 \\ 0 & -\beta \end{bmatrix} + \widehat{\mathcal{E}}_g - \frac{1}{2n_0} \beta_g \widehat{\mathcal{O}}_g^2$$

$$= -n_0 \begin{bmatrix} \beta & 0 \\ 0 & \beta \end{bmatrix} \beta_g + E_g + \frac{1}{2n_0} \lambdabar^2 \boldsymbol{w} \cdot \boldsymbol{w} \begin{bmatrix} \beta & 0 \\ 0 & -\beta \end{bmatrix} \beta_g. \tag{31}$$

We drop the β_g, as before and then get the following

$$i\lambdabar \frac{\partial}{\partial z} \psi(\boldsymbol{r}) = \widehat{H} \psi(\boldsymbol{r})$$

$$\widehat{H} = -n_0 \beta + \widehat{\mathcal{E}} + \widehat{\mathcal{O}}$$

$$\widehat{\mathcal{E}} = -(n(\boldsymbol{r}) - n_0) \beta - i\lambdabar \beta \boldsymbol{\Sigma} \cdot \boldsymbol{u} + \frac{1}{2n_0} \lambdabar^2 w^2 \beta$$

$$\widehat{\mathcal{O}} = i (M_y p_x - M_x p_y)$$

$$= \beta (\boldsymbol{M}_\perp \cdot \widehat{\boldsymbol{p}}_\perp), \tag{32}$$

where, $w^2 = \boldsymbol{w} \cdot \boldsymbol{w}$, the square of the logarithmic gradient of the resistance function. This is how the basic beam optical Hamiltonian (Eq. (24)) gets modified. The next degree of accuracy is achieved by going a step further in the Foldy–Wouthuysen iteration and obtaining the $\widehat{\mathcal{H}}_g^{(4)}$. Then, this would be the higher refined starting beam optical Hamiltonian, further modifying the basic beam optical Hamiltonian in Eq. (24). This way, we can apply the Foldy–Wouthuysen in *cascade* to obtain the higher order contributions coming from the logarithmic gradient of the resistance function, to any desired degree of accuracy. Thus, it is possible to incorporate the resistance function into the beam optics formalism, knowing well that we are unlikely to need the contributions arising from it.

4. APPLICATIONS

In the preceding sections, we presented the specially developed exact eight-dimensional matrix representation of Maxwell's equations in a medium with varying permittivity and permeability following the recipe in Khan (2005b). This matrix representation formed the basis for an exact beam-optical Hamiltonian, which was in a close algebraic correspondence with the Dirac equation. This correspondence enabled us to use quantum methodologies, the Foldy–Wouthuysen transformation in particular. We obtained expressions for the paraxial and leading order aberrating Hamiltonians and developed a procedure to obtain the Hamiltonians to still higher orders. The formal expressions for the Hamiltonians were obtained without making any assumptions on the form of the varying refractive index. It was seen that even the paraxial Hamiltonian has a wavelength-dependent matrix term. This matrix term is very similar to the spin term in the Dirac equation. The aberrating Hamiltonian contains numerous wavelength-dependent terms in two guises: one of these is the explicit wavelength-dependent terms coming from the commutators inbuilt in the formalism with $\bar{\lambda}$ playing the role played by $\hbar$ in quantum mechanics. The other set arises from the matrix terms.

Now, we consider the application of the formalism to specific systems. The first example is that of a medium with a constant refractive index. This system is exactly solvable just like the free particle in relativistic quantum mechanics. This example enables us to have a closer look at the aberration expansion, and compare it with the exact result. As a second example, we shall consider the axially symmetric graded-index medium. This example enables us to demonstrate the power of the formalism, reproducing the familiar results from the traditional approaches and further giving rise to new results, dependent on the wavelength. We derive the three new aberrations permitted by the axial symmetry and predict an image rotation. Next, we examine the general case of thick and thin lenses. Explicit formulae relating the incident and emergent light rays for the general case of a thick lens are presented for this system (Khan, 2016g).

4.1 Medium with Constant Refractive Index

Medium with a constant refractive index constitutes an ideal system. This is exactly solvable as the Hamiltonian can be exactly diagonalized. This is similar to the exact diagonalization of the free particle Hamiltonians in relativistic quantum mechanics. Other systems are diagonalized approximately using the Foldy–Wouthuysen iterative procedure. Hence, we have extensively used the Foldy–Wouthuysen scheme in our optics formalism.

For any medium with a constant refractive index, $n(\boldsymbol{r}) = n_c$, we have

$$\widehat{H}_c = -n_c\beta + i\left(M_y p_x - M_x p_y\right),\tag{33}$$

which is exactly diagonalized by the following transform,

$$
\begin{aligned}
T^{\pm} &= \exp\left[i\left(\pm i\beta\right)\widehat{\mathcal{O}}\theta\right] \\
&= \exp\left[\mp i\beta\left(M_y p_x - M_x p_y\right)\theta\right] \\
&= \cosh\left(\left|\widehat{\boldsymbol{p}}_{\perp}\right|\theta\right) \mp i\frac{\beta\left(M_y p_x - M_x p_y\right)}{\left|\widehat{\boldsymbol{p}}_{\perp}\right|}\sinh\left(\left|\widehat{\boldsymbol{p}}_{\perp}\right|\theta\right).
\end{aligned}
\tag{34}
$$

We choose

$$\tanh\left(2\left|\widehat{\boldsymbol{p}}_{\perp}\right|\theta\right) = \frac{\left|\widehat{\boldsymbol{p}}_{\perp}\right|}{n_c},\tag{35}$$

then

$$T^{\pm} = \frac{\left(n_c + P_z\right)\mp i\beta\left(M_y p_x - M_x p_y\right)}{\sqrt{2P_z\left(n_c + P_z\right)}},\tag{36}$$

where $P_z = +\sqrt{\left(n_c^2 - \widehat{\boldsymbol{p}}_{\perp}^2\right)}$. Then we obtain

$$
\begin{aligned}
\widehat{H}_c^{\text{diagonal}} &= T^{+}\widehat{H}_c T^{-} \\
&= T^{+}\left\{-n_c\beta + i\left(M_y p_x - M_x p_y\right)\right\}T^{-} \\
&= -\left\{n_c^2 - \widehat{\boldsymbol{p}}_{\perp}^2\right\}^{\frac{1}{2}}\beta.
\end{aligned}
\tag{37}
$$

This exact result is to be compared with the series expansion obtained using the Foldy–Wouthuysen procedure:

$$\widehat{\mathcal{H}}_c^{(4)} = -n_c \left\{ 1 - \frac{1}{2n_c^2}\widehat{\boldsymbol{p}}_\perp^2 - \frac{1}{8n_c^4}\widehat{\boldsymbol{p}}_\perp^4 - \cdots \right\} \beta$$

$$\approx -n_c \left\{ 1 - \frac{1}{n_c^2}\widehat{\boldsymbol{p}}_\perp^2 \right\}^{\frac{1}{2}} \beta$$

$$= -\left\{ n_c^2 - \widehat{\boldsymbol{p}}_\perp^2 \right\}^{\frac{1}{2}} \beta$$

$$= \widehat{\mathrm{H}}_c^{\mathrm{diagonal}} . \tag{38}$$

The transfer maps between any pair of points $\{(z'', z') | z'' > z'\}$ on the z-axis is obtained through the transfer operator

$$\left| \psi(z'', z') \right| = \widehat{\mathcal{T}}(z'', z') \left| \psi(z'', z') \right\rangle , \tag{39}$$

with

$$\mathrm{i}\lambdabar \frac{\partial}{\partial z}\widehat{\mathcal{T}}(z'', z') = \widehat{\mathcal{H}}\widehat{\mathcal{T}}(z'', z') , \quad \widehat{\mathcal{T}}(z'', z') = \widehat{\mathcal{I}} ,$$

$$\widehat{\mathcal{T}}(z'', z') = \wp \left\{ \exp\left[-\frac{\mathrm{i}}{\lambdabar}\int_{z'}^{z''} dz\,\widehat{\mathcal{H}}(z) \right] \right\}$$

$$= \widehat{\mathcal{I}} - \frac{\mathrm{i}}{\lambdabar}\int_{z'}^{z''} dz\,\widehat{\mathcal{H}}(z)$$

$$+ \left(-\frac{\mathrm{i}}{\lambdabar} \right)^2 \int_{z'}^{z''} dz \int_{z'}^{z} dz'\,\widehat{\mathcal{H}}(z)\widehat{\mathcal{H}}(z')$$

$$+ \cdots , \tag{40}$$

where $\widehat{\mathcal{I}}$ is the identity operator and $\wp$ represents the path-ordered exponential. In general, $\widehat{\mathcal{T}}(z'', z')$ does not have a closed form expression. Following is the most useful way to express the z-evolution operator $\widehat{\mathcal{T}}(z'', z')$, or the z-propagator

$$\widehat{\mathcal{T}}(z'', z') = \exp\left[-\frac{\mathrm{i}}{\lambdabar}\widehat{T}(z'', z') \right] , \tag{41}$$

with

$$\widehat{T}(z'', z') = \int_{z'}^{z''} dz \widehat{\mathcal{H}}(z)$$
$$+ \frac{1}{2}\left(-\frac{i}{\lambda}\right) \int_{z'}^{z''} dz \int_{z'}^{z} dz' \left[\widehat{\mathcal{H}}(z), \widehat{\mathcal{H}}(z')\right]$$
$$+ \cdots . \tag{42}$$

The series expression in Eq. (42) is the Magnus formula (Blanes, Casas, Oteo, & Ros, 2009; Magnus, 1954; Mananga & Charpentier, 2016; Wilcox, 1967). Additional details are covered in Appendix D. The Magnus formula is required for the next example, where the refractive index is varying.

For a constant refractive index the transformer operator is obtained exactly as

$$\widehat{U}_c\left(z_{\text{out}}, z_{\text{in}}\right) = \exp\left[-\frac{i}{\lambda}\Delta z \widehat{\mathcal{H}}_c\right]$$
$$= \exp\left[+\frac{i}{\lambda} n_c \Delta z \left\{1 - \frac{1}{2}\frac{\widehat{p}_\perp^2}{n_c^2} - \frac{1}{8}\left(\frac{\widehat{p}_\perp^2}{n_c^2}\right)^2 - \cdots\right\}\right], \tag{43}$$

where, $\Delta z = (z_{\text{out}}, z_{\text{in}})$. Using Eq. (43), we obtain the transfer maps

$$\begin{pmatrix} \langle \mathbf{r}_\perp \rangle \\ \langle \mathbf{p}_\perp \rangle \end{pmatrix}_{\text{out}} = \begin{pmatrix} 1 & \frac{1}{\sqrt{n_c^2 - \mathbf{p}_\perp^2}}\Delta z \\ 0 & 1 \end{pmatrix} \begin{pmatrix} \langle \mathbf{r}_\perp \rangle \\ \langle \mathbf{p}_\perp \rangle \end{pmatrix}_{\text{in}}. \tag{44}$$

It is to be noted that the beam-optical Hamiltonian is inherently aberrating. Even the ideal system of a constant refractive index has aberrations to all orders (Dragt et al., 1986).

4.2 Axially Symmetric Graded-Index Medium

In the first example, we considered the ideal system of constant refractive index, which we did both exactly and approximately. The second example is that of the axially symmetric graded-index medium, which cannot be done exactly. The refractive index for this system is expressed as an infinite series (see p. 117 in Dragt et al., 1986)

$$n(\mathbf{r}) = n_0 + \alpha_2(z)\mathbf{r}_\perp^2 + \alpha_4(z)\mathbf{r}_\perp^4 + \cdots . \tag{45}$$

We assume that the optic-axis is along the z-axis. The axial symmetry does not permit the other powers of $\boldsymbol{r}_\perp$. We note

$$\widehat{\mathcal{E}} = -\left\{\alpha_2(z)r_\perp^2 + \alpha_4(z)r_\perp^4 + \cdots\right\}\beta - i\lambdabar\beta\boldsymbol{\Sigma}\cdot\boldsymbol{u}$$
$$\widehat{\mathcal{O}} = i\left(M_y p_x - M_x p_y\right)$$
$$= \beta\left(\boldsymbol{M}_\perp\cdot\widehat{\boldsymbol{p}}_\perp\right), \tag{46}$$

where

$$\boldsymbol{\Sigma}\cdot\boldsymbol{u} = -\frac{1}{n_0}\alpha_2(z)\boldsymbol{\Sigma}_\perp\cdot\boldsymbol{r}_\perp - \frac{1}{2n_0}\left(\frac{d}{dz}\alpha_2(z)\right)\Sigma_z r_\perp^2. \tag{47}$$

To simplify the formal expression for the beam-optical Hamiltonian $\widehat{\mathcal{H}}^{(4)}$ given in Eqs. (28)–(29), we make use of the following

$$\left(\boldsymbol{M}_\perp\cdot\widehat{\boldsymbol{p}}_\perp\right)^2 = \widehat{\boldsymbol{p}}_\perp^2, \qquad \widehat{\mathcal{O}}^2 = -\widehat{\boldsymbol{p}}_\perp^2, \qquad \frac{\partial}{\partial z}\widehat{\mathcal{O}} = 0,$$

$$\left(\boldsymbol{M}_\perp\cdot\widehat{\boldsymbol{p}}_\perp\right)r_\perp^2\left(\boldsymbol{M}_\perp\cdot\widehat{\boldsymbol{p}}_\perp\right) = \frac{1}{2}\left(r_\perp^2\widehat{\boldsymbol{p}}_\perp^2 + \widehat{\boldsymbol{p}}_\perp^2 r_\perp^2\right) + 2\lambdabar\beta\widehat{L}_z + 2\lambdabar^2, \tag{48}$$

where, $\widehat{L}_z$ is the angular momentum operator. Finally, the beam-optical Hamiltonian to order $\left(\widehat{\boldsymbol{p}}_\perp^2/n_0^2\right)$ is

$$\widehat{\mathcal{H}} = \widehat{H}_{0,p} + \widehat{H}_{0,(4)}$$
$$+ \widehat{H}_{0,(2)}^{(\lambdabar)} + \widehat{H}_{0,(4)}^{(\lambdabar)}$$
$$+ \widehat{H}^{(\lambdabar,\sigma)}$$

$$\widehat{H}_{0,p} = -n_0 + \frac{1}{2n_0}\widehat{\boldsymbol{p}}_\perp^2 - \alpha_2(z)r_\perp^2$$

$$\widehat{H}_{0,(4)} = \frac{1}{8n_0^3}\widehat{\boldsymbol{p}}_\perp^4$$
$$- \frac{\alpha_2(z)}{4n_0^2}\left(r_\perp^2\widehat{\boldsymbol{p}}_\perp^2 + \widehat{\boldsymbol{p}}_\perp^2 r_\perp^2\right)$$
$$- \alpha_4(z)r_\perp^4$$

$$\widehat{H}_{0,(2)}^{(\lambdabar)} = -\frac{\lambdabar^2}{2n_0^2}\alpha_2(z) - \frac{\lambdabar}{2n_0^2}\alpha_2(z)\widehat{L}_z + \frac{\lambdabar^2}{2n_0^3}\alpha_2^2(z)r_\perp^2$$

$$\widehat{H}_{0,(4)}^{(\lambdabar)} = \frac{\lambdabar}{4n_0^3}\alpha_2^2(z)\left(r_\perp^2\widehat{L}_z + \widehat{L}_z r_\perp^2\right) + \frac{\lambdabar^2}{2n_0^3}\alpha_2(z)\alpha_4(z)r_\perp^4$$

$$\widehat{H}^{(\lambdabar,\sigma)} = \frac{i\lambdabar^3}{2n_0^3}\left\{\frac{d}{dz}\alpha_2(z)\right\}\beta\Sigma_z$$

$$+ \frac{i\hbar^2}{4n_0^3} \alpha_2(z) \left(\Sigma_x p_y - \Sigma_y p_x \right)$$

$$+ \frac{i\hbar^3}{2n_0^3} \left\{ \frac{d}{dz} \alpha_2(z) \right\} \Sigma_z \widehat{L}_z$$

$$+ \frac{i\hbar}{4n_0^3} \alpha_2(z)\beta \left[\mathbf{\Sigma}_\perp \cdot \mathbf{r}_\perp, \widehat{\mathbf{p}}_\perp^2 \right]_+$$

$$+ \frac{i\hbar}{8n_0^3} \left\{ \frac{d}{dz} \alpha_2(z) \right\} \beta \Sigma_z \left[\mathbf{r}_\perp^2, \widehat{\mathbf{p}}_\perp^2 \right]_+$$

$$+ \cdots \tag{49}$$

where $[A, B]_+ = (AB + BA)$ and '$\cdots$' are the numerous other terms arising from the matrix terms. We have retained only the leading order of such terms above for an illustration.

The reasons for partitioning the beam-optical Hamiltonian $\widehat{\mathcal{H}}$ in the above manner are as follows. The paraxial Hamiltonian, $\widehat{H}_{0,p}$, describes the ideal behavior. $\widehat{H}_{0,(4)}$ is responsible for the third-order aberrations. Both of these Hamiltonians are modified by the wavelength-dependent contributions given in $\widehat{H}_{0,(2)}^{(\hbar)}$ and $\widehat{H}_{0,(4)}^{(\hbar)}$ respectively. Lastly, we have $\widehat{H}^{(\hbar,\sigma)}$, which is associated with the polarization.

4.2.1 The Paraxial Hamiltonian

The modified paraxial Hamiltonian for an axially symmetric graded-index medium is

$$\widehat{\mathcal{H}} = \widehat{H}_{0,p} + + \widehat{H}_{0,(2)}^{(\hbar)} + \cdots$$

$$= -n_0 + \frac{1}{2n_0} \widehat{\mathbf{p}}_\perp^2 - \alpha_2(z) \mathbf{r}_\perp^2$$

$$- \frac{\hbar^2}{2n_0^2} \alpha_2(z) - \frac{\hbar}{2n_0^2} \alpha_2(z) \widehat{L}_z + \frac{\hbar^2}{2n_0^3} \alpha_2^2(z) \mathbf{r}_\perp^2 + \cdots \tag{50}$$

where '$\cdots$' are the contributions arising from the matrix terms. These extra terms are in the Hamiltonian are sure to influence the beam optics of the system.

The paraxial transfer maps are expressed as

$$\begin{pmatrix} \langle \mathbf{r}_\perp \rangle \\ \langle \mathbf{p}_\perp \rangle \end{pmatrix}_{\text{out}} = \begin{pmatrix} P & Q \\ R & S \end{pmatrix} \begin{pmatrix} \langle \mathbf{r}_\perp \rangle \\ \langle \mathbf{p}_\perp \rangle \end{pmatrix}_{\text{in}}. \tag{51}$$

The functions, P, Q, R and S are the solutions of the paraxial Hamiltonian (49). The four functions are related as $PS - QR = 1$, by the virtue of the symplecticity. We have an additional relation originating from the structure of the paraxial equations: $R = P'$ and $S = Q'$ where $(\ )'$ is the z-derivative.

4.2.2 Image Rotation

The traditional paraxial Hamiltonian of light optics (Dragt et al., 1986) gets modified by several wavelength-dependent terms. The term $\frac{\lambdabar}{2n_0^2}\alpha_2(z)\widehat{L}_z$ is of extra interest to us. In charged-particle optics the $\widehat{L}_z$ term in the Hamiltonian of the axially symmetric magnetic lens, is responsible for the *image rotation* (Hawkes & Kasper, 1989a, 1989b; Ryne & Dragt, 1991; Jagannathan & Khan, 1996; Khan, 1997). With similar reasoning, we conclude that there exists an image rotation in the axially symmetric graded-index medium and its magnitude is (Khan, 2006c, 2010, 2014a, 2016f, 2017c)

$$\theta(z'', z') = \frac{\lambdabar}{2n_0^2}\int_{z'}^{z''} dz\,\alpha_2(z)\,. \tag{52}$$

This wavelength-dependent image rotation (which need not be small) has no analogue in the *square-root approach* (Dragt et al., 1986; Dragt, 1988) and the recently developed non-traditional formalism of Helmholtz optics (Khan et al., 2002; Khan, 2005a, 2016g).

4.2.3 Aberrations

The Hamiltonian $\widehat{H}_{0,(4)}$ in Eq. (49) is the familiar one from the traditional prescriptions and is responsible for the six aberrations. The term, $\widehat{H}_{0,(4)}^{(\lambdabar)}$ modifies the above six aberrations by wavelength-dependent contributions. Moreover, this term leads to the remaining three aberrations permitted by the axial symmetry. The axial symmetry permits *exactly* nine third-order aberrations, which are enumerated in Table 1. The name *POCUS* is used in Dragt et al. (1986) on page 137.

The axial symmetry allows only the terms (in the Hamiltonian) which are produced out of, $\widehat{\boldsymbol{p}}_\perp^2$, $r_\perp^2$, $(\widehat{\boldsymbol{p}}_\perp \cdot \boldsymbol{r}_\perp + \boldsymbol{r}_\perp \cdot \widehat{\boldsymbol{p}}_\perp)$ and $\widehat{L}_z$. Combinatorially, to fourth-order one would get ten terms including $\widehat{L}_z^2$. We have listed nine of them in Table 1. The tenth one is

$$\widehat{L}_z^2 = \frac{1}{2}\left(\widehat{\boldsymbol{p}}_\perp^2 r_\perp^2 + r_\perp^2 \widehat{\boldsymbol{p}}_\perp^2\right) - \frac{1}{4}\left(\widehat{\boldsymbol{p}}_\perp \cdot \boldsymbol{r}_\perp + \boldsymbol{r}_\perp \cdot \widehat{\boldsymbol{p}}_\perp\right)^2 + \lambdabar^2\,. \tag{53}$$

Table 1 The nine third-order aberrations permitted by the axial symmetry

Symbol	Polynomial	Name
C	$\widehat{\boldsymbol{p}}_{\perp}^4$	Spherical aberration
K	$[\widehat{\boldsymbol{p}}_{\perp}^2, (\widehat{\boldsymbol{p}}_{\perp} \cdot \boldsymbol{r}_{\perp} + \boldsymbol{r}_{\perp} \cdot \widehat{\boldsymbol{p}}_{\perp})]_+$	Coma
k	$\widehat{\boldsymbol{p}}_{\perp}^2 \widehat{L}_z$	Anisotropic coma
A	$(\widehat{\boldsymbol{p}}_{\perp} \cdot \boldsymbol{r}_{\perp} + \boldsymbol{r}_{\perp} \cdot \widehat{\boldsymbol{p}}_{\perp})^2$	Astigmatism
a	$(\widehat{\boldsymbol{p}}_{\perp} \cdot \boldsymbol{r}_{\perp} + \boldsymbol{r}_{\perp} \cdot \widehat{\boldsymbol{p}}_{\perp}) \widehat{L}_z$	Anisotropic astigmatism
F	$(\widehat{\boldsymbol{p}}_{\perp}^2 \boldsymbol{r}_{\perp}^2 + \boldsymbol{r}_{\perp}^2 \widehat{\boldsymbol{p}}_{\perp}^2)$	Curvature of field
D	$[\boldsymbol{r}_{\perp}^2, (\widehat{\boldsymbol{p}}_{\perp} \cdot \boldsymbol{r}_{\perp} + \boldsymbol{r}_{\perp} \cdot \widehat{\boldsymbol{p}}_{\perp})]_+$	Distortion
d	$\boldsymbol{r}_{\perp}^2 \widehat{L}_z$	Anisotropic distortion
E	$\boldsymbol{r}_{\perp}^4$	Nameless or $POCUS$

So, $\widehat{L}_z^2$ is not listed separately. Hence, we have only nine third-order aberrations permitted by axial symmetry, as stated earlier.

Next, we have the task of obtaining the transfer operator. This is done elegantly using the paraxial solutions, P, Q, R, and S, through the *interaction picture* from the Lie algebraic formulation of light beam optics and charged-particle beam optics respectively (Dragt & Forest, 1986):

$$
\begin{aligned}
\widehat{\mathcal{T}}(z, z_0) &= \exp\left[-\frac{\mathrm{i}}{\lambda} \widehat{T}(z, z_0)\right] \\
&= \exp\left[-\frac{\mathrm{i}}{\lambda}\left\{ C\left(z'', z'\right) \widehat{\boldsymbol{p}}_{\perp}^4 \right.\right. \\
&\quad + K\left(z'', z'\right) \left[\widehat{\boldsymbol{p}}_{\perp}^2, (\widehat{\boldsymbol{p}}_{\perp} \cdot \boldsymbol{r}_{\perp} + \boldsymbol{r}_{\perp} \cdot \widehat{\boldsymbol{p}}_{\perp})\right]_+ \\
&\quad + k\left(z'', z'\right) \widehat{\boldsymbol{p}}_{\perp}^2 \widehat{L}_z \\
&\quad + A\left(z'', z'\right) (\widehat{\boldsymbol{p}}_{\perp} \cdot \boldsymbol{r}_{\perp} + \boldsymbol{r}_{\perp} \cdot \widehat{\boldsymbol{p}}_{\perp})^2 \\
&\quad + a\left(z'', z'\right) (\widehat{\boldsymbol{p}}_{\perp} \cdot \boldsymbol{r}_{\perp} + \boldsymbol{r}_{\perp} \cdot \widehat{\boldsymbol{p}}_{\perp}) \widehat{L}_z \\
&\quad + F\left(z'', z'\right) (\widehat{\boldsymbol{p}}_{\perp}^2 \boldsymbol{r}_{\perp}^2 + \boldsymbol{r}_{\perp}^2 \widehat{\boldsymbol{p}}_{\perp}^2) \\
&\quad + D\left(z'', z'\right) \left[\boldsymbol{r}_{\perp}^2, (\widehat{\boldsymbol{p}}_{\perp} \cdot \boldsymbol{r}_{\perp} + \boldsymbol{r}_{\perp} \cdot \widehat{\boldsymbol{p}}_{\perp})\right]_+ \\
&\quad + d\left(z'', z'\right) \boldsymbol{r}_{\perp}^2 \widehat{L}_z \\
&\quad \left.\left. + E\left(z'', z'\right) \boldsymbol{r}_{\perp}^4 \right\}\right].
\end{aligned}
\tag{54}
$$

The nine aberration coefficients are

$$
C\left(z'',z'\right) = \int_{z'}^{z''} dz \left\{ \frac{1}{8n_0^3} S^4 - \frac{\alpha_2(z)}{2n_0^2} Q^2 S^2 - \alpha_4(z) Q^4 \right. \\
\left. + \frac{\hbar^2}{2n_0^3} \alpha_2(z)\alpha_4(z) Q^4 \right\}
$$

$$
K\left(z'',z'\right) = \int_{z'}^{z''} dz \left\{ \frac{1}{8n_0^3} RS^3 - \frac{\alpha_2(z)}{4n_0^2} QS(PS+QR) - \alpha_4(z) PQ^3 \right. \\
\left. + \frac{\hbar^2}{2n_0^3} \alpha_2(z)\alpha_4(z) PQ^3 \right\}
$$

$$
k\left(z'',z'\right) = \frac{\hbar}{2n_0^3} \int_{z'}^{z''} dz \alpha_2^2(z) Q^2
$$

$$
A\left(z'',z'\right) = \int_{z'}^{z''} dz \left\{ \frac{1}{8n_0^3} R^2 S^2 - \frac{\alpha_2(z)}{2n_0^2} PQRS - \alpha_4(z) P^2 Q^2 \right. \\
\left. + \frac{\hbar^2}{2n_0^3} \alpha_2(z)\alpha_4(z) P^2 Q^2 \right\}
$$

$$
a\left(z'',z'\right) = \frac{\hbar}{2n_0^3} \int_{z'}^{z''} dz \alpha_2^2(z) PQ
$$

$$
F\left(z'',z'\right) = \int_{z'}^{z''} dz \left\{ \frac{1}{8n_0^3} R^2 S^2 - \frac{\alpha_2(z)}{4n_0^2} (P^2 S^2 + Q^2 R^2) - \alpha_4(z) P^2 Q^2 \right. \\
\left. + \frac{\hbar^2}{2n_0^3} \alpha_2(z)\alpha_4(z) P^2 Q^2 \right\}
$$

$$
D\left(z'',z'\right) = \int_{z'}^{z''} dz \left\{ \frac{1}{8n_0^3} R^3 S - \frac{\alpha_2(z)}{4n_0^2} PR(PS+QR) - \alpha_4(z) P^3 Q \right. \\
\left. + \frac{\hbar^2}{2n_0^3} \alpha_2(z)\alpha_4(z) P^3 Q \right\}
$$

$$
d\left(z'',z'\right) = \frac{\hbar}{2n_0^3} \int_{z'}^{z''} dz \alpha_2^2(z) P^2
$$

$$
E\left(z'',z'\right) = \int_{z'}^{z''} dz \left\{ \frac{1}{8n_0^3} R^4 - \frac{\alpha_2(z)}{2n_0^2} P^2 R^2 - \alpha_4(z) P^4 \right. \\
\left. + \frac{\hbar^2}{2n_0^3} \alpha_2(z)\alpha_4(z) P^4 \right\} . \tag{55}
$$

Thus, we see that the current approach gives rise to all the nine permissible aberrations. The six aberrations, familiar from the traditional prescriptions get modified by the wavelength-dependent contributions. The extra three (k, a and d, all anisotropic!) are all pure wavelength-dependent aberrations and totally absent in the traditional *square-root approach* (Dragt et al., 1986; Dragt, 1988) and the recently developed non-traditional formalism of Helmholtz optics (Khan et al., 2002; Khan, 2005a, 2016g).

4.3 General Thick Lens

Let us consider a thick lens of thickness $t = t_1 + t_2$ as shown in Fig. 1. Let the refractive index of the lens be n and that of the surrounding medium to be unity without loss of generality. The optic-axis is along z-axis; x and y are along the face of the lens. The following equations of the curved surfaces of the lens describe the lens system

$$z_1(\mathbf{r}_\perp) = \gamma_2(r_\perp^2) + \gamma_4(r_\perp^4) + \gamma_6(r_\perp^6) + \cdots,$$
$$z_2(\mathbf{r}_\perp) = \beta_2(r_\perp^2) + \beta_4(r_\perp^4) + \beta_6(r_\perp^6) + \cdots. \tag{56}$$

The other powers of $\mathbf{r}_\perp$ are absent due to the assumed axial symmetry.

The transfer operator for this lens is given by the product of three transfer operators corresponding to the two surfaces and the transit through the lens respectively. The transfer operator within the paraxial approximation leads to the transfer map

$$\begin{pmatrix} \langle \mathbf{r}_\perp \rangle \\ \langle \mathbf{p}_\perp \rangle \end{pmatrix}_{\text{out}} = S \begin{pmatrix} \langle \mathbf{r}_\perp \rangle \\ \langle \mathbf{p}_\perp \rangle \end{pmatrix}_{\text{in}}, \tag{57}$$

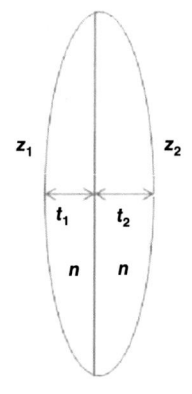

Figure 1 Lens of thickness $t = t_1 + t_2$ and uniform refractive index n.

where

$$S = \begin{bmatrix} 1 & 0 \\ 2\beta_2(1-n) & 1 \end{bmatrix} \begin{bmatrix} 1 & t/n \\ 0 & 1 \end{bmatrix} \begin{bmatrix} 1 & 0 \\ 2\gamma_2(1-n) & 1 \end{bmatrix}$$

$$= \begin{bmatrix} 1 + 2\gamma_2(1-n)t/n & t/n \\ 2(1-n)\left[(\beta_2+\gamma_2) + 2(1-n)\gamma_2\beta_2 t/n\right] & 1 + 2\beta_2(1-n)t/n \end{bmatrix}. \tag{58}$$

Let the spherical lens have surfaces with radii of curvature R_1 and R_2 respectively. Then, the constants can be identified as $\gamma_2 = 1/2R_1$ and $\beta_2 = -1/2R_2$. Then the transfer map S simplifies to

$$S = \begin{bmatrix} 1 + (n-1)t/nR_1 & t/n \\ (1-n)\left[\left(\frac{1}{R_1} - \frac{1}{R_2}\right) + \frac{(n-1)t}{nR_1R_2}\right] & 1 - (n-1)t/nR_2 \end{bmatrix}. \tag{59}$$

This is the well known $ABCD$ matrix or $\begin{bmatrix} A & B \\ C & D \end{bmatrix}$ also known as the Kogelnik's ABCD-law (Kogelnik, 1965). The dioptric power F is

$$F = -C = \frac{1}{f} = (n-1)\left[\left(\frac{1}{R_1} - \frac{1}{R_2}\right) + \frac{(n-1)t}{nR_1R_2}\right]. \tag{60}$$

This is the familiar relation giving the power of a general thick lens along with the Lensmaker's equation. The example of the thick lens has been done in detail using Lie algebraic methods in Lakshminarayanan et al. (1998, 2002) and Rangarajan and Sachidanand (1997). The same system has been worked out using quantum methodologies (Khan, 2016g).

5. POLARIZATION

In the preceding section, we saw the development of beam optics using a matrix formulation of Maxwell's equations. The scalar optics was generalized to wave optics by Mukunda, Simon and Sudarshan for paraxial systems. The Mukunda–Simon–Sudarshan (MSS) theory for transition from scalar to vector wave optics is based on a group theoretic approach and takes care of the polarization. The matrix formulation of Maxwell optics, we have developed describes both beam optics and polarization. The MSS theory is obtained as a leading order approximation of our formalism. Our formalism is an appropriate candidate to extend the MSS-theory beyond the paraxial approximation (Khan, 2016f). In this section, we shall derive the elegant MSS-rule for transition from scalar to vector optics. We shall then apply the MSS-rule to the Gaussian beams.

5.1 Mukunda–Simon–Sudarshan Rule

The beam-optical Hamiltonian in our formalism is exact and does not have any assumptions on the form of the spatially varying refractive index. In the initial studies of polarization, it suffices to focus on the restricted case of a *constant* refractive index. The MSS-rule is obtained even with this simplification! Consequently, in this section, we shall confine our derivations to the case of a constant refractive index. When the refractive index has a constant value, $n(\mathbf{r}) = n_c$, the beam-optical Hamiltonian in Eq. (24) reduces to

$$\widehat{\mathsf{H}}_c = -n_c\beta + \beta\left(\mathbf{M}_\perp \cdot \widehat{\mathbf{p}}_\perp\right)$$
$$= -n_c\beta + \mathrm{i}\left(M_y p_x - M_x p_y\right). \tag{61}$$

The Dirac Hamiltonian can be exactly diagonalized for a free particle. Similar is the situation, when the refractive index is a constant. The required transform in the Foldy–Wouthuysen procedure is

$$\psi \quad \longrightarrow \quad \psi_0 = T^+\psi = \frac{(n_c + P_z) - \mathrm{i}\beta\left(M_y p_x - M_x p_y\right)}{\sqrt{2P_z(n_c + P_z)}}\psi, \tag{62}$$

where $P_z = +\sqrt{n_c^2 - \widehat{\mathbf{p}}_\perp^2}$. This transforms the basic beam-optical equation in Eq. (61) to the exactly diagonalized form

$$\mathrm{i}\hbar\frac{\partial}{\partial z}\psi_0 = -\beta\left\{n_c^2 - \widehat{\mathbf{p}}_\perp^2\right\}^{\frac{1}{2}}\psi_0. \tag{63}$$

The square-roots in Eq. (62) are to expanded in a power series, in $(\widehat{\mathbf{p}}_\perp^2/n_c^2)$, to the required degree of accuracy. From Eq. (63), we conclude that the components of ψ_0 behave as 'scalars' as the beam propagates along the z-axis. In scalar optics, it is assumed that the components of $\mathbf{E}$ and $\mathbf{B}$, which behave as scalars; this is not the case. It is the components of the defined vector ψ_0 in Eq. (8) and Eq. (62), which behave as scalars.

In the input free-space region, Eq. (63) is

$$\mathrm{i}\hbar\frac{\partial}{\partial z}\psi_{0(\mathrm{in})} = -\beta P_z\psi_{0(\mathrm{in})},$$

$$\psi_{0(\mathrm{in})} = T^+\psi_{(\mathrm{in})} = \frac{(n_c + P_z) - \mathrm{i}\beta\left(M_y p_x - M_x p_y\right)}{\sqrt{2P_z(n_c + P_z)}}\psi_{(\mathrm{in})}. \tag{64}$$

The input beam is in the forward direction along the z-axis (optic axis). So,

$$i \hbar \frac{\partial}{\partial z} \psi_{(in)} = -P_z \psi_{(in)} . \tag{65}$$

We note, that p_x, p_y and $\partial/\partial z$ commute with each other and also with β. So, P_z commutes with β. We also note, that $\partial/\partial z$, P_z, T^+ and β commute among themselves. The mutual consistency of Eq. (62) and Eq. (65) requires that

$$\beta \psi_{0(in)} = \psi_{0(in)} = \begin{bmatrix} \psi_{0u(in)} \\ \psi_{0L(in)} \end{bmatrix} . \tag{66}$$

Consequently, the lower part of $\psi_{0(in)}$

$$\psi_{0L(in)} = 0 . \tag{67}$$

The vanishing of the lower part has consequences. The z-evolution described by Eq. (63) for ψ_0 does not couple the upper and lower components of ψ_0. So, the lower part of $\psi_0(z)$ is identically zero for all values of z, as the beam propagates. So, in all generality,

$$\psi_{0L}(z) = 0 . \tag{68}$$

The z-evolution of the two-component upper part $\psi_{0u}(z)$ is given by

$$i \hbar \frac{\partial}{\partial z} \psi_{0u} = - \left\{ n^2 - \widehat{\boldsymbol{p}}_\perp^2 \right\}^{\frac{1}{2}} \psi_{0u} = \widehat{H} \psi_{0u} . \tag{69}$$

The beam is now completely characterized jointly by the constraints in Eq. (68) and the z-evolution in Eq. (69).

The constraints in Eq. (68) has interesting implications on the beam fields. In the region of free space, we have

$$\psi_0 = T^+ \psi = \frac{1}{\sqrt{2P_z (n_c + P_z)}} \begin{bmatrix} (n_c + P_z)(-F_x + iF_y) + (-p_x + ip_y)F_z \\ (n_c + P_z)F_z + (-p_x + ip_y)(F_x + iF_y) \\ (-p_x - ip_y)(-F_x + iF_y) + (n_c + P_z)F_z \\ (-p_x - ip_y)F_z + (n_c + P_z)(F_x + iF_y) \end{bmatrix} . \tag{70}$$

The constraints $\psi_{0L}(z) = 0$ in Eq. (68) in the region of free space translates to the following relations

$$E_z = \frac{i\lambdabar}{(n_c + P_z)} \{(\mathbf{\nabla}_\perp \cdot \mathbf{E}_\perp) + v(\mathbf{\nabla} \times \mathbf{B})_z\},$$

$$B_z = +\frac{i\lambdabar}{(n_c + P_z)} \left\{(\mathbf{\nabla}_\perp \cdot \mathbf{B}_\perp) - \frac{1}{v}(\mathbf{\nabla} \times \mathbf{E})_z\right\},$$

$$B_x = -\frac{1}{v}E_y - \frac{i\lambdabar}{(n_c + P_z)} \left\{\frac{1}{v}\frac{\partial}{\partial y}E_z + \frac{\partial}{\partial x}B_z\right\},$$

$$B_y = +\frac{1}{v}E_x + \frac{i\lambdabar}{(n_c + P_z)} \left\{\frac{1}{v}\frac{\partial}{\partial x}E_z - \frac{\partial}{\partial y}B_z\right\}. \tag{71}$$

These relations hold for any beam moving in the direction of $+z$-axis. The aforementioned relations in Eq. (71) are consistent with the results of Mukunda, Simon and Sudarshan, which were obtained using a group theoretical analysis (Mukunda et al., 1983a, 1983b, 1985a, 1985b; Simon et al., 1986, 1987).

Let us now examine the implications of the constraints in Eq. (68) on the z-evolution of the fields $(\mathbf{E}, \mathbf{B})$. The values of the fields in the output plane at z'' and the input plane at z' is formally given by integrating the Hamiltonian in Eq. (63)

$$\psi_{0u}(z'') = \wp \left\{\exp\left[-\frac{i}{\lambdabar}\int_{z'}^{z''} dz\widehat{H}_0(z)\right]\right\}\psi_0(z')$$

$$= \widehat{G}_0(z'', z')\psi_{0u}(z')$$

$$= \exp\left[-\frac{i}{\lambdabar}\phi(z'', z')\right]\psi_{0u}(z'), \tag{72}$$

where $\wp$ signifies the z-ordering (Blanes et al., 2009; Magnus, 1954; Mananga & Charpentier, 2016; Wilcox, 1967). We write the z-evolution of the fields as

$$\begin{bmatrix} \sqrt{\epsilon}\mathbf{E} \\ \frac{1}{\sqrt{\mu}}\mathbf{B} \end{bmatrix}(z'') = \widehat{G}_0(z'', z')\begin{bmatrix} \sqrt{\epsilon}\mathbf{E} \\ \frac{1}{\sqrt{\mu}}\mathbf{B} \end{bmatrix}(z'). \tag{73}$$

Now, using Eq. (73) along with the conditions on the fields in Eq. (71), we express the output fields $(\mathbf{E}(z''), \mathbf{B}(z''))$, in terms of the input fields $(\mathbf{E}(z'), \mathbf{B}(z'))$. After a straightforward algebra, we obtain the relations given below

$$E_x(z'') = \widehat{G}_0 E_x(z') - \widehat{G}_0 \frac{1}{n_c + P_z}\left\{E_z(z')\frac{\partial\phi}{\partial x} - vB_z(z')\frac{\partial\phi}{\partial y}\right\},$$

$$E_y(z'') = \widehat{G}_0 E_y(z') - \widehat{G}_0 \frac{1}{n_c + P_z} \left\{ E_z(z') \frac{\partial \phi}{\partial y} + v B_z(z') \frac{\partial \phi}{\partial x} \right\},$$

$$E_z(z'') = \widehat{G}_0 \mathbf{\nabla}_\perp \cdot \mathbf{E}_\perp(z')$$
$$+ \widehat{G}_0 \frac{1}{n_c + P_z} \left\{ E_x(z') \frac{\partial \phi}{\partial x} + E_y(z') \frac{\partial \phi}{\partial y} \right.$$
$$+ v \left(B_y(z') \frac{\partial \phi}{\partial x} - B_x(z') \frac{\partial \phi}{\partial y} \right) \right\},$$

$$B_x(z'') = \widehat{G}_0 B_x(z') - \widehat{G}_0 \frac{1}{n_c + P_z} \left\{ \frac{1}{v} E_z(z') \frac{\partial \phi}{\partial y} + B_z(z') \frac{\partial \phi}{\partial x} \right\},$$

$$B_y(z'') = \widehat{G}_0 B_y(z') + \widehat{G}_0 \frac{1}{n_c + P_z} \left\{ \frac{1}{v} E_z(z') \frac{\partial \phi}{\partial x} - B_z(z') \frac{\partial \phi}{\partial y} \right\},$$

$$B_z(z'') = \widehat{G}_0 B_z(z')$$
$$+ \widehat{G}_0 \frac{1}{n_c + P_z} \left\{ B_x(z') \frac{\partial \phi}{\partial x} + B_y(z') \frac{\partial \phi}{\partial y} \right.$$
$$- \frac{1}{v} \left(E_y(z') \frac{\partial \phi}{\partial x} - E_x(z') \frac{\partial \phi}{\partial y} \right) \right\}. \tag{74}$$

The aforementioned six relations for the fields in Eq. (74) are summarized in the matrix below

$$\begin{bmatrix} \sqrt{\epsilon} E_x \\ \sqrt{\epsilon} E_y \\ \sqrt{\epsilon} E_z \\ \frac{1}{\sqrt{\mu}} B_x \\ \frac{1}{\sqrt{\mu}} B_y \\ \frac{1}{\sqrt{\mu}} B_z \end{bmatrix} (z'') = \widehat{G}_0(z'', z') \mathcal{M} \begin{bmatrix} \sqrt{\epsilon} E_x \\ \sqrt{\epsilon} E_y \\ \sqrt{\epsilon} E_z \\ \frac{1}{\sqrt{\mu}} B_x \\ \frac{1}{\sqrt{\mu}} B_y \\ \frac{1}{\sqrt{\mu}} B_z \end{bmatrix} (z'), \tag{75}$$

where

$$\mathcal{M} = \begin{bmatrix} 1 & 0 & -\frac{1}{2n_c}\left(\frac{\partial \phi}{\partial x}\right) & 0 & 0 & \frac{1}{2n_c}\left(\frac{\partial \phi}{\partial y}\right) \\ 0 & 1 & -\frac{1}{2n_c}\left(\frac{\partial \phi}{\partial y}\right) & 0 & 0 & -\frac{1}{2n_c}\left(\frac{\partial \phi}{\partial x}\right) \\ \frac{1}{2n_c}\left(\frac{\partial \phi}{\partial x}\right) & \frac{1}{2n_c}\left(\frac{\partial \phi}{\partial y}\right) & 1 & -\frac{1}{2n_c}\left(\frac{\partial \phi}{\partial y}\right) & \frac{1}{2n_c}\left(\frac{\partial \phi}{\partial x}\right) & 0 \\ 0 & 0 & -\frac{1}{2n_c}\left(\frac{\partial \phi}{\partial y}\right) & 1 & 0 & -\frac{1}{2n_c}\left(\frac{\partial \phi}{\partial x}\right) \\ 0 & 0 & \frac{1}{2n_c}\left(\frac{\partial \phi}{\partial x}\right) & 0 & 1 & -\frac{1}{2n_c}\left(\frac{\partial \phi}{\partial y}\right) \\ \frac{1}{2n_c}\left(\frac{\partial \phi}{\partial y}\right) & -\frac{1}{2n_c}\left(\frac{\partial \phi}{\partial x}\right) & 0 & \frac{1}{2n_c}\left(\frac{\partial \phi}{\partial x}\right) & \frac{1}{2n_c}\left(\frac{\partial \phi}{\partial y}\right) & 1 \end{bmatrix}. \tag{76}$$

We have kept only the leading order terms in the matrix $\mathcal{M}$ in Eq. (76). This is in view of the paraxial approximation, when approximating

$$\frac{1}{n_c + P_z} = \frac{1}{2n_c} + \frac{1}{8n_c^3}\widehat{\boldsymbol{p}}_\perp^2 + \ldots \approx \frac{1}{2n_c}. \tag{77}$$

We follow the notation in Mukunda et al. (1983a) and have

$$G_x = \frac{1}{2}\begin{bmatrix} -S_2 & S_1 \\ -S_1 & -S_2 \end{bmatrix}, \quad G_y = \frac{1}{2}\begin{bmatrix} S_1 & S_2 \\ -S_2 & S_1 \end{bmatrix},$$

$$S_1 = \begin{bmatrix} 0 & 0 & 0 \\ 0 & 0 & -i \\ 0 & i & 0 \end{bmatrix}, \quad S_2 = \begin{bmatrix} 0 & 0 & i \\ 0 & 0 & 0 \\ -i & 0 & 0 \end{bmatrix}. \tag{78}$$

The matrices G_x and G_y satisfy the following identities

$$G_x^2 + G_y^2 = 0, \quad G_x^3 = G_y^3 = 0, \quad G_x G_y = G_y G_x. \tag{79}$$

If $\phi(\boldsymbol{r}_\perp)$ is a general polynomial in $(\boldsymbol{r}_\perp)$, then to the leading order in $\lambdabar$

$$\mathcal{M} \approx \exp\left[-\frac{i}{\lambdabar}\{\phi(\boldsymbol{Q}_\perp) - \phi(\boldsymbol{r}_\perp)\}\right],$$

$$\boldsymbol{Q}_\perp = \boldsymbol{r}_\perp + \frac{\lambdabar}{n_0}\boldsymbol{G}_\perp. \tag{80}$$

This enables us to express Eq. (73) as

$$\begin{bmatrix} \sqrt{\epsilon}\boldsymbol{E} \\ \frac{1}{\sqrt{\mu}}\boldsymbol{B} \end{bmatrix}(z'') = \widehat{G}_0\left(\widehat{\boldsymbol{p}}_\perp, \boldsymbol{r}_\perp \to \boldsymbol{Q}_\perp; z'', z'\right)\begin{bmatrix} \sqrt{\epsilon}\boldsymbol{E} \\ \frac{1}{\sqrt{\mu}}\boldsymbol{B} \end{bmatrix}(z'). \tag{81}$$

Eq. (81) represents the elegant Mukunda–Simon–Sudarshan rule for the transition from the scalar theory to the vector theory of wave optics. Whenever, any ideal linear optical system is described by an operator $\widehat{G}_0\left(\boldsymbol{r}_\perp, \widehat{\boldsymbol{p}}_\perp\right)$

$$\psi_{(\text{out})} = \widehat{G}_0\left(\boldsymbol{r}_\perp, \widehat{\boldsymbol{p}}_\perp\right)\psi_{(\text{in})}, \tag{82}$$

in scalar optics, then the same system is described completely in vector wave optics by a matrix function of $\boldsymbol{r}_\perp$ and $\widehat{\boldsymbol{p}}_\perp$ obtained from $\widehat{G}_0\left(\boldsymbol{r}_\perp, \widehat{\boldsymbol{p}}_\perp\right)$ by the simple replacement $\boldsymbol{r}_\perp \to \boldsymbol{Q}_\perp$.

As mentioned earlier, our formalism is based on an exact matrix representation and works to all orders for beam optics as well as polarization.

The MSS-rule is readily obtained within the paraxial approximation and the simplification of a constant refractive index (Khan, 2016f). So, our formalism of Maxwell optics is an appropriate candidate for generalizing the MSS-theory far beyond the paraxial approximation.

5.2 Gaussian Beams in Maxwell Optics

As a first application of the Mukunda–Simon–Sudarshan rule, let us consider the Gaussian beams. The beam is propagating along the z-axis, with its waist positioned at the transverse plane at $z = 0$. The field in the waist plane is

$$F(\mathbf{r}_\perp; 0) = \sqrt{\frac{2}{\pi}} \frac{1}{\sigma_0} \exp\left[-\frac{1}{\sigma_0^2} r_\perp^2\right] \begin{bmatrix} \sqrt{\epsilon} \mathbf{E} \\ \frac{1}{\sqrt{\mu}} \mathbf{B} \end{bmatrix}. \tag{83}$$

We note, that $k = 1/\lambdabar$. The complex radius of curvature, $q(z)$ is related to the radius of curvature of the phase front, $R(z)$ as

$$\frac{1}{q(z)} = \frac{1}{R(z)} + \frac{2i}{k\sigma^2(z)}, \tag{84}$$

where $\sigma(z)$ is the beam width. Let us consider a plane wave polarized in the x-direction (without any loss of generality and $n_0 = 1$). Then the column vector $\begin{bmatrix} \sqrt{\epsilon} \mathbf{E} \\ \frac{1}{\sqrt{\mu}} \mathbf{B} \end{bmatrix}$ for the fields simplifies to

$$F(\mathbf{r}_\perp) = \begin{bmatrix} \sqrt{\epsilon} E_{x_0} \\ 0 \\ 0 \\ 0 \\ \frac{1}{\sqrt{\mu}} B_{y_0} \\ 0 \end{bmatrix}. \tag{85}$$

Using the MSS-rule, the fields in the plane at $z = 0$ are

$$F(\mathbf{r}_\perp; 0) = \sqrt{\frac{2}{\pi}} \frac{1}{\sigma_0} \exp\left[-\frac{1}{\sigma_0^2} (\mathbf{r}_\perp + \lambdabar \mathbf{G}_\perp)^2\right] \begin{bmatrix} \sqrt{\epsilon} E_{x_0} \\ 0 \\ 0 \\ 0 \\ \frac{1}{\sqrt{\mu}} B_{y_0} \\ 0 \end{bmatrix}. \tag{86}$$

The fields at a later transverse plane at $z = z''$ are

$$F(\boldsymbol{r}_\perp; z'') = \sqrt{\frac{2}{\pi}} \frac{1}{\sigma(z)} \exp\left[\frac{ik}{2q}(\boldsymbol{r}_\perp + \lambdabar \boldsymbol{G}_\perp)^2\right] \begin{bmatrix} \sqrt{\epsilon}E_{x0} \\ 0 \\ 0 \\ 0 \\ \frac{1}{\sqrt{\mu}}B_{y0} \\ 0 \end{bmatrix}. \qquad (87)$$

Using the properties of the matrices G_x and G_y in Eq. (79), the exponential in Eq. (87) is readily obtained in a closed form. So, the expression in Eq. (87) simplifies to

$$F(\boldsymbol{r}_\perp; z'') = \sqrt{\frac{2}{\pi}} \frac{1}{\sigma(z)} \exp\left[\frac{ik}{2q(z)}r_\perp^2\right]\mathcal{M} \begin{bmatrix} \sqrt{\epsilon}E_{x0} \\ 0 \\ 0 \\ 0 \\ \frac{1}{\sqrt{\mu}}B_{y0} \\ 0 \end{bmatrix}, \qquad (88)$$

where $\mathcal{M}$ is the matrix

$$\mathcal{M} = \begin{bmatrix} 1 + \frac{y^2-x^2}{8q^2} & -\frac{xy}{4q^2} & \frac{x}{2q} & \frac{xy}{4q^2} & \frac{y^2-x^2}{8q^2} & -\frac{y}{2q} \\ -\frac{xy}{4q^2} & 1 - \frac{y^2-x^2}{8q^2} & \frac{y}{2q} & \frac{y^2-x^2}{8q^2} & -\frac{xy}{4q^2} & \frac{x}{2q} \\ -\frac{x}{2q} & -\frac{y}{2q} & 1 & \frac{y}{2q} & -\frac{x}{2q} & 0 \\ -\frac{xy}{4q^2} & -\frac{y^2-x^2}{8q^2} & \frac{y}{2q} & 1 + \frac{y^2-x^2}{8q^2} & -\frac{xy}{4q^2} & \frac{x}{2q} \\ -\frac{y^2-x^2}{8q^2} & \frac{xy}{4q^2} & -\frac{x}{2q} & -\frac{xy}{4q^2} & 1 - \frac{y^2-x^2}{8q^2} & \frac{y}{2q} \\ -\frac{y}{2q} & \frac{x}{2q} & 0 & -\frac{x}{2q} & -\frac{y}{2q} & 1 \end{bmatrix}. \qquad (89)$$

In order to express the fields in component form, we shall use the field amplitudes given by

$$\Omega(\boldsymbol{r}_\perp; z) = \sqrt{\frac{2}{\pi}} \frac{1}{\sigma(z)} \exp\left[\frac{ik}{2q(z)}r_\perp^2\right]. \qquad (90)$$

In term of components, we have

$$E_x(\boldsymbol{r}_\perp; z) = \Omega(\boldsymbol{r}_\perp; z)\left[\sqrt{\epsilon}E_{x0} + \frac{y^2-x^2}{8q^2}\left(\sqrt{\epsilon}E_{x0} + \frac{1}{\sqrt{\mu}}B_{y0}\right)\right]$$

$$E_y(\boldsymbol{r}_\perp; z) = \Omega(\boldsymbol{r}_\perp; z) \left[-\frac{xy}{4q^2} \left(\sqrt{\epsilon} E_{x0} + \frac{1}{\sqrt{\mu}} B_{y0} \right) \right]$$

$$E_z(\boldsymbol{r}_\perp; z) = \Omega(\boldsymbol{r}_\perp; z) \left[-\frac{x}{2q} \left(\sqrt{\epsilon} E_{x0} + \frac{1}{\sqrt{\mu}} B_{y0} \right) \right]$$

$$B_x(\boldsymbol{r}_\perp; z) = \Omega(\boldsymbol{r}_\perp; z) \left[-\frac{xy}{4q^2} \left(\sqrt{\epsilon} E_{x0} + \frac{1}{\sqrt{\mu}} B_{y0} \right) \right]$$

$$B_y(\boldsymbol{r}_\perp; z) = \Omega(\boldsymbol{r}_\perp; z) \left[-\frac{y^2 - x^2}{8q^2} \left(\sqrt{\epsilon} E_{x0} + \frac{1}{\sqrt{\mu}} B_{y0} \right) + \frac{1}{\sqrt{\mu}} B_{y0} \right]$$

$$B_z(\boldsymbol{r}_\perp; z) = \Omega(\boldsymbol{r}_\perp; z) \left[-\frac{y}{2q} \left(\sqrt{\epsilon} E_{x0} + \frac{1}{\sqrt{\mu}} B_{y0} \right) \right]. \tag{91}$$

We note, that $E_y(\boldsymbol{r}_\perp; z) = cB_x(\boldsymbol{r}_\perp; z)$. Eq. (91) has the electric and magnetic fields of a Gaussian beam in Maxwell optics. It is interesting to note, that the electric field $\boldsymbol{E}(\boldsymbol{r}_\perp; z)$ has a component $E_z(\boldsymbol{r}_\perp; z)$ along the beam axis. Additionally, $E_y(\boldsymbol{r}_\perp; z)$ and $B_x(\boldsymbol{r}_\perp; z)$ have the cross-polarization components. These components vanish only along the two planes defined by $x = 0$ and $y = 0$ respectively. It is to be noted, that the intersection of these two planes is the z-axis, which is the optics axis. The MSS-rule leading to the Gaussian Maxwell beams was derived using a group theoretical analysis (Mukunda et al., 1983a, 1983b, 1985a, 1985b; Simon et al., 1986, 1987). Here, the same rule is obtained using an entirely different framework.

6. CONCLUSION

We have developed an exact matrix representation of Maxwell's equations taking into account the spatial and temporal variations of the permittivity and permeability. This representation, using 8×8 matrices is the basis for an exact formalism of Maxwell optics presented here. The exact beam optical Hamiltonian, derived from this representation has an algebraic structure in direct correspondence with the Dirac equation of the electron. We exploit this correspondence to adopt the standard machinery, namely the Foldy–Wouthuysen transformation technique of the Dirac theory, to the beam optical formalism. This enabled us to develop a systematic procedure to obtain the aberration expansion from the beam-optical Hamiltonian to any desired degree of accuracy. We further get the wavelength-dependent contributions at each order, starting with the lowest-order paraxial Hamiltonian. Formal expressions were obtained for the paraxial and leading order aberrating Hamiltonians, without making any assumption on the form of the varying refractive index.

The beam-optical Hamiltonians, we derived also have the wavelength-dependent matrix terms, which are related to the polarization. In our approach, we have been able to derive a Hamiltonian which contains both the beam-optics and the polarization.

We applied the formalism to the specific examples and saw how the beam-optics (paraxial behavior and the aberrations) gets modified by the wavelength-dependent contributions. First of the two examples is the *medium with a constant refractive index*. This case is exactly solvable. This example enables us to compare the expansion of the beam-optical Hamiltonian with the exact result. Moreover, we could acquaint ourselves with the techniques needed for the realistic examples.

As a second example, we considered the *axially symmetric graded-index medium*. For this example, in the traditional approaches one gets only six aberrations. In our formalism we get all the nine aberrations permitted by the axial symmetry. The six aberration coefficients of the traditional approaches get modified by the wavelength-dependent contributions. It is very interesting to note that apart from the wavelength-dependent modifications of the aberrations, this approach also gives rise to the *image rotation* (see Khan, 2006c, 2010, 2014a, 2016f, 2017c). This image rotation is proportional to the wavelength and we have derived an explicit relationship for the angle in Eq. (52). Such, an image rotation has no analogue/counterpart in any of the traditional prescriptions. It would be worthwhile to experimentally look for the predicted image rotation (Khan, 2006c, 2010, 2014a, 2016f, 2017c). The presence of the nine third-order aberrations and image rotation are well-known in *axially symmetric magnetic electron lenses*, even when treated classically. The quantum treatment of this system leads to their wavelength-dependent modifications (Jagannathan & Khan, 1996; Khan, 1997).

The optical Hamiltonian has two components: *Beam-Optics* and *Polarization* respectively. We have addressed both in detail. The formalism presented in this article provides a natural framework for the study of light polarization. This provides a unified treatment for the beam-optics and the light polarization. The new formalism is a suitable candidate to extend traditional theory of polarization beyond the paraxial approximation. This is so, because the paraxial limit of our formalism readily leads to the elegant Mukunda–Simon–Sudarshan rule (Simon et al., 1987) for passage from scalar to vector wave optics in the case of paraxial systems (Khan, 2016f).

The close analogy between geometrical optics and charged-particle beam optics has been known for a very long time. Until now, it was possible to see this analogy only between the geometrical optics and the classical prescriptions of charged-particle optics. A quantum theory of charged-particle optics was presented in recent years (Conte et al., 1996; Jagannathan et al., 1989; Jagannathan, 1990, 1993, 1999, 2002, 2003; Jagannathan & Khan, 1995, 1996, 1997, 2018; Khan, 1997, 1999a, 1999b, 2001, 2002a, 2002b, 2002c, 2003d, 2016d; Khan & Jagannathan, 1993, 1994, 1995). With the current development of the non-traditional prescriptions of Helmholtz optics (Khan et al., 2002; Khan, 2005a, 2016g) and the matrix formulation of Maxwell optics presented here, using the rich algebraic machinery of quantum mechanics, it is now possible to see a parallel of the analogies at each level. The non-traditional prescription of the Helmholtz optics is in close analogy with the quantum theory of charged-particles based on the Klein–Gordon equation. The matrix formulation of Maxwell optics presented here is in close analogy with the quantum theory of charged-particles based on the Dirac equation. The use of the Foldy–Wouthuysen transformation technique in optics has been able to shed light on the deeper connections in the wavelength-dependent regime between the light optics and charged-particle optics (Khan, 2002c, 2017a). The parallel of these analogies is described in Appendix F.

We have presented an alternate and exact way of deriving the beam optical Hamiltonian, which reproduces the established results. Furthermore, we have derived the extra wavelength-dependent contributions. In the low wavelength limit our formalism reproduces the Lie algebraic formalism of optics (Dragt, 1982, 1988; Dragt et al., 1986; Lakshminarayanan et al., 1998). The Foldy–Wouthuysen technique employed by us is ideally suited for the Lie algebraic approach to optics.

Quantum methodologies are a powerful tool to handle a variety of systems. Quantum methodologies have been also used to study multiparticle effects in charged-particle beam optics. In passing, we note the models developed by Fedele et al. (the *thermal wave model*; Fedele, Miele, Palumbo, & Vaccaro, 1993; Fedele & Man'ko, 1999; Fedele, Man'ko, & Man'ko, 2000; Fedele, Jovanovi, De Nicola, Mannan, & Tanjia, 2014; Fedele, Tanjia, Jovanović, De Nicola, & Ronsivalle, 2014). Cufaro Petroni et al. have developed *stochastic collective dynamical model* for treating the beam phenomenologically as a quasiclassical many-body system (Petroni, De Martino, De Siena, & Illuminati, 2000). In these models, the basic equation is a Schrödinger-*like* equation with the *beam emittance* playing the role of $\hbar$, the Planck's constant. The quantum-like approach has been also used to develop a *Diffraction Model for the beam halo* (Khan & Pusterla, 1999, 2000a, 2000b, 2001).

ACKNOWLEDGMENTS

I am grateful to Professor Ramaswamy Jagannathan for all my training in the very exciting field of *quantum theory of charged-particle beam optics*, which was the topic of my doctoral thesis. He elegantly supervised my doctoral thesis. Naturally, we were dealing with the relativistic wave equations; he taught me the related techniques including the Foldy–Wouthuysen transformation. He gave the brilliant suggestion to use the Foldy–Wouthuysen transformation technique to investigate the scalar optics (Helmholtz optics). Later, on Professor Jagannathan guided me to the logical continuation from scalar optics to the vector optics leading to the matrix formulation of Maxwell optics, again employing the Foldy–Wouthuysen transformation. During the course of the above investigations, I had the privilege to enjoy the warm hospitality on several occasions at the Institute of Mathematical Sciences (MatScience/IMSc) in the coastal city of Chennai (Madras), India. I would also like to acknowledge thankfully the benefit of collaboration, during the initial work on Helmholtz optics, with Professor Rajiah Simon (IMSc). Also I thank Professor Hawkes for showing a keen interest in the quantum theory of charged-particle beam optics, which led Jagannathan and me to write a long and comprehensive chapter two decades ago. This was followed by a chapter on the Foldy–Wouthuysen transformation technique in optics. Professor Hawkes' continued encouragement has resulted in this chapter. The new formalism of optics using quantum methodologies described in this article was selected for the "Year in Optics 2016" by the Optical Society [of America]. The summary of the two selected papers (Khan, 2016f, 2016g) is available in the special issue of the Optics & Photonics News (Khan, 2016h).

APPENDIX A. RIEMANN–SILBERSTEIN VECTOR

The Riemann–Silberstein complex vector (Silberstein, 1907a, 1907b) $F(r, t)$ built from the electric field $D(r, t)$ and the magnetic filed $B(r, t)$ is given by

$$
\begin{aligned}
F(r, t) &= \frac{1}{\sqrt{2}} \left(\frac{1}{\sqrt{\epsilon(r)}} D(r, t) + i\frac{1}{\sqrt{\mu(r)}} B(r, t) \right) \\
&= \frac{1}{\sqrt{2}} \left(\sqrt{\epsilon(r)} E(r, t) + i\frac{1}{\sqrt{\mu(r)}} B(r, t) \right),
\end{aligned} \tag{A.1}
$$

where $\epsilon(r)$ is the permittivity of the medium and $\mu(r)$ is the permeability of the medium. In vacuum, we have $\epsilon_0 = 8.85 \times 10^{-12}$ C^2/N.m^2 and $\mu_0 = 4\pi \times 10^{-7}$ N/A^2. The Riemann–Silberstein complex vector $F(r, t)$ can also be derived from the potential $Z(r, t)$ (for example, see Panofsky & Phillips, 1962),

$$F(r, t) = \nabla \times \left\{ \frac{i}{v} \frac{\partial}{\partial t} Z(r, t) + \nabla \times Z(r, t) \right\}. \qquad (A.2)$$

$Z(r, t)$ is the *superpotential* and is commonly known as the *polarization potential* or the *Hertz Vector* (see Panofsky & Phillips, 1962). This further leads to the wave-equation

$$\left\{ \nabla^2 - \frac{1}{v^2} \frac{\partial^2}{\partial t^2} \right\} Z(r, t) = 0. \qquad (A.3)$$

The Riemann–Silberstein vector is a mixture of both a vector and a pseudovector. But it is to be noted, that $\begin{pmatrix} F^+ \\ F^- \end{pmatrix}$ transforms according to the Lorentz group extended by parity (Wang, Qiu, Wang, & Shi, 2015). It is advantageous to use the Riemann–Silberstein vector for obtaining the matrix-representations of Maxwell's equations (see Bialynicki-Birula, 1994, 1996a, 1996b; Bialynicki-Birula & Bialynicka-Birula, 2013; Barnett, 2014; Mehrafarin & Balajany, 2010; Red'kov, Tokarevskaya, & Spix, 2012). Riemann–Silberstein vector can be used to express many of the quantities associated with the electromagnetic field

$$\text{Poynting Vector: } S = \frac{1}{\mu} E \times B$$

$$= -iv \left(F^\dagger \times F \right)$$

$$\text{Energy Density: } u = \frac{1}{2} \left(\epsilon E \cdot E + \frac{1}{\mu} B \cdot B \right)$$

$$= F^\dagger \cdot F$$

$$\text{Momentum Density: } p_{EB} = \epsilon (E \times B)$$

$$= -\frac{i}{v} \left(F^\dagger \times F \right)$$

$$\text{Angular Momentum Density: } L_{EB} = \epsilon \left\{ r \times (E \times B) \right\}$$

$$= -\frac{i}{v} \left\{ r \times \left(F^\dagger \times F \right) \right\}, \qquad (A.4)$$

where † is the Hermitian conjugate. The other quantities are

$$\text{Total Energy: } E = \frac{1}{2} \int d^3r \left\{ \epsilon \boldsymbol{E} \cdot \boldsymbol{E} + \frac{1}{\mu} \boldsymbol{B} \cdot \boldsymbol{B} \right\}$$

$$= \int d^3r \left\{ \boldsymbol{F}^\dagger \cdot \boldsymbol{F} \right\}$$

$$\text{Total Momentum: } \boldsymbol{P} = \epsilon \int d^3r \left\{ \boldsymbol{E} \times \boldsymbol{B} \right\}$$

$$= -\frac{i}{v} \int d^3r \left\{ \boldsymbol{F}^\dagger \times \boldsymbol{F} \right\}$$

$$\text{Total Angular Momentum: } \boldsymbol{M} = \epsilon \int d^3r \left\{ \boldsymbol{r} \times (\boldsymbol{E} \times \boldsymbol{B}) \right\}$$

$$= -\frac{i}{v} \int d^3r \left\{ \boldsymbol{r} \times \left(\boldsymbol{F}^\dagger \times \boldsymbol{F} \right) \right\}$$

$$\text{Moment of Energy: } \boldsymbol{N} = \frac{1}{2} \int d^3r \left\{ \boldsymbol{r} \left(\epsilon \boldsymbol{E} \cdot \boldsymbol{E} + \frac{1}{\mu} \boldsymbol{B} \cdot \boldsymbol{B} \right) \right\}$$

$$= \int d^3r \left\{ \boldsymbol{r} \left(\boldsymbol{F}^\dagger \cdot \boldsymbol{F} \right) \right\}. \tag{A.5}$$

In this form these quantities look like the quantum-mechanical expectation values! The use of the Riemann–Silberstein vector as a possible candidate for the *photon wavefunction* has been advocated for a long time (Bialynicki-Birula, 1994, 1996a, 1996b).

APPENDIX B. MATRIX REPRESENTATION OF MAXWELL'S EQUATIONS

Matrix representations of Maxwell's equations are very well-known (Laporte & Uhlenbeck, 1931; Moses, 1959; Majorana, 1974). However, all these representations lack an exactness or/and are given in terms of a *pair* of matrix equations (Bialynicki-Birula, 1994, 1996a, 1996b). Some of these representations are in free space. Such a representation is an approximation in a medium with space- and time-dependent permittivity $\epsilon(\boldsymbol{r}, t)$ and permeability $\mu(\boldsymbol{r}, t)$ respectively. Even this approximation is often expressed through a pair of equations using 3×3 matrices: one for the curl and one for the divergence, which occur in Maxwell's equations. This practice of writing the divergence condition separately is completely avoidable by using 4×4 matrices for Maxwell's equations in free-space (Moses, 1959). A single equation using 4×4 matrices is necessary and sufficient when $\epsilon(\boldsymbol{r}, t)$ and $\mu(\boldsymbol{r}, t)$ are treated as 'local' constants (Moses, 1959; Bialynicki-Birula, 1996b).

A treatment taking into account the variations of $\epsilon(\boldsymbol{r}, t)$ and $\mu(\boldsymbol{r}, t)$ has been presented in Bialynicki-Birula (1996b). This treatment uses the Riemann–Silberstein vectors, $\boldsymbol{F}^{\pm}(\boldsymbol{r}, t)$ to reexpress Maxwell's equations as four equations: two equations are for the curl and two are for the divergences and there is mixing in $\boldsymbol{F}^{+}(\boldsymbol{r}, t)$ and $\boldsymbol{F}^{-}(\boldsymbol{r}, t)$. This mixing is very neatly expressed through the two derived functions of $\epsilon(\boldsymbol{r}, t)$ and $\mu(\boldsymbol{r}, t)$. These four equations are then expressed as a pair of matrix equations using 6×6 matrices: again one for the curl and one for the divergence. Even though this treatment is exact it involves a pair of matrix equations.

Here, we present a treatment which enables us to express Maxwell's equations in a single matrix equation instead of a *pair* of matrix equations. Our approach is a logical continuation of the treatment in Bialynicki-Birula (1996b). We use the linear combination of the components of the Riemann–Silberstein vectors, $\boldsymbol{F}^{\pm}(\boldsymbol{r}, t)$ and the final matrix representation is a single equation using 8×8 matrices. This representation contains all the four Maxwell's equations in presence of sources taking into account the spatial and temporal variations of the permittivity $\epsilon(\boldsymbol{r}, t)$ and the permeability $\mu(\boldsymbol{r}, t)$.

In Section B.1 we shall summarize the treatment for a homogeneous medium and introduce the required functions and notation. In Section B.2 we shall present the matrix representation in an inhomogeneous medium, in presence of sources.

B.1 Homogeneous Medium

We shall start with Maxwell's equations (Jackson, 1998; Panofsky & Phillips, 1962) in an inhomogeneous medium with sources,

$$\nabla \cdot \boldsymbol{D}(\boldsymbol{r}, t) = \rho \,,$$

$$\nabla \times \boldsymbol{H}(\boldsymbol{r}, t) - \frac{\partial}{\partial t}\boldsymbol{D}(\boldsymbol{r}, t) = \boldsymbol{J} \,,$$

$$\nabla \times \boldsymbol{E}(\boldsymbol{r}, t) + \frac{\partial}{\partial t}\boldsymbol{B}(\boldsymbol{r}, t) = 0 \,,$$

$$\nabla \cdot \boldsymbol{B}(\boldsymbol{r}, t) = 0 \,. \tag{B.1}$$

We assume the media to be linear, that is $\boldsymbol{D} = \epsilon \boldsymbol{E}$, and $\boldsymbol{B} = \mu \boldsymbol{H}$, where ϵ is the permittivity of the medium and μ is the permeability of the medium. In general $\epsilon = \epsilon(\boldsymbol{r}, t)$ and $\mu = \mu(\boldsymbol{r}, t)$. In this section, we treat them as 'local' constants in the various derivations. The magnitude of the velocity of light in the medium is given by $v(\boldsymbol{r}, t) = |\boldsymbol{v}(\boldsymbol{r}, t)| = 1/\sqrt{\epsilon(\boldsymbol{r}, t)\mu(\boldsymbol{r}, t)}$. In vacuum we have, $\epsilon_0 = 8.85 \times 10^{-12}$ C^2/N.m^2 and $\mu_0 = 4\pi \times 10^{-7}$ N/A^2.

One possible way to obtain the required matrix representation is to use the Riemann–Silberstein vector (Bialynicki-Birula, 1996b) given by

$$\boldsymbol{F}^+ (\boldsymbol{r}, t) = \frac{1}{\sqrt{2}} \left(\sqrt{\epsilon(\boldsymbol{r}, t)}\boldsymbol{E}(\boldsymbol{r}, t) + \mathrm{i}\frac{1}{\sqrt{\mu(\boldsymbol{r}, t)}}\boldsymbol{B}(\boldsymbol{r}, t) \right)$$

$$\boldsymbol{F}^- (\boldsymbol{r}, t) = \frac{1}{\sqrt{2}} \left(\sqrt{\epsilon(\boldsymbol{r}, t)}\boldsymbol{E}(\boldsymbol{r}, t) - \mathrm{i}\frac{1}{\sqrt{\mu(\boldsymbol{r}, t)}}\boldsymbol{B}(\boldsymbol{r}, t) \right). \qquad \text{(B.2)}$$

For any homogeneous medium it is equivalent to use either $\boldsymbol{F}^+ (\boldsymbol{r}, t)$ or $\boldsymbol{F}^- (\boldsymbol{r}, t)$. The two differ by the sign before 'i' and are not the complex conjugate of one another. We have not assumed any form for $\boldsymbol{E}(\boldsymbol{r}, t)$ and $\boldsymbol{B}(\boldsymbol{r}, t)$. We will be needing both of them in an inhomogeneous medium, to be considered in detail in Section B.2.

If for a certain medium $\epsilon(\boldsymbol{r}, t)$ and $\mu(\boldsymbol{r}, t)$ are constants (or can be treated as 'local' constants under certain approximations), then the vectors $\boldsymbol{F}^\pm (\boldsymbol{r}, t)$ satisfy

$$\mathrm{i}\frac{\partial}{\partial t}\boldsymbol{F}^\pm (\boldsymbol{r}, t) = \pm v\boldsymbol{\nabla} \times \boldsymbol{F}^\pm (\boldsymbol{r}, t) - \frac{1}{\sqrt{2\epsilon}}(\mathrm{i}\boldsymbol{J})$$

$$\boldsymbol{\nabla} \cdot \boldsymbol{F}^\pm (\boldsymbol{r}, t) = \frac{1}{\sqrt{2\epsilon}}(\rho). \qquad \text{(B.3)}$$

Thus, by using the Riemann–Silberstein vector it has been possible to re-express the four Maxwell's equations (for a medium with constant ϵ and μ) as two equations. The first one contains the two Maxwell's equations with curl and the second one contains the two Maxwell's equations with divergences. The first of the two equations in Eq. (B.3) can be immediately converted into a 3×3 matrix representation. However, this representation does not contain the divergence conditions (the first and the fourth Maxwell's equations) contained in the second equation in Eq. (B.3). A further compactification is possible only by expressing Maxwell's equations in a 4×4 matrix representation. To this end, using the components of the Riemann–Silberstein vector, we define

$$\Psi^+(\boldsymbol{r}, t) = \begin{bmatrix} -F_x^+ + \mathrm{i}F_y^+ \\ F_z^+ \\ F_z^+ \\ F_x^+ + \mathrm{i}F_y^+ \end{bmatrix}, \quad \Psi^-(\boldsymbol{r}, t) = \begin{bmatrix} -F_x^- - \mathrm{i}F_y^- \\ F_z^- \\ F_z^- \\ F_x^- - \mathrm{i}F_y^- \end{bmatrix}. \qquad \text{(B.4)}$$

The vectors for the sources are

$$W^+ = \left(\frac{1}{\sqrt{2\epsilon}}\right) \begin{bmatrix} -J_x + iJ_y \\ J_z - v\rho \\ J_z + v\rho \\ J_x + iJ_y \end{bmatrix}, \quad W^- = \left(\frac{1}{\sqrt{2\epsilon}}\right) \begin{bmatrix} -J_x - iJ_y \\ J_z - v\rho \\ J_z + v\rho \\ J_x - iJ_y \end{bmatrix}. \quad \text{(B.5)}$$

Then we obtain

$$\frac{\partial}{\partial t}\Psi^+ = -v\{\boldsymbol{M} \cdot \boldsymbol{\nabla}\}\Psi^+ - W^+$$

$$\frac{\partial}{\partial t}\Psi^- = -v\{\boldsymbol{M}^* \cdot \boldsymbol{\nabla}\}\Psi^- - W^-, \quad \text{(B.6)}$$

where ( )* denotes complex-conjugation and the triplet $\boldsymbol{M} = (M_x, M_y, M_z)$ is expressed in terms of

$$\Omega = \begin{bmatrix} 0 & -\mathbb{1} \\ \mathbb{1} & 0 \end{bmatrix}, \quad \beta = \begin{bmatrix} \mathbb{1} & 0 \\ 0 & -\mathbb{1} \end{bmatrix}, \quad \mathbb{1} = \begin{bmatrix} 1 & 0 \\ 0 & 1 \end{bmatrix}. \quad \text{(B.7)}$$

Alternately, we may use the matrix $J = -\Omega$. Both differ by a sign. For our purpose, it is fine to use either Ω or J. However, they have a different meaning: J is *contravariant* and Ω is *covariant*; the matrix Ω corresponds to the Lagrange brackets of classical mechanics and J corresponds to the Poisson brackets. An important relation is $\Omega = J^{-1}$. The M-matrices are

$$M_x = \begin{bmatrix} 0 & 0 & 1 & 0 \\ 0 & 0 & 0 & 1 \\ 1 & 0 & 0 & 0 \\ 0 & 1 & 0 & 0 \end{bmatrix} = -\beta\Omega,$$

$$M_y = \begin{bmatrix} 0 & 0 & -i & 0 \\ 0 & 0 & 0 & -i \\ i & 0 & 0 & 0 \\ 0 & i & 0 & 0 \end{bmatrix} = i\Omega,$$

$$M_z = \begin{bmatrix} 1 & 0 & 0 & 0 \\ 0 & 1 & 0 & 0 \\ 0 & 0 & -1 & 0 \\ 0 & 0 & 0 & -1 \end{bmatrix} = \beta. \quad \text{(B.8)}$$

Each of the four Maxwell's equations are easily obtained from the matrix representation in Eq. (B.6). This is done by taking the sums and differences of row-I with row-IV and row-II with row-III respectively. The first three give the y, x and z components of the curl and the last one gives the divergence conditions present in the evolution equation (B.3).

It is to be noted that the matrices M are all non-singular and all are Hermitian. Moreover, they satisfy the usual algebra of the Dirac matrices, including

$$M_x \beta = -\beta M_x ,$$
$$M_y \beta = -\beta M_y ,$$
$$M_x^2 = M_y^2 = M_z^2 = I ,$$
$$M_x M_y = -M_y M_x = i M_z ,$$
$$M_y M_z = -M_z M_y = i M_x ,$$
$$M_z M_x = -M_x M_z = i M_y . \tag{B.9}$$

Before proceeding further we note the following: The pair ($\Psi^{\pm}$, M) are **not** unique. Different choices of $\Psi^{\pm}$ would give rise to different M, such that the triplet M continues to satisfy the algebra of the Dirac matrices in Eq. (B.9). We have preferred $\Psi^{\pm}$ *via* the Riemann–Silberstein vector in Eq. (B.2) in Bialynicki-Birula (1996b). This vector has certain advantages over the other possible choices. The Riemann–Silberstein vector is well-known in classical electrodynamics and has certain interesting properties and uses (Bialynicki-Birula, 1996b).

In deriving the above 4×4 matrix representation of Maxwell's equations we have ignored the spatial and temporal derivatives of $\epsilon(r, t)$ and $\mu(r, t)$ in the first two of Maxwell's equations. We have treated ϵ and μ as 'local' constants.

B.2 Inhomogeneous Medium

In the previous section, we wrote the evolution equations for the Riemann–Silberstein vector in Eq. (B.3), for a medium, treating $\epsilon(r, t)$ and $\mu(r, t)$ as 'local' constants. From these pairs of equations we wrote the matrix form of Maxwell's equations. In this section, we shall write the exact equations taking into account the spatial and temporal variations of $\epsilon(r, t)$ and $\mu(r, t)$. It is very much possible to write the required evolution equations using $\epsilon(r, t)$ and $\mu(r, t)$. But we shall follow the procedure in Bialynicki-Birula (1996b) of using the two derived *laboratory functions*

$$\text{Velocity Function: } v(r, t) = \frac{1}{\sqrt{\epsilon(r, t) \mu(r, t)}}$$

$$\text{Resistance Function: } h(r, t) = \sqrt{\frac{\mu(r, t)}{\epsilon(r, t)}} . \tag{B.10}$$

The function, $v(r, t)$ has the dimensions of velocity and the function, $h(r, t)$ has the dimensions of resistance (measured in Ohms). We can equivalently use the *Conductance Function*, $\kappa(r, t) = 1/h(r, t) = \sqrt{\epsilon(r, t)/\mu(r, t)}$ (measured in Ohms^{-1} or Mhos!) in place of the resistance function, $h(r, t)$. These derived functions enable us to understand the dependence of the variations more transparently (Bialynicki-Birula, 1996b). Moreover the derived functions are the ones which are measured experimentally. In terms of these functions, $\epsilon = 1/\sqrt{vh}$ and $\mu = \sqrt{h/v}$. Using these functions the exact equations satisfied by $F^{\pm}(r, t)$ are

$$i\frac{\partial}{\partial t}F^{+}(r, t) = v(r, t)\left(\nabla \times F^{+}(r, t)\right) + \frac{1}{2}\left(\nabla v(r, t) \times F^{+}(r, t)\right)$$

$$+ \frac{v(r, t)}{2h(r)}\left(\nabla h(r, t) \times F^{-}(r, t)\right) - \frac{i}{\sqrt{2}}\sqrt{v(r, t)h(r, t)}J$$

$$+ \frac{i}{2}\frac{\dot{v}(r, t)}{v(r, t)}F^{+}(r, t) + \frac{i}{2}\frac{\dot{h}(r, t)}{h(r, t)}F^{-}(r, t) ,$$

$$i\frac{\partial}{\partial t}F^{-}(r, t) = -v(r, t)\left(\nabla \times F^{-}(r, t)\right) - \frac{1}{2}\left(\nabla v(r, t) \times F^{-}(r, t)\right)$$

$$- \frac{v(r, t)}{2h(r, t)}\left(\nabla h(r, t) \times F^{+}(r, t)\right) - \frac{i}{\sqrt{2}}\sqrt{v(r, t)h(r, t)}J$$

$$+ \frac{i}{2}\frac{\dot{v}(r, t)}{v(r, t)}F^{-}(r, t) + \frac{i}{2}\frac{\dot{h}(r, t)}{h(r, t)}F^{+}(r, t) ,$$

$$\nabla \cdot F^{+}(r, t) = \frac{1}{2v(r, t)}\left(\nabla v(r, t) \cdot F^{+}(r, t)\right)$$

$$+ \frac{1}{2h(r, t)}\left(\nabla h(r, t) \cdot F^{-}(r, t)\right)$$

$$+ \frac{1}{\sqrt{2}}\sqrt{v(r, t)h(r, t)}\,\rho ,$$

$$\nabla \cdot F^{-}(r, t) = \frac{1}{2v(r, t)}\left(\nabla v(r, t) \cdot F^{-}(r, t)\right)$$

$$+ \frac{1}{2h(r, t)}\left(\nabla h(r, t) \cdot F^{+}(r, t)\right)$$

$$+ \frac{1}{\sqrt{2}}\sqrt{v(r, t)h(r, t)}\,\rho , \tag{B.11}$$

where $\dot{v} = \frac{\partial v}{\partial t}$ and $\dot{h} = \frac{\partial h}{\partial t}$. The evolution equations in Eq. (B.11) are exact (for a linear media) and the dependence on the variations of $\epsilon(\boldsymbol{r}, t)$ and $\mu(\boldsymbol{r}, t)$ has been neatly expressed through the two derived functions. The *coupling* between $\boldsymbol{F}^+(\boldsymbol{r}, t)$ and $\boldsymbol{F}^-(\boldsymbol{r}, t)$ is *via* the gradient and time-derivative of only one derived function namely, $h(\boldsymbol{r}, t)$ or equivalently $\kappa(\boldsymbol{r}, t)$. Either of these can be used and both are the directly measured quantities. We further note that the dependence of the coupling is logarithmic

$$\frac{1}{h(\boldsymbol{r}, t)} \nabla h(\boldsymbol{r}, t) = \nabla \left\{ \ln \left(h(\boldsymbol{r}, t) \right) \right\} ,$$

$$\frac{1}{h(\boldsymbol{r}, t)} \dot{h}(\boldsymbol{r}, t) = \frac{\partial}{\partial t} \left\{ \ln \left(h(\boldsymbol{r}, t) \right) \right\} , \tag{B.12}$$

where 'ln' is the natural logarithm.

The coupling can be best summarized by expressing the equations in Eq. (B.11) in a (block) matrix form. For this we introduce the following logarithmic function

$$\mathcal{L}(\boldsymbol{r}, t) = \frac{1}{2} \left\{ \mathbb{1} \ln \left(v(\boldsymbol{r}, t) \right) + \sigma_x \ln \left(h(\boldsymbol{r}, t) \right) \right\} , \tag{B.13}$$

where σ_x is one of the Pauli triplet matrices

$$\boldsymbol{\sigma} = \left(\sigma_x = \begin{bmatrix} 0 & 1 \\ 1 & 0 \end{bmatrix} , \ \sigma_y = \begin{bmatrix} 0 & -i \\ i & 0 \end{bmatrix} , \ \sigma_z = \begin{bmatrix} 1 & 0 \\ 0 & -1 \end{bmatrix} \right) . \tag{B.14}$$

Using the above notation, the matrix form of the equations in Eq. (B.11) is

$$i \left\{ \mathbb{1} \frac{\partial}{\partial t} - \frac{\partial}{\partial t} \mathcal{L} \right\} \begin{bmatrix} \boldsymbol{F}^+(\boldsymbol{r}, t) \\ \boldsymbol{F}^-(\boldsymbol{r}, t) \end{bmatrix} = v(\boldsymbol{r}) \sigma_z \left\{ \mathbb{1} \nabla + \nabla \mathcal{L} \right\} \times \begin{bmatrix} \boldsymbol{F}^+(\boldsymbol{r}, t) \\ \boldsymbol{F}^-(\boldsymbol{r}, t) \end{bmatrix}$$

$$- \frac{i}{\sqrt{2}} \sqrt{v(\boldsymbol{r}, t) h(\boldsymbol{r}, t)} \boldsymbol{J} ,$$

$$\left\{ \mathbb{1} \nabla - \nabla \mathcal{L} \right\} \cdot \begin{bmatrix} \boldsymbol{F}^+(\boldsymbol{r}, t) \\ \boldsymbol{F}^-(\boldsymbol{r}, t) \end{bmatrix} = + \frac{1}{\sqrt{2}} \sqrt{v(\boldsymbol{r}, t) h(\boldsymbol{r}, t)} \rho , \tag{B.15}$$

where the dot-product and the cross-product are to be understood as

$$\begin{bmatrix} A & B \\ C & D \end{bmatrix} \cdot \begin{bmatrix} u \\ v \end{bmatrix} = \begin{bmatrix} A \cdot u + B \cdot v \\ C \cdot u + D \cdot v \end{bmatrix}$$

$$\begin{bmatrix} A & B \\ C & D \end{bmatrix} \times \begin{bmatrix} u \\ v \end{bmatrix} = \begin{bmatrix} A \times u + B \times v \\ C \times u + D \times v \end{bmatrix}. \qquad (B.16)$$

It is to be noted that the 6×6 matrices in the evolution equations in Eq. (B.15) are either Hermitian or antihermitian. Any dependence on the variations of $\epsilon(r, t)$ and $\mu(r, t)$ is at best 'weak'. We further note, $\nabla \left(\ln \left(v(r, t) \right) \right) = -\nabla \left(\ln \left(n(r, t) \right) \right)$ and $\frac{\partial}{\partial t} \left(\ln \left(v(r, t) \right) \right) = -\frac{\partial}{\partial t} \left(\ln \left(n(r, t) \right) \right)$. In some media, the coupling may vanish ($\nabla h(r, t) = 0$ and $\dot{h}(r, t) = 0$) and in the same medium the refractive index, $n(r, t) = c/v(r, t)$ may vary ($\nabla n(r, t) \neq 0$ or/and $\dot{n}(r, t) \neq 0$). It may be further possible to use the approximations $\nabla \left(\ln \left(h(r, t) \right) \right) \approx 0$ and $\frac{\partial}{\partial t} \left(\ln \left(h(r, t) \right) \right) \approx 0$.

We shall be using the following matrices to express the exact representation

$$\Sigma = \begin{bmatrix} \sigma & 0 \\ 0 & \sigma \end{bmatrix}, \qquad \alpha = \begin{bmatrix} 0 & \sigma \\ \sigma & 0 \end{bmatrix}, \qquad I = \begin{bmatrix} \mathbb{1} & 0 \\ 0 & \mathbb{1} \end{bmatrix}, \qquad (B.17)$$

where Σ are the Dirac spin matrices and α are the matrices used in the Dirac equation. Then,

$$\begin{aligned}
\frac{\partial}{\partial t} & \begin{bmatrix} I & 0 \\ 0 & I \end{bmatrix} \begin{bmatrix} \Psi^+ \\ \Psi^- \end{bmatrix} - \frac{\dot{v}(r, t)}{2v(r, t)} \begin{bmatrix} I & 0 \\ 0 & I \end{bmatrix} \begin{bmatrix} \Psi^+ \\ \Psi^- \end{bmatrix} \\
& + \frac{\dot{h}(r, t)}{2h(r, t)} \begin{bmatrix} 0 & i\beta\alpha_y \\ i\beta\alpha_y & 0 \end{bmatrix} \begin{bmatrix} \Psi^+ \\ \Psi^- \end{bmatrix} \\
& = -v(r, t) \begin{bmatrix} \{M \cdot \nabla + \Sigma \cdot u\} & -i\beta \left(\Sigma \cdot w \right) \alpha_y \\ -i\beta \left(\Sigma^* \cdot w \right) \alpha_y & \{M^* \cdot \nabla + \Sigma^* \cdot u\} \end{bmatrix} \begin{bmatrix} \Psi^+ \\ \Psi^- \end{bmatrix} \\
& \quad - \begin{bmatrix} I & 0 \\ 0 & I \end{bmatrix} \begin{bmatrix} W^+ \\ W^- \end{bmatrix}, \qquad (B.18)
\end{aligned}$$

where

$$\begin{aligned}
u(r, t) &= \frac{1}{2v(r, t)} \nabla v(r, t) = \frac{1}{2} \nabla \left\{ \ln v(r, t) \right\} = -\frac{1}{2} \nabla \left\{ \ln n(r, t) \right\} \\
w(r, t) &= \frac{1}{2h(r, t)} \nabla h(r, t) = \frac{1}{2} \nabla \left\{ \ln h(r, t) \right\}. \qquad (B.19)
\end{aligned}$$

Our matrix representation has thirteen 8×8 matrices. Among these, ten are Hermitian. The remaining three are antihermitian. These three antihermitian matrices couple the upper and lower components (Ψ^+ and Ψ^-) through the logarithmic gradient of the resistance function ($w(r, t)$).

We have been able to express Maxwell's equations in a matrix form in a medium with varying permittivity $\epsilon(r, t)$ and permeability $\mu(r, t)$, in presence of sources. We have been able to do so using a single equation instead of a *pair* of matrix equations. We have used 8×8 matrices and have been able to separate the dependence of the coupling between the upper components (Ψ^+) and the lower components (Ψ^-) through the two laboratory functions. Significantly, the exact matrix representation has an algebraic structure of the Dirac equation. We feel that this 8×8 representation would be more suitable for some of the studies related to the *photon wave function* (Bialynicki-Birula, 1996b). In passing, we note that it is possible to derive Maxwell's equations from the Fermat's principle of geometrical optics by the process of 'wavization' analogous to the quantization of classical mechanics (Pradhan, 1987). This derivation emphasizes the significance of quantum methodologies.

APPENDIX C. FOLDY–WOUTHUYSEN TRANSFORMATION

The Foldy–Wouthuysen transform is widely used in high energy physics. It was historically formulated by Leslie Lawrence Foldy and Siegfried Adolf Wouthuysen in 1949 to understand the nonrelativistic limit of the Dirac equation, the equation for the spin-1/2 particles (Foldy & Wouthuysen, 1950; Foldy, 1952; see also Pryce, 1948; Tani, 1951; see Acharya & Sudarshan, 1960 for a detailed general discussion of the Foldy–Wouthuysen-type transformations in particle interpretation of relativistic wave equations). The approach of the Foldy and Wouthuysen made use of a canonical transform that has now come to be known as the Foldy–Wouthuysen transformation (a brief account of the history of the transformation is to be found in the obituaries of Foldy and Wouthuysen, Brown, Krauss, & Taylor, 2001 and Leopold, 1997; and the biographical memoir of Foldy, 2006). Before their work, there was some difficulty in understanding and gathering all the interaction terms of a given order, such as those for a Dirac particle immersed in an external field. With their procedure, the physical interpretation of the terms was clear, and it became possible to apply their work in a systematic way to a number of problems that had previously defied solution (see Bjorken & Drell, 1964; Costella & McKellar, 1995 for technical details). The Foldy–Wouthuysen transform was extended to the physically important cases of the spin-0 and the spin-1 particles (Case, 1954), and even generalized to the case of arbitrary spins (Jayaraman, 1975; Silenko, 2016). The powerful machinery of the Foldy–Wouthuysen transform has found applications in very diverse areas such as atomic systems (Asaga, Fujita, & Hiramoto, 2000; Pachucki, 2005) synchrotron radiation (Lippert, Brückel, Köhler, & Schneider, 1994) and derivation of the Bloch equation for polarized beams (Heinemann & Barber, 1999). The applications of the Foldy–Wouthuysen transformation in acoustics is very natural; comprehensive and mathematically rigorous accounts can be found by Fishman (1992, 2004), Leviandier (2009), Orris and Wurmser (1995), and Wurmser (2001, 2004). For ocean acoustic see Patton (1986).

In the traditional scheme the purpose of expanding the *light optics* Hamiltonian $\widehat{H} = -(n^2(\mathbf{r}) - \widehat{\boldsymbol{p}}_\perp^2)^{1/2}$ in a series using $(\widehat{\boldsymbol{p}}_\perp^2/n_0^2)$ as the expansion parameter is to understand the propagation of the quasiparaxial beam in terms of a series of approximations (paraxial + nonparaxial). Similar is the situation in the case of the *charged-particle optics*. Let us recall that in relativistic quantum mechanics too one has a similar problem of understanding the relativistic wave equations as the nonrelativistic approximation plus the relativistic correction terms in the quasirelativistic regime. For the Dirac equation (which is first order in time) this is done most conveniently using the Foldy–Wouthuysen transformation leading to an iterative diagonalization technique.

The main framework of the newly developed formalisms of optics (both light optics and the charged-particle optics) is based on the transformation technique of the Foldy–Wouthuysen theory which casts the Dirac equation in a form displaying the different interaction terms between the Dirac particle and an applied electromagnetic field in a nonrelativistic and easily interpretable form. In the Foldy–Wouthuysen theory the Dirac equation is decoupled through a canonical transformation into two two-component equations: one reduces to the Pauli equation (Osche, 1977) in the nonrelativistic limit and the other describes the negative-energy states. There is a close algebraic analogy between the Helmholtz equation (governing scalar optics) and the Klein–Gordon equation; and the matrix-form of Maxwell's equations (governing vector optics) and the Dirac equation. So, it is natural to use the powerful machinery of standard quantum mechanics (particularly, the Foldy–Wouthuysen transform) in analyzing these systems.

The Foldy–Wouthuysen technique is ideally suited for the Lie algebraic approach to optics. With all these plus points, the powerful and ambiguity-free expansion, the Foldy–Wouthuysen Transformation is still little used in optics. The technique of the Foldy–Wouthuysen Transformation results in what we call as the non-traditional prescriptions of Helmholtz optics (Khan, 2005a, 2016g) and Maxwell optics (Khan, 2006c, 2010, 2014a, 2016f, 2017c) respectively. The non-traditional approaches give rise to very interesting wavelength-dependent modifications of the paraxial and aberrating behavior. The non-traditional formalism of Maxwell optics provides a unified framework of light beam optics and polarization. The *non-traditional prescriptions of light optics* are in close analogy with the *quantum theory of charged-particle beam optics* (Jagannathan et al., 1989; Jagannathan, 1990, 1993, 1999, 2002, 2003; Jagannathan & Khan, 1995, 1996, 1997, 2018; Khan & Jagannathan, 1993, 1994, 1995; Khan, 1997, 1999a, 1999b, 2002a, 2002b; Conte et al., 1996). In the following sections, we shall look at the standard Foldy–Wouthuysen transform in detail. An outline of the quantum theory of charged-particle beam optics is presented in Appendix E. A comprehensive account can be found in the references. An exact matrix representation of Maxwell's equations is presented in Appendix B.

The standard Foldy–Wouthuysen theory is described briefly to clarify its use for the purpose of the above studies in optics. Let us consider a charged-particle of rest-mass m_0, charge q in the presence of an electromagnetic field characterized by $\boldsymbol{E} = -\nabla\phi - \frac{\partial}{\partial t}\boldsymbol{A}$ and $\boldsymbol{B} = \nabla \times \boldsymbol{A}$. Then the Dirac equation is

$$i\hbar\frac{\partial}{\partial t}\Psi(\boldsymbol{r}, t) = \widehat{H}_D\Psi(\boldsymbol{r}, t) \tag{C.1}$$

$$\widehat{H}_D = m_0c^2\beta + q\phi + c\boldsymbol{\alpha} \cdot \widehat{\boldsymbol{\pi}}$$
$$= m_0c^2\beta + \widehat{\mathcal{E}} + \widehat{\mathcal{O}}$$
$$\widehat{\mathcal{E}} = q\phi$$
$$\widehat{\mathcal{O}} = c\boldsymbol{\alpha} \cdot \widehat{\boldsymbol{\pi}}, \tag{C.2}$$

where

$$\boldsymbol{\alpha} = \begin{bmatrix} 0 & \sigma \\ \sigma & 0 \end{bmatrix}, \quad \beta = \begin{bmatrix} \mathbb{1} & 0 \\ 0 & -\mathbb{1} \end{bmatrix}, \quad \mathbb{1} = \begin{bmatrix} 1 & 0 \\ 0 & 1 \end{bmatrix},$$

$$\boldsymbol{\sigma} = \begin{bmatrix} \sigma_x = \begin{bmatrix} 0 & 1 \\ 1 & 0 \end{bmatrix}, & \sigma_y = \begin{bmatrix} 0 & -i \\ i & 0 \end{bmatrix}, & \sigma_z = \begin{bmatrix} 1 & 0 \\ 0 & -1 \end{bmatrix} \end{bmatrix}, \tag{C.3}$$

with $\widehat{\boldsymbol{\pi}} = \widehat{\boldsymbol{p}} - q\boldsymbol{A}$, $\widehat{\boldsymbol{p}} = -i\hbar\nabla$, and $\widehat{\pi}^2 = \left(\widehat{\pi}_x^2 + \widehat{\pi}_y^2 + \widehat{\pi}_z^2\right)$.

In the nonrelativistic situation, the upper pair of components of the Dirac spinor Ψ are large compared to the lower pair of components. The operator $\widehat{\mathcal{E}}$ which does not couple the large and small components of Ψ is called 'even' and $\widehat{\mathcal{O}}$ is called an 'odd' operator which couples the large to the small components. Note, that

$$\beta\widehat{\mathcal{O}} = -\widehat{\mathcal{O}}\beta, \qquad \beta\widehat{\mathcal{E}} = \widehat{\mathcal{E}}\beta. \tag{C.4}$$

Now, the search is for a unitary transformation, $\Psi' = \Psi \longrightarrow \widehat{U}\Psi$, such that the equation for Ψ' does not contain any odd operator.

In the free particle case (with $\phi = 0$ and $\widehat{\boldsymbol{\pi}} = \widehat{\boldsymbol{p}}$) such a Foldy–Wouthuysen transformation is given by

$$\Psi \longrightarrow \Psi' = \widehat{U}_F\Psi$$

$$\widehat{U}_F = e^{i\widehat{S}} = e^{\beta\boldsymbol{\alpha}\cdot\widehat{\boldsymbol{p}}\theta}, \qquad \tan 2|\widehat{\boldsymbol{p}}|\theta = \frac{|\widehat{\boldsymbol{p}}|}{m_0 c}. \tag{C.5}$$

This transformation eliminates the odd part completely from the free particle Dirac Hamiltonian reducing it to the diagonal form:

$$\begin{aligned}
i\hbar\frac{\partial}{\partial t}\Psi' &= e^{i\widehat{S}}\left(m_0 c^2\beta + c\boldsymbol{\alpha}\cdot\widehat{\boldsymbol{p}}\right)e^{-i\widehat{S}}\Psi' \\
&= \left(\cos|\widehat{\boldsymbol{p}}|\theta + \frac{\beta\boldsymbol{\alpha}\cdot\widehat{\boldsymbol{p}}}{|\widehat{\boldsymbol{p}}|}\sin|\widehat{\boldsymbol{p}}|\theta\right)\left(m_0 c^2\beta + c\boldsymbol{\alpha}\cdot\widehat{\boldsymbol{p}}\right) \\
&\qquad\qquad \times\left(\cos|\widehat{\boldsymbol{p}}|\theta - \frac{\beta\boldsymbol{\alpha}\cdot\widehat{\boldsymbol{p}}}{|\widehat{\boldsymbol{p}}|}\sin|\widehat{\boldsymbol{p}}|\theta\right)\Psi' \\
&= \left(m_0 c^2\cos 2|\widehat{\boldsymbol{p}}|\theta + c|\widehat{\boldsymbol{p}}|\sin 2|\widehat{\boldsymbol{p}}|\theta\right)\beta\Psi' \\
&= \left(\sqrt{m_0^2 c^4 + c^2\widehat{\boldsymbol{p}}^2}\right)\beta\,\Psi'.
\end{aligned} \tag{C.6}$$

Generally, when the electron is in a time-dependent electromagnetic field it is not possible to construct an $\exp(i\widehat{S})$ which removes the odd operators from the transformed Hamiltonian completely. Therefore, one has to be content with a nonrelativistic expansion of the transformed Hamiltonian in a power series in $1/m_0 c^2$ keeping through any desired order. Note, that in the nonrelativistic case, when $|\boldsymbol{p}| \ll m_0 c$, the transformation operator $\widehat{U}_F = \exp(i\widehat{S})$ with $\widehat{S} \approx -i\beta\widehat{\mathcal{O}}/2m_0 c^2$, where $\widehat{\mathcal{O}} = c\boldsymbol{\alpha}\cdot\widehat{\boldsymbol{p}}$ is the odd part of the free Hamiltonian. So, in the general case we can start with the transformation

$$\Psi^{(1)} = e^{i\widehat{S}_1}\Psi, \qquad \widehat{S}_1 = -\frac{i\beta\widehat{\mathcal{O}}}{2m_0c^2} = -\frac{i\beta\boldsymbol{\alpha}\cdot\widehat{\boldsymbol{\pi}}}{2m_0c}. \qquad (C.7)$$

Then, the equation for $\Psi^{(1)}$ is

$$
\begin{aligned}
i\hbar\frac{\partial}{\partial t}\Psi^{(1)} &= i\hbar\frac{\partial}{\partial t}\left(e^{i\widehat{S}_1}\Psi\right) = i\hbar\frac{\partial}{\partial t}\left(e^{i\widehat{S}_1}\right)\Psi + e^{i\widehat{S}_1}\left(i\hbar\frac{\partial}{\partial t}\Psi\right) \\
&= \left[i\hbar\frac{\partial}{\partial t}\left(e^{i\widehat{S}_1}\right) + e^{i\widehat{S}_1}\widehat{H}_D\right]\Psi \\
&= \left[i\hbar\frac{\partial}{\partial t}\left(e^{i\widehat{S}_1}\right)e^{-i\widehat{S}_1} + e^{i\widehat{S}_1}\widehat{H}_De^{-i\widehat{S}_1}\right]\Psi^{(1)} \\
&= \left[e^{i\widehat{S}_1}\widehat{H}_De^{-i\widehat{S}_1} - i\hbar e^{i\widehat{S}_1}\frac{\partial}{\partial t}\left(e^{-i\widehat{S}_1}\right)\right]\Psi^{(1)} \\
&= \widehat{H}_D^{(1)}\Psi^{(1)} \qquad (C.8)
\end{aligned}
$$

where we have used the identity $\frac{\partial}{\partial t}\left(e^{\widehat{A}}\right)e^{-\widehat{A}} + e^{\widehat{A}}\frac{\partial}{\partial t}\left(e^{-\widehat{A}}\right) = \frac{\partial}{\partial t}\widehat{I} = 0$.

Now, using the identities

$$e^{\widehat{A}}\widehat{B}e^{-\widehat{A}} = \widehat{B} + [\widehat{A},\widehat{B}] + \frac{1}{2!}[\widehat{A},[\widehat{A},\widehat{B}]] + \frac{1}{3!}[\widehat{A},[\widehat{A},[\widehat{A},\widehat{B}]]] + \dots$$

$$
\begin{aligned}
e^{\widehat{A}(t)}&\frac{\partial}{\partial t}\left(e^{-\widehat{A}(t)}\right) \\
&= \left(1 + \widehat{A}(t) + \frac{1}{2!}\widehat{A}(t)^2 + \frac{1}{3!}\widehat{A}(t)^3\cdots\right) \\
&\quad \times \frac{\partial}{\partial t}\left(1 - \widehat{A}(t) + \frac{1}{2!}\widehat{A}(t)^2 - \frac{1}{3!}\widehat{A}(t)^3\cdots\right) \\
&= \left(1 + \widehat{A}(t) + \frac{1}{2!}\widehat{A}(t)^2 + \frac{1}{3!}\widehat{A}(t)^3\cdots\right) \\
&\quad \times \left(-\frac{\partial\widehat{A}(t)}{\partial t} + \frac{1}{2!}\left\{\frac{\partial\widehat{A}(t)}{\partial t}\widehat{A}(t) + \widehat{A}(t)\frac{\partial\widehat{A}(t)}{\partial t}\right\}\right. \\
&\qquad -\frac{1}{3!}\left\{\frac{\partial\widehat{A}(t)}{\partial t}\widehat{A}(t)^2 + \widehat{A}(t)\frac{\partial\widehat{A}(t)}{\partial t}\widehat{A}(t)\right. \\
&\qquad\qquad \left.\left. + \widehat{A}(t)^2\frac{\partial\widehat{A}(t)}{\partial t}\right\}\cdots\right) \\
&\approx -\frac{\partial\widehat{A}(t)}{\partial t} - \frac{1}{2!}\left[\widehat{A}(t),\frac{\partial\widehat{A}(t)}{\partial t}\right]
\end{aligned}
$$

$$-\frac{1}{3!}\left[\widehat{A}(t),\left[\widehat{A}(t),\frac{\partial\widehat{A}(t)}{\partial t}\right]\right]$$

$$-\frac{1}{4!}\left[\widehat{A}(t),\left[\widehat{A}(t),\left[\widehat{A}(t),\frac{\partial\widehat{A}(t)}{\partial t}\right]\right]\right], \tag{C.9}$$

with $\widehat{A}=i\widehat{S}_1$, we find

$$\widehat{H}_D^{(1)}\approx\widehat{H}_D-\hbar\frac{\partial\widehat{S}_1}{\partial t}+i\left[\widehat{S}_1,\widehat{H}_D-\frac{\hbar}{2}\frac{\partial\widehat{S}_1}{\partial t}\right]$$

$$-\frac{1}{2!}\left[\widehat{S}_1,\left[\widehat{S}_1,\widehat{H}_D-\frac{\hbar}{3}\frac{\partial\widehat{S}_1}{\partial t}\right]\right]$$

$$-\frac{i}{3!}\left[\widehat{S}_1,\left[\widehat{S}_1,\left[\widehat{S}_1,\widehat{H}_D-\frac{\hbar}{4}\frac{\partial\widehat{S}_1}{\partial t}\right]\right]\right]. \tag{C.10}$$

Substituting in Eq. (C.10), $\widehat{H}_D=m_0c^2\beta+\widehat{\mathcal{E}}+\widehat{\mathcal{O}}$, simplifying the right hand side using the relations $\beta\widehat{\mathcal{O}}=-\widehat{\mathcal{O}}\beta$ and $\beta\widehat{\mathcal{E}}=\widehat{\mathcal{E}}\beta$ and collecting everything together, we have

$$\widehat{H}_D^{(1)}\approx m_0c^2\beta+\widehat{\mathcal{E}}_1+\widehat{\mathcal{O}}_1$$

$$\widehat{\mathcal{E}}_1\approx\widehat{\mathcal{E}}+\frac{1}{2m_0c^2}\beta\widehat{\mathcal{O}}^2-\frac{1}{8m_0^2c^4}\left[\widehat{\mathcal{O}},\left([\widehat{\mathcal{O}},\widehat{\mathcal{E}}]+i\hbar\frac{\partial\widehat{\mathcal{O}}}{\partial t}\right)\right]$$

$$-\frac{1}{8m_0^3c^6}\beta\widehat{\mathcal{O}}^4$$

$$\widehat{\mathcal{O}}_1\approx\frac{\beta}{2m_0c^2}\left([\widehat{\mathcal{O}},\widehat{\mathcal{E}}]+i\hbar\frac{\partial\widehat{\mathcal{O}}}{\partial t}\right)-\frac{1}{3m_0^2c^4}\widehat{\mathcal{O}}^3, \tag{C.11}$$

with $\widehat{\mathcal{E}}_1$ and $\widehat{\mathcal{O}}_1$ obeying the relations $\beta\widehat{\mathcal{O}}_1=-\widehat{\mathcal{O}}_1\beta$ and $\beta\widehat{\mathcal{E}}_1=\widehat{\mathcal{E}}_1\beta$ exactly like $\widehat{\mathcal{E}}$ and $\widehat{\mathcal{O}}$. It is seen that while the term $\widehat{\mathcal{O}}$ in $\widehat{H}_D$ is of order zero with respect to the expansion parameter $1/m_0c^2$ (i.e., $\widehat{\mathcal{O}}=O((1/m_0c^2)^0)$) the odd part of $\widehat{H}_D^{(1)}$, namely $\widehat{\mathcal{O}}_1$, contains only terms of order $1/m_0c^2$ and higher powers of $1/m_0c^2$ (i.e., $\widehat{\mathcal{O}}_1=O((1/m_0c^2))$).

To reduce the strength of the odd terms further in the transformed Hamiltonian a second Foldy–Wouthuysen transformation is applied with the same prescription:

$$\Psi^{(2)}=e^{i\widehat{S}_2}\Psi^{(1)},$$

$$\widehat{S}_2=-\frac{i\beta\widehat{\mathcal{O}}_1}{2m_0c^2}$$

$$= -\frac{i\beta}{2m_0c^2}\left[\frac{\beta}{2m_0c^2}\left([\hat{\mathcal{O}},\hat{\mathcal{E}}]+i\hbar\frac{\partial\hat{\mathcal{O}}}{\partial t}\right)-\frac{1}{3m_0^2c^4}\hat{\mathcal{O}}^3\right]. \qquad \text{(C.12)}$$

After this transformation,

$$i\hbar\frac{\partial}{\partial t}\Psi^{(2)}=\hat{H}_D^{(2)}\Psi^{(2)},\quad \hat{H}_D^{(2)}=m_0c^2\beta+\hat{\mathcal{E}}_2+\hat{\mathcal{O}}_2$$

$$\hat{\mathcal{E}}_2\approx\hat{\mathcal{E}}_1,\quad \hat{\mathcal{O}}_2\approx\frac{\beta}{2m_0c^2}\left([\hat{\mathcal{O}}_1,\hat{\mathcal{E}}_1]+i\hbar\frac{\partial\hat{\mathcal{O}}_1}{\partial t}\right), \qquad \text{(C.13)}$$

where, now, $\hat{\mathcal{O}}_2=O\big((1/m_0c^2)^2\big)$. After the third transformation

$$\Psi^{(3)}=e^{i\hat{S}_3}\Psi^{(2)},\qquad \hat{S}_3=-\frac{i\beta\hat{\mathcal{O}}_2}{2m_0c^2} \qquad \text{(C.14)}$$

we have

$$i\hbar\frac{\partial}{\partial t}\Psi^{(3)}=\hat{H}_D^{(3)}\Psi^{(3)},\quad \hat{H}_D^{(3)}=m_0c^2\beta+\hat{\mathcal{E}}_3+\hat{\mathcal{O}}_3$$

$$\hat{\mathcal{E}}_3\approx\hat{\mathcal{E}}_2\approx\hat{\mathcal{E}}_1,\quad \hat{\mathcal{O}}_3\approx\frac{\beta}{2m_0c^2}\left([\hat{\mathcal{O}}_2,\hat{\mathcal{E}}_2]+i\hbar\frac{\partial\hat{\mathcal{O}}_2}{\partial t}\right), \qquad \text{(C.15)}$$

where $\hat{\mathcal{O}}_3=O\big((1/m_0c^2)^3\big)$. So, neglecting $\hat{\mathcal{O}}_3$,

$$\hat{H}_D^{(3)}\approx m_0c^2\beta+\hat{\mathcal{E}}+\frac{1}{2m_0c^2}\beta\hat{\mathcal{O}}^2$$

$$-\frac{1}{8m_0^2c^4}\left[\hat{\mathcal{O}},\left([\hat{\mathcal{O}},\hat{\mathcal{E}}]+i\hbar\frac{\partial\hat{\mathcal{O}}}{\partial t}\right)\right]$$

$$-\frac{1}{8m_0^3c^6}\beta\left\{\hat{\mathcal{O}}^4+\left([\hat{\mathcal{O}},\hat{\mathcal{E}}]+i\hbar\frac{\partial\hat{\mathcal{O}}}{\partial t}\right)^2\right\} \qquad \text{(C.16)}$$

It may be noted that starting with the second transformation successive $(\hat{\mathcal{E}},\hat{\mathcal{O}})$ pairs can be obtained recursively using the rule

$$\hat{\mathcal{E}}_j=\hat{\mathcal{E}}_1\left(\hat{\mathcal{E}}\to\hat{\mathcal{E}}_{j-1},\hat{\mathcal{O}}\to\hat{\mathcal{O}}_{j-1}\right)$$

$$\hat{\mathcal{O}}_j=\hat{\mathcal{O}}_1\left(\hat{\mathcal{E}}\to\hat{\mathcal{E}}_{j-1},\hat{\mathcal{O}}\to\hat{\mathcal{O}}_{j-1}\right),\quad j>1, \qquad \text{(C.17)}$$

and retaining only the relevant terms of desired order at each step.

With $\hat{\mathcal{E}}=q\phi$ and $\hat{\mathcal{O}}=c\boldsymbol{\alpha}\cdot\hat{\boldsymbol{\pi}}$, the final reduced Hamiltonian in Eq. (C.16) is, to the order calculated,

$$\hat{H}_D^{(3)}=\beta\left(m_0c^2+\frac{\hat{\pi}^2}{2m_0}-\frac{\hat{p}^4}{8m_0^3c^6}\right)+q\phi-\frac{q\hbar}{2m_0c}\beta\boldsymbol{\Sigma}\cdot\boldsymbol{B}$$

$$- \frac{iq\hbar^2}{8m_0^2 c^2} \mathbf{\Sigma} \cdot \text{curl}\, \mathbf{E} - \frac{q\hbar}{4m_0^2 c^2} \mathbf{\Sigma} \cdot \mathbf{E} \times \widehat{\boldsymbol{p}}$$

$$- \frac{q\hbar^2}{8m_0^2 c^2} \text{div}\,\mathbf{E}, \tag{C.18}$$

with the individual terms having direct physical interpretations. The terms in the first parenthesis result from the expansion of $\sqrt{m_0^2 c^4 + c^2 \widehat{\pi}^2}$ showing the effect of the relativistic mass increase. The second and third terms are the electrostatic and magnetic dipole energies. The next two terms, taken together (for hermiticity), contain the spin–orbit interaction. The last term, the so-called Darwin term, is attributed to the *zitterbewegung* (trembling motion) of the Dirac particle: because of the rapid coordinate fluctuations over distances of the order of the Compton wavelength $(2\pi \hbar/m_0 c)$ the particle sees a somewhat smeared out electric potential.

The Foldy–Wouthuysen transformation technique clearly expands the Dirac Hamiltonian as a power series in the parameter $1/m_0 c^2$ enabling the use of a systematic approximation procedure for studying the deviations from the nonrelativistic situation. We note the analogy between the non-relativistic particle dynamics and paraxial optics in Table 2.

Noting the above analogy, the idea of Foldy–Wouthuysen form of the Dirac theory has been adopted to study the paraxial optics and deviations from it. The Helmholtz equation governing scalar optics is first linearized in a procedure similar to the way in which the Klein–Gordon equation is writ-ten in the Feshbach–Villars form (linear in $\partial/\partial t$), unlike the Klein–Gordon equation (quadratic in $\partial/\partial t$). This enables us to use the Foldy–Wouthuysen transformation technique. In the case of the vector optics, Maxwell's equa-tions are cast in a spinor form resembling exactly the Dirac equation (C.1), (C.2) in all respects: i.e., a multicomponent Ψ having the upper half of its components large compared to the lower components and the Hamiltonian

Table 2 The correspondence between particle dynamics and paraxial optics

Standard Dirac equation	Beam optical form
$m_0 c^2 \beta + \widehat{\mathcal{E}}_D + \widehat{\mathcal{O}}_D$	$-n_0 \beta + \widehat{\mathcal{E}} + \widehat{\mathcal{O}}$
$m_0 c^2$	$-n_0$
Positive energy	Forward propagation
Nonrelativistic, $\lvert \widehat{\pi} \rvert \ll m_0 c$	Paraxial beam, $\lvert \widehat{\boldsymbol{p}}_\perp \rvert \ll n_0$
Non-relativistic motion	Paraxial behavior
+ relativistic corrections	+ aberration corrections

having an even part ($\widehat{\mathcal{E}}$), an odd part ($\widehat{\mathcal{O}}$), a suitable expansion parameter ($|\widehat{\boldsymbol{p}}_\perp|/n_0 \ll 1$) characterizing the dominant forward propagation, and a leading term with a β coefficient commuting with $\widehat{\mathcal{E}}$ and anticommuting with $\widehat{\mathcal{O}}$. It has to be borne in mind that the Dirac field and the electromagnetic field are two entirely different entities. But the striking resemblance in the underlying algebraic structure can be exploited to carry out some useful calculations with meaningful results. See Appendix B for the derivation of an exact matrix representation of Maxwell's equations and how it differs from numerous other representations. The additional feature of our formalism is to return finally to the original representation after making an extra approximation, dropping β from the final reduced optical Hamiltonian, taking into account the fact that we are primarily interested only in the forward-propagating beam. The Foldy–Wouthuysen transformation has enabled entirely new approaches to light optics and charged particle optics respectively (Khan, 2006b, 2008).

APPENDIX D. THE MAGNUS FORMULA

The Magnus formula is the continuous analogue of the famous Baker–Campbell–Hausdorff (BCH) formula

$$e^{\widehat{A}}e^{\widehat{B}} = e^{\widehat{A}+\widehat{B}+\frac{1}{2}[\widehat{A},\widehat{B}]+\frac{1}{12}\{[[\widehat{A},\widehat{A}],\widehat{B}]+[[\widehat{A},\widehat{B}],\widehat{B}]\}+\cdots}. \tag{D.1}$$

Let it be required to solve the differential equation

$$\frac{\partial}{\partial t}u(t) = \widehat{A}(t)u(t) \tag{D.2}$$

to get $u(T)$ at $T > t_0$, given the value of $u(t_0)$; the operator $\widehat{A}$ can represent any linear operation. For an infinitesimal Δt, we can write

$$u(t_0 + \Delta t) = e^{\Delta t \widehat{A}(t_0)}u(t_0). \tag{D.3}$$

Iterating this solution we have

$$u(t_0 + 2\Delta t) = e^{\Delta t\widehat{A}(t_0+\Delta t)}e^{\Delta t\widehat{A}(t_0)}u(t_0)$$

$$u(t_0 + 3\Delta t) = e^{\Delta t\widehat{A}(t_0+2\Delta t)}e^{\Delta t\widehat{A}(t_0+\Delta t)}e^{\Delta t\widehat{A}(t_0)}u(t_0)$$

$$\cdots \quad \text{and so on.} \tag{D.4}$$

If $T = t_0 + N\Delta t$ we would have

$$u(T) = \left\{ \prod_{n=0}^{N-1} e^{\Delta t \widehat{A}(t_0 + n\Delta t)} \right\} u(t_0). \tag{D.5}$$

Thus, $u(T)$ is given by computing the product in Eq. (D.5) using successively the BCH-formula (D.1) and considering the limit $\Delta t \longrightarrow 0, N \longrightarrow \infty$ such that $N\Delta t = T - t_0$. The resulting expression is the Magnus formula (Blanes et al., 2009; Magnus, 1954; Mananga & Charpentier, 2016; Wilcox, 1967):

$$u(T) = \widehat{\mathcal{T}}(T, t_0) u(t_0)$$

$$\mathcal{T}(T, t_0) = \exp \left\{ \int_{t_0}^{T} dt_1 \, \widehat{A}(t_1) \right.$$

$$+ \frac{1}{2} \int_{t_0}^{T} dt_2 \int_{t_0}^{t_2} dt_1 \, [\widehat{A}(t_2), \widehat{A}(t_1)]$$

$$+ \frac{1}{6} \int_{t_0}^{T} dt_3 \int_{t_0}^{t_3} dt_2 \int_{t_0}^{t_2} dt_1 \, ([[\widehat{A}(t_3), \widehat{A}(t_2)], \widehat{A}(t_1)]$$

$$\left. + [[\widehat{A}(t_1), \widehat{A}(t_2)], \widehat{A}(t_3)]) + \dots \right\}. \tag{D.6}$$

To see how Eq. (D.6) is obtained, let us substitute the assumed form of the solution, $u(t) = \widehat{\mathcal{T}}(t, t_0) u(t_0)$, in Eq. (D.2). Then, it is seen that $\widehat{\mathcal{T}}(t, t_0)$ obeys the equation

$$\frac{\partial}{\partial t} \widehat{\mathcal{T}}(t, t_0) = \widehat{A}(t) \mathcal{T}(t, t_0), \qquad \widehat{\mathcal{T}}(t_0, t_0) = \widehat{I}. \tag{D.7}$$

Introducing an iteration parameter λ in Eq. (D.7), let

$$\frac{\partial}{\partial t} \widehat{\mathcal{T}}(t, t_0; \lambda) = \lambda \widehat{A}(t) \widehat{\mathcal{T}}(t, t_0; \lambda), \tag{D.8}$$

$$\widehat{\mathcal{T}}(t_0, t_0; \lambda) = \widehat{I}, \quad \widehat{\mathcal{T}}(t, t_0; 1) = \widehat{\mathcal{T}}(t, t_0). \tag{D.9}$$

Assume a solution of Eq. (D.8) to be of the form

$$\widehat{\mathcal{T}}(t, t_0; \lambda) = e^{\Omega(t, t_0; \lambda)} \tag{D.10}$$

with

$$\Omega(t, t_0; \lambda) = \sum_{n=1}^{\infty} \lambda^n \Delta_n(t, t_0), \qquad \Delta_n(t_0, t_0) = 0 \quad \text{for all } n. \tag{D.11}$$

Now, using the identity (see Wilcox, 1967)

$$\frac{\partial}{\partial t} e^{\Omega(t,t_0;\lambda)} = \left\{ \int_0^1 ds\, e^{s\Omega(t,t_0;\lambda)} \frac{\partial}{\partial t} \Omega(t,t_0;\lambda) e^{-s\Omega(t,t_0;\lambda)} \right\} e^{\Omega(t,\lambda)}, \qquad (D.12)$$

one has

$$\int_0^1 ds\, e^{s\Omega(t,t_0;\lambda)} \frac{\partial}{\partial t} \Omega(t,t_0;\lambda) e^{-s\Omega(t,t_0;\lambda)} = \lambda \widehat{A}(t). \qquad (D.13)$$

Substituting in Eq. (D.13) the series expression for $\Omega(t,t_0;\lambda)$ (D.11), expanding the left hand side using the first identity in Eq. (D.8), integrating and equating the coefficients of λ^j on both sides, we get, recursively, the equations for $\Delta_1(t,t_0), \Delta_2(t,t_0), \ldots$, etc. For $j=1$

$$\frac{\partial}{\partial t} \Delta_1(t,t_0) = \widehat{A}(t), \qquad \Delta_1(t_0,t_0) = 0 \qquad (D.14)$$

and hence

$$\Delta_1(t,t_0) = \int_{t_0}^t dt_1 \widehat{A}(t_1). \qquad (D.15)$$

For $j=2$

$$\frac{\partial}{\partial t} \Delta_2(t,t_0) + \frac{1}{2}\left[\Delta_1(t,t_0), \frac{\partial}{\partial t}\Delta_1(t,t_0)\right] = 0, \qquad \Delta_2(t_0,t_0) = 0 \qquad (D.16)$$

and hence

$$\Delta_2(t,t_0) = \frac{1}{2}\int_{t_0}^t dt_2 \int_{t_0}^{t_2} dt_1 \left[\widehat{A}(t_2), \widehat{A}(t_1)\right]. \qquad (D.17)$$

Similarly,

$$\Delta_3(t,t_0) = \frac{1}{6}\int_{t_0}^t dt_1 \int_{t_0}^{t_1} dt_2 \int_{t_0}^{t_2} dt_3 \left\{ \left[\left[\widehat{A}(t_1), \widehat{A}(t_2)\right], \widehat{A}(t_3)\right] \right.$$
$$\left. + \left[\left[\widehat{A}(t_3), \widehat{A}(t_2)\right], \widehat{A}(t_1)\right]\right\}. \qquad (D.18)$$

Then, the Magnus formula in Eq. (D.6) follows from (D.9)–(D.11). Eq. (42) in the context of z-evolution follows from the above discussion with the identification $t \longrightarrow z$, $t_0 \longrightarrow z^{(1)}$, $T \longrightarrow z^{(2)}$ and $\widehat{A}(t) \longrightarrow -\frac{i}{\hbar}\widehat{\mathcal{H}}_0(z)$.

For more details on the exponential solutions of linear differential equations, related operator techniques and applications to physical problems the reader is referred to Wilcox (1967), Bellman and Vasudevan (1986), Dattoli, Renieri, and Torre (1993), and references therein.

APPENDIX E. QUANTUM THEORY OF CHARGED-PARTICLE BEAM OPTICS

The classical mechanics is obtained from the quantum mechanics, by taking $\hbar \longrightarrow 0$, where $\hbar = h/2\pi$ is the reduced Planck's constant. This naturally leads to the question, *where is the $\hbar$ in charged-particle beam optics?* This question has been addressed in substantial detail by developing a quantum theory of charged-particle beam optics.

The designing and operating of a variety of small devices such as electron microscopes and large-scale accelerator facilities is based on classical mechanics of charged-particle beam optics (Hawkes & Kasper, 1989a, 1989b). Owing to this success, there have been very few attempts to develop quantum theories of charged-particle beam optics (Hawkes & Kasper, 1994). The following are noteworthy:

1. 1930, Glaser: A semiclassical approach to image formation in an electron microscope, based on the non-relativistic Schrödinger equation (Glaser, 1956).

2. 1934, 1957, 1963, 1965, 1966, Rubinovicz; 1953, Durand; and 1953, 1954, 1955, 1958, 1960, Phan-Van-Loc: Used the Dirac equation to understand the electron diffraction (Durand, 1953; Phan-Van-Loc, 1953, 1954, 1955, 1958a, 1958b, 1960; Rubinowicz, 1934, 1957, 1963, 1965).

3. 1986, Ferwerda, Hoenders, and Slump: Application of the Klein–Gordon equation to practical electron microscopes operating at relativistic energies (Ferwerda, Hoenders, & Slump, 1986a, 1986b).

4. 1989, Jagannathan, Mukunda, Simon, and Sudarshan: Focusing action of a magnetic round lens *ab initio* using the Dirac equation (Jagannathan et al., 1989).

5. 1990, Jagannathan: Derived the focusing theory for electron lenses, and in particular for magnetic and electrostatic round lenses and quadrupole lenses, from the Dirac equation; examined curved optical axes. Prepared the complete *blueprint* of the *Quantum Theory of Charged-Particle Beam Optics* (Jagannathan, 1990, 1993).

6. 1995, Jagannathan and Khan: Following the Jagannathan's 1990 blueprint, the quantum theory of aberrations is developed using the Klein–Gordon equation and the Dirac equation respectively (Khan & Jagannathan, 1995; Jagannathan & Khan, 1996, 1997).

7. 1996, Conte, Jagannathan, Khan, and Pusterla: A unified theory of the beam-optics and the spin dynamics of a Dirac particle beam is presented, considering the anomalous magnetic moment (Conte et al., 1996). The paraxial limit of this approach leads to the beam-optical version of the Thomas–Bargmann–Michel–Telegdi (Thomas–BMT) equation.

8. 1997, Khan: *First Doctoral Thesis* on the *Quantum Theory of Charged-Particle Beam Optics* (Khan, 1997) under the supervision of Jagannathan (Institute of Mathematical Sciences, Chennai, India).

See Hawkes and Kasper (1994) for a detailed historical account. Thereafter, the field of quantum approaches to beam optics has been under active research (Conte et al., 1996; Jagannathan et al., 1989; Jagannathan, 1990, 1993, 1999, 2002, 2003; Jagannathan & Khan, 1995, 1996, 1997, 2018; Khan, 1997, 1999a, 1999b, 2001, 2002a, 2002b, 2002c, 2003d, 2016d; Khan & Jagannathan, 1993, 1994, 1995). A diffractive quantum limit has been also considered for particle beams (Hill, 2000). There have been a series of meetings on the emerging field of QABP: the *Quantum Aspects of Beam Physics*, whose proceedings are in Chen (1999, 2002), Chen and Reil (2003), and their reports (Chen, 1998, 2000, 2003a, 2003b).

Let us recall that there are several equations in quantum mechanics governing specific situations: Schrödinger equation (for non-relativistic spin-less particles); Klein–Gordon equation (for relativistic spin-less particles); Dirac equation (for relativistic spin-1/2 particles); and so forth. The quantum prescriptions of charged-particle beam optics originate from one of the aforementioned time-dependent equations applicable to the system being studied. In beam optics, the focus is in the evolution of the relevant beam parameters along the optic axis of the system, as the beam propagates through the optical elements (quadrupoles, sextupoles, bending magnets, ...). For a system having a straight optic axis (say along z-axis), it would suffice to study the z-evolution of the beam parameters. In general, we can have a curvilinear coordinate system, $(x, y; s)$ depending on the geometry of the system, leading to the s-evolution. Consequently, we first transform the time-dependent equations from quantum mechanics to the following form

$$i\hbar\frac{\partial}{\partial s}\psi\left(\boldsymbol{r}_\perp,s\right)=\widehat{\mathcal{H}}\left(\boldsymbol{r}_\perp,s\right)\psi\left(\boldsymbol{r}_\perp,s\right),\tag{E.1}$$

where $\boldsymbol{r}_\perp = (x, y)$ are the transverse coordinates. Eq. (E.1) is the fundamental equation governing the quantum mechanics of charged-particle beam optics for the system under study. Hence, we name it as the *beam-optical equation*. Accordingly, $\mathcal{H}$ is the *beam-optical Hamiltonian* and ψ is the *beam wavefunction*. It would be incomplete to assume that the beam-optical equation is obtained by just eliminating the time 't' in the basic equations of quantum mechanics in preference to the spatial variable 's' along the optic axis. One needs to make suitable transforms ensuring that the derived beam-optical equation has both the physical as well as the mathematical correspondence with the parent time-dependent equation from quantum mechanics. Otherwise, it would be difficult to proceed further with the beam-optical equation. The aforementioned correspondence enables us to apply the rich and powerful techniques developed to handle each of the equations of quantum mechanics. Now, we are equipped to obtain the required transfer maps completely characterizing the optical system.

Let $\{\langle O\rangle(s_{in})\}$ be a relevant observable at the transverse plane at some input reference point 's_{in}'. Its evolution along the optic axis at the transverse-plane at 's' is obtained by integrating the beam-optical equation in Eq. (E.1)

$$\psi\left(\boldsymbol{r}_\perp,s\right)=\widehat{U}\left(s,s_{in}\right)\psi\left(\boldsymbol{r}_\perp,s_{in}\right).\tag{E.2}$$

Then, the required transfer maps are obtained as follows

$$\langle O\rangle\left(s_{in}\right)\longrightarrow\langle O\rangle\left(s\right)=\langle\psi\left(\boldsymbol{r}_\perp,s\right)\,|O|\,\psi\left(\boldsymbol{r}_\perp,s\right)\rangle$$
$$=\langle\psi\left(\boldsymbol{r}_\perp,s_{in}\right)\,|\widehat{U}^\dagger O\widehat{U}|\,\psi\left(\boldsymbol{r}_\perp,s_{in}\right)\rangle.\tag{E.3}$$

To illustrate the algorithm, let us consider the transport of a quasiparaxial beam of charged-particles of mean momentum p_0 through a magnetic lens, using the Dirac equation. A magnetic lens is characterized by a magnetic field $\boldsymbol{B}(\boldsymbol{r}) = (B_x, B_y, B_z)$ corresponding to the vector potential $\boldsymbol{A}(\boldsymbol{r}) = (A_x, A_y, A_z)$. For an axially symmetric magnetic lens the vector potential is

$$\boldsymbol{A}(\boldsymbol{r})=\left(-\frac{y}{2}\Pi\left(\boldsymbol{r}_\perp,z\right),\frac{x}{2}\Pi\left(\boldsymbol{r}_\perp,z\right),0\right)$$

$$\Pi\left(\mathbf{r}_\perp, z\right) = \sum_{n=0}^{\infty} \frac{1}{n!(n+1)!}\left(-\frac{\mathbf{r}_\perp^2}{4}\right)^n B^{(2n)}(z)$$

$$= B(z) - \frac{1}{8}r_\perp^2 B''(z) + \frac{1}{192}r_\perp^4 B''''(z) - \cdots, \qquad \text{(E.4)}$$

where $B^0(z) = B(z)$, $B'(z) = dB(z)/dz$, $B''(z) = d^2 B(z)/dz^2, \cdots$. The magnetic field corresponding to this vector potential is axially symmetric with respect to the z-axis, which is the optic axis. It is given by

$$\mathbf{B}_\perp = -\frac{1}{2}\left(B'(z) - \frac{1}{8}r_\perp^2 B'''(z) + \cdots\right)\mathbf{r}_\perp$$

$$B_z = B(z) - \frac{1}{4}r_\perp^2 B''(z) + \frac{1}{64}r_\perp^4 B''''(z) - \cdots. \qquad \text{(E.5)}$$

The axially symmetric magnetic round lenses are of interest for electron microscopes. In accelerator optics, the primary interest is in the quadrupoles, bending magnets and sextupoles. The quadrupole is characterized by the following magnetic field

$$\mathbf{B} = (-Q_m y, -Q_m x, 0), \qquad \text{(E.6)}$$

corresponding to the vector potential

$$\mathbf{A}(\mathbf{r}) = \left(0, 0, \frac{1}{2}Q_m(x^2 - y^2)\right), \qquad \text{(E.7)}$$

where Q_m is the quadrupole strength. The sextupole is characterized by the following magnetic field

$$\mathbf{B} = \left(-6S_m xy, -3S_m(x^2 - y^2), 0\right), \qquad \text{(E.8)}$$

corresponding to the vector potential

$$\mathbf{A}(\mathbf{r}) = \left(0, 0, \frac{1}{2}S_m(x^3 - 3xy^2)\right), \qquad \text{(E.9)}$$

where S_m is the sextupole strength. In all types of lenses, the field is assumed to be confined to the lens region, (z_{in}, z_{out}) and zero outside this region. The beam is assumed to move along a straight optical axis chosen to be along the z-axis in the positive direction.

The Dirac equation for the magnetic lens system is

$$i\hbar\frac{\partial\Psi(\mathbf{r}, t)}{\partial t} = \left[\left(mc^2\beta + c\boldsymbol{\alpha}_\perp \cdot \hat{\boldsymbol{\pi}}_\perp\right) + c\alpha_z\left(p_z - qA_z\right)\right]\Psi(\mathbf{r}, t). \qquad \text{(E.10)}$$

The beam wavefunction is assumed to be in the following form

$$\Psi(\mathbf{r}, t) = \int_{p_0-\Delta p}^{p_0+\Delta p} dp \, \exp\left[-\frac{i}{\hbar}E(p)t\right]\psi(\mathbf{r}; \mathbf{p}), \qquad (E.11)$$

where $\Delta p \ll p_0$, $p = |\mathbf{p}|$, $E(p) = +(m^2c^4 + c^2p^2)^{1/2}$, and $\psi(\mathbf{r}; \mathbf{p})$ obeys the time-independent Dirac equation

$$\left[E(p) - mc^2\beta - c\boldsymbol{\alpha}_\perp \cdot \widehat{\boldsymbol{\pi}}_\perp + c\alpha_z\left(i\hbar\frac{\partial}{\partial z} + qA_z\right)\right]\psi(\mathbf{r}; \mathbf{p}) = 0. \qquad (E.12)$$

Eq. (E.12) can be integrated in the form

$$\psi(\mathbf{r}_\perp, z_{\text{out}}; p) = \widehat{G}(z_{\text{out}}, z_{\text{in}}; p)\psi(\mathbf{r}_\perp, z_{\text{in}}; p). \qquad (E.13)$$

Here, $\widehat{G}(z_{\text{out}}, z_{\text{in}}; p)$ is the z-propagator for the transport of the Dirac wavefunction through the magnetic lens. Now, we rewrite Eq. (E.12) as

$$i\hbar\frac{\partial\psi}{\partial z} = \left[-qA_z - p_0\chi\beta\alpha_z + \alpha_z\boldsymbol{\alpha}_\perp \cdot \widehat{\boldsymbol{\pi}}_\perp\right]\psi,$$

$$\chi = \begin{pmatrix} \xi I & 0 \\ 0 & -\xi^{-1}I \end{pmatrix}, \qquad \xi = \frac{(E(p) + mc^2)}{cp_0}. \qquad (E.14)$$

Now, a transformation $\psi \rightarrow \psi' = M\psi = (1/\sqrt{2})(I + \chi\alpha_z)\psi$ converts (E.14) into the required beam-optical representation

$$i\hbar\frac{\partial\psi'}{\partial z} = \left[-p_0\beta - qA_z + \chi\boldsymbol{\alpha}_\perp \cdot \widehat{\boldsymbol{\pi}}_\perp\right]\psi'. \qquad (E.15)$$

Eq. (E.15) has the desired beam-optical form as in Eq. (E.1). Now, it is possible to apply the standard machinery of relativistic quantum mechanics to Eq. (E.15) in order to calculate the transfer maps.

This completes the central scheme for constructing quantum prescriptions of charged-particle beam optics. The calculation of the transfer maps is accomplished by using the standard and powerful techniques of quantum mechanics, accompanied with systematic approximations. In most computations, the beam-optical Hamiltonian is expanded in a power series of $|\widehat{\boldsymbol{\pi}}_\perp/p_0|$. Lets us note, that $\widehat{\boldsymbol{\pi}}_\perp$ is the transverse kinetic momentum, obeying the paraxial condition $|\widehat{\boldsymbol{\pi}}_\perp/p_0| \ll 1$, where p_0 is the design (or mean) momentum of beam particles propagating predominantly along the direction of the optic axis. The leading order approximation (up to quadratic in $\hat{\boldsymbol{\pi}}_\perp$) describes the paraxial behavior. Terms of higher orders in the expansion are responsible for the aberrations, order-by-order. Compared to the classical prescriptions, the paraxial and the aberrating quantum Hamiltonians contain extra terms, which appear in the powers of the reduced de Broglie wavelength ($\lambdabar_0 = \hbar/p_0$). These terms are responsible for the quantum modification of the classical behavior. The aforementioned wavelength-dependent contributions are the chief outcome of the quantum prescriptions. In the limit of low wavelength, the quantum prescriptions reduce to the familiar 'Lie algebraic formalism of charged-particle beam optics' (Dragt & Forest, 1986; Dragt et al., 1988; Forest et al., 1989; Forest & Hirata, 1992; Radlička, 2008; Rangarajan et al., 1990; Rangarajan & Sachidanand, 1997; Rangarajan & Sridharan, 2010; Ryne & Dragt, 1991; Turchetti et al., 1989; Todesco, 1999).

APPENDIX F. HAMILTON'S OPTICAL-MECHANICAL ANALOGY IN THE WAVELENGTH-DEPENDENT REGIME

The analogy between geometrical light optics and the classical theories of charged-particle beam optics has been known and exploited for a very long time. For comprehensive and historical accounts see Hawkes and Kasper (1989a, 1989b, 1994), Born and Wolf (1999), and Forbes (2001). Hamilton's optical-mechanical analogy played a significant role in the initial development of Schrödinger's wave mechanics (Masters, 2014). It is also called as 'optico-mechanical analogy' and 'opto-mechanical analogy'. Many of the equations of classical physics can be derived from the *variational principles*. In the context of optics, it is *Fermat's principle of least time* and in mechanics, it is *Maupertuis' principle* (Lakshminarayanan et al., 2002; Gloge & Marcuse, 1969).

Usually, René Descartes (1596–1650) is credited as the originator of the analogy. The analogy has a much older history and its beginning can be attributed to Ibn al-Haytham (965–1037) also known by Latinized name: Alhazen (Ambrosini, Ponticiello, Schirripa Spagnolo, Borghi, & Gori, 1997; Wolf & Krötzsch, 1995; Wolf, 2004; Rashed, 1990, 1993; Azzedine, Rashed, & Lakshminarayanan, 2017; Khan, 2006a, 2006b, 2007, 2015a, 2015b, 2016a, 2016b, 2016c, 2017a, 2017d; Al-Amri, El-Gomati, & Zubairy, 2017). But it was William Rowan Hamilton (1805–1865), who in 1831, closely examined the trajectories of material particles in various potential fields and compared them with the paths of rays of light in media with spatially varying refractive index. Thus, it was Hamilton, who laid a rigorous mathematical foundation to the optical-mechanical analogy. In the 1920s, the analogy was the guiding force for the development of both theoretical and practical electron optics. Based on the analogy, in 1926, Hans Busch used the techniques from geometrical light beam optics to derive the focusing action (lens-*like* behavior) of a short axially symmetric magnetic field. Busch was able to show that the action of a short axially symmetric magnetic field on charged-particle rays is very similar to that of glass lens on rays of light, bringing electrons from an object point together at an image point. Such fields are easily produced using solenoids. The work of Hans Busch was used by Ernst Ruska, resulting in the invention of the electron microscope in 1931. The optical-mechanical analogy goes much further as it also influenced the development of quantum mechanics. In 1924, Louis de Broglie proposed the wave–particle duality and assigned a wavelength to all moving particles. Using the de Broglie wavelength, Schrödinger penned down his now famous equation and thereby extended the analogy. Some historical aspects of the analogy were also addressed during the *2015 International Year of Light and Light-Based Technologies* (Khan, 2014b, 2014c, 2014d, 2015a, 2015b, 2016a, 2016b, 2016c, 2016e, 2017d; Dudley, Rivero González, Niemela, & Plenkovich, 2016; Al-Amri et al., 2017; Azzedine et al., 2017). The Hamilton's optical-mechanical analogy predates quantum mechanics and Maxwell's theory of electromagnetism. Both quantum mechanics and the Maxwell optics lead to wavelength modifications of the classical charged-particle beam optics and geometrical light beam optics respectively. So, it is natural to examine the status of the analogy in the wavelength-dependent regime.

Until very recently, it was possible to see this analogy only between the geometrical optics and classical prescriptions of electron optics. The reasons being that the quantum theories of charged-particle beam optics have been under development only for about two decades (Conte et al., 1996; Jagannathan et al., 1989; Jagannathan, 1990, 1993, 1999, 2002, 2003; Jagannathan & Khan, 1995, 1996, 1997, 2018; Khan, 1997, 1999a, 1999b, 2001, 2002a, 2002b, 2002c, 2003d, 2016d; Khan & Jagannathan, 1993, 1994, 1995). The quantum prescriptions are outlined in Appendix E. The quantum prescriptions have the very expected feature of wavelength-dependent effects, which have no analogue in the traditional descriptions of light beam optics. With the current development of the non-traditional prescriptions of Helmholtz optics (Khan et al., 2002; Khan, 2005a, 2016g) and the matrix formulation of Maxwell optics

Table 3 Hamiltonians in the different prescriptions

Light beam optics	Charged-particle beam optics
Fermat's principle	**Maupertuis' principle**
$\mathcal{H} = -\left\{ n^2(r) - p_\perp^2 \right\}^{1/2}$	$\mathcal{H} = -\left\{ p_0^2 - \pi_\perp^2 \right\}^{1/2} - qA_z$
Non-traditional Helmholtz	**Klein–Gordon formalism**
$\widehat{H}_{0,p} = -n(r) + \frac{1}{2n_0}\widehat{p}_\perp^2$ $\quad - \frac{i\hbar}{16n_0^3}\left[\widehat{p}_\perp^2, \frac{\partial}{\partial z}n(r) \right]$	$\widehat{H}_{0,p} = -p_0 - qA_z + \frac{1}{2p_0}\widehat{\pi}_\perp^2$ $\quad + \frac{i\hbar}{16p_0^4}\left[\widehat{\pi}_\perp^2, \frac{\partial}{\partial z}\widehat{\pi}_\perp^2 \right]$
Maxwell, matrix	**Dirac formalism**
$\widehat{H}_{0,p} = -n(r) + \frac{1}{2n_0}\widehat{p}_\perp^2$ $\quad - i\hbar\beta\boldsymbol{\Sigma}\cdot\boldsymbol{u}$ $\quad + \frac{1}{2n_0}\hbar^2 w^2 \beta$	$\widehat{H}_{0,p} = -p_0 - qA_z + \frac{1}{2p_0}\widehat{\pi}_\perp^2$ $\quad - \frac{\hbar}{2p_0}\left\{ \mu\gamma\boldsymbol{\Sigma}_\perp\cdot\boldsymbol{B}_\perp + (q+\mu)\Sigma_z B_z \right\}$ $\quad + \frac{i\hbar}{m_0 c}\epsilon B_z$
Notation	
Variable Refractive Index,	$\widehat{\pi}_\perp = \widehat{p}_\perp - q\boldsymbol{A}_\perp$
$n(r) = c\sqrt{\epsilon(r)\mu(r)}$	μ_a, anomalous magnetic moment
Resistance, $h(r) = \sqrt{\mu(r)/\epsilon(r)}$	ϵ_a, anomalous electric moment
$\boldsymbol{u}(r) = -\frac{1}{2n(r)}\boldsymbol{\nabla}n(r)$	$\mu = 2m_0\mu_a/\hbar, \epsilon = 2m_0\epsilon_a/\hbar$
$\boldsymbol{w}(r) = \frac{1}{2h(r)}\boldsymbol{\nabla}h(r)$	$\gamma = E/m_0 c^2$
$\boldsymbol{\Sigma}$ and β are the Dirac matrices	

(Khan, 2006c, 2010, 2014a, 2016f, 2017c), accompanied with wavelength-dependent effects, it is seen that the analogy between the two systems persists. The non-traditional prescription of Helmholtz optics is in close analogy with the quantum theory of charged-particle beam optics based on the Klein–Gordon equation. The matrix formulation of Maxwell optics is in close analogy with the quantum theory of charged-particle beam optics based on the Dirac equation. This analogy is summarized in the table of Hamiltonians (see Khan, 2002c, 2017a for details). In Table 3, we list the Hamiltonians, in the different prescriptions of light beam optics and charged-particle beam optics for magnetic systems. $\widehat{H}_{0,p}$ are the paraxial Hamiltonians, with the lowest order wavelength-dependent contributions.

REFERENCES

Acharya, R., & Sudarshan, E. C. G. (1960). Front description in relativistic quantum mechanics. *J. Math. Phys.*, *1*, 532–536. http://dx.doi.org/10.1063/1.1703689.

Al-Amri, M. D., El-Gomati, M. M., & Zubairy, M. S. (2017). *Optics in our time*. Springer. http://dx.doi.org/10.1007/978-3-319-31903-2.

Ambrosini, D., Ponticiello, A., Schirripa Spagnolo, G., Borghi, R., & Gori, F. (1997). Bouncing light beams and the Hamiltonian analogy. *Eur. J. Phys.*, *18*, 284–289. http://dx.doi.org/10.1088/0143-0807/18/4/008.

Asaga, T., Fujita, T., & Hiramoto, M. (2000). EDM operator free from Schiff's theorem. Retrieved from http://arXiv.org/abs/hep-ph/0005314/.

Azzedine, B., Rashed, R., & Lakshminarayanan, V. (2017). *Light-based science: Technology and sustainable development*. CRC Press, Taylor & Francis. Retrieved from http://isbn.nu/9781498779388.

Barnett, S. M. (2014). Optical Dirac equation. *New J. Phys.*, *16*, 093008. http://dx.doi.org/10.1088/1367-2630/16/9/093008.

Bellman, R., & Vasudevan, R. (1986). *Wave propagation: An invariant imbedding approach.* Dordrecht: Reidel. http://dx.doi.org/10.1007/978-94-009-5227-0.

Bialynicki-Birula, I. (1994). On the wave function of the photon. *Acta Phys. Pol. A*, *86*, 97–116. Retrieved from http://przyrbwn.icm.edu.pl/APP/ABSTR/86/a86-1-8.html.

Bialynicki-Birula, I. (1996a). The photon wave function. In H. H. Eberly, L. Mandel, & E. Wolf (Eds.), *Coherence and quantum optics VII* (pp. 313–322). New York, USA: Plenum Press. http://dx.doi.org/10.1007/978-1-4757-9742-8_38.

Bialynicki-Birula, I. (1996b). Photon wave function. In E. Wolf (Ed.), *Progress in optics: Vol. XXXVI* (pp. 245–294). Amsterdam: Elsevier. http://dx.doi.org/10.1016/S0079-6638(08)70316-0.

Bialynicki-Birula, I., & Bialynicka-Birula, Z. (2013). The role of the Riemann–Silberstein vector in classical and quantum theories of electromagnetism. *J. Phys. A, Math. Theor.*, *46*, 053001. http://dx.doi.org/10.1088/1751-8113/46/5/053001.

Bjorken, J. D., & Drell, S. D. (1964). *Relativistic quantum mechanics.* New York, San Francisco, USA: McGraw-Hill.

Blanes, S., Casas, F., Oteo, J. A., & Ros, J. (2009). The Magnus expansion and some of its applications. *Phys. Rep.*, *470*(5), 151–238. http://dx.doi.org/10.1016/j.physrep.2008.11.001.

Born, M., & Wolf, E. (1999). *Principles of optics*. Cambridge, UK: Cambridge University Press.

Brown, R. W., Krauss, L. M., & Taylor, P. L. (2001). Obituary of Leslie Lawrence Foldy. *Phys. Today*, *54*(12), 75–76. http://dx.doi.org/10.1063/1.1445566.

Case, K. M. (1954). Some generalizations of the Foldy–Wouthuysen transformation. *Phys. Rev.*, *95*, 1323–1328. http://dx.doi.org/10.1103/PhysRev.95.1323.

Chen, P. (1998). Workshop report. *ICFA Beam Dyn. Newslett.*, *16*, 22–25. Retrieved from http://icfa-usa.jlab.org/archive/newsletter/icfa_bd_nl_16.pdf.

Chen, P. (Ed.). (1999). *Proceedings of the 15th advanced ICFA beam dynamics workshop on quantum aspects of beam physics*. Singapore: World Scientific. Retrieved from http://www.slac.stanford.edu/grp/ara/qabp/qabp.html.

Chen, P. (2000). Workshop report. *ICFA Beam Dyn. Newslett.*, *23*, 13–14. Retrieved from http://icfa-usa.jlab.org/archive/newsletter/icfa_bd_nl_23.pdf.

Chen, P. (Ed.). (2002). *Proceedings of the 18th advanced ICFA beam dynamics workshop on quantum aspects of beam physics*. Singapore: World Scientific. Retrieved from http://qabp2k.sa.infn.it/.

Chen, P. (2003a). Workshop report. *ICFA Beam Dyn. Newslett.*, *30*, 72–75. Retrieved from http://icfa-usa.jlab.org/archive/newsletter/icfa_bd_nl_30.pdf.

Chen, P. (2003b). Workshop report. *Bull. Assoc. Asia Pacific Phys. Soc.*, *13*(1), 34–37.

Chen, P., & Reil, K. (Eds.). (2003). *Proceedings of the joint 28th ICFA advanced beam dynamics and advanced & novel accelerators workshop on quantum aspects of beam physics and other critical issues of beams in physics and astrophysics.* Singapore: World Scientific. Retrieved from http://home.hiroshima-u.ac.jp/ogata/qabp/home.html, http://www.slac.stanford.edu/pubs/slacreports/slac-r-630.html.

Conte, M., Jagannathan, R., Khan, S. A., & Pusterla, M. (1996). Beam optics of the Dirac particle with anomalous magnetic moment. *Part. Accel., 56,* 99–126. Retrieved from http://cds.cern.ch/record/307931/files/p99.pdf.

Costella, J. P., & McKellar, B. H. J. (1995). The Foldy–Wouthuysen transformation. *Am. J. Phys., 63,* 1119–1121. http://dx.doi.org/10.1119/1.18017. Retrieved from http://arXiv.org/abs/hep-ph/9503416/.

Dattoli, G., Renieri, A., & Torre, A. (1993). *Lectures on the free electron laser theory and related topics.* Singapore: World Scientific. http://dx.doi.org/10.1142/1334.

Dragt, A. J. (1982). A Lie algebraic theory of geometrical optics and optical aberrations. *J. Opt. Soc. Am., 72*(3), 372–379. http://dx.doi.org/10.1364/JOSA.72.000372.

Dragt, A. J. (1988). *Lie algebraic method for ray and wave optics.* University of Maryland Physics Department report.

Dragt, A. J., & Forest, E. (1986). Lie algebraic theory of charged particle optics and electron microscopes. In P. W. Hawkes (Ed.), *Advances in imaging and electron physics: Vol. 67* (pp. 65–120). San Diego: Academic Press. http://dx.doi.org/10.1016/S0065-2539(08)60329-7.

Dragt, A. J., Forest, E., & Wolf, K. B. (1986). Foundations of a Lie algebraic theory of geometrical optics. In J. Sánchez Mondragón, & K. B. Wolf (Eds.), *Lecture notes in physics: Vol. 250. Lie methods in optics* (pp. 105–157). Berlin, Germany: Springer-Verlag. http://dx.doi.org/10.1007/3-540-16471-5_4.

Dragt, A. J., Neri, F., Rangarajan, G., Douglas, D. R., Healy, L. M., & Ryne, R. D. (1988). Lie algebraic treatment of linear and nonlinear beam dynamics. *Annu. Rev. Nucl. Part. Sci., 38,* 455–496. http://dx.doi.org/10.1146/annurev.ns.38.120188.002323.

Dudley, J., Rivero González, J., Niemela, J., & Plenkovich, K. (2016). *The international year of light and light-based technologies 2015.* A Successful Community Partnership for Global Outreach final report (October 2016). Document code: SC/2016/IYL; catalog number: 246088. Retrieved from http://unesdoc.unesco.org/images/0024/002460/246088e.pdf, http://www.light2015.org/dam/About/IYL2015-Final-Report.pdf.

Durand, E. (1953). *C. R. Acad. Sci. Paris, 236,* 1337–1339.

Dvoeglazov, V. V. (1993). Electrodynamics with Weinberg's photons. *Hadron. J., 16,* 423–428. Retrieved from http://arxiv.org/abs/hep-th/9306108.

Esposito, S. (1998). Covariant Majorana formulation of electrodynamics. *Found. Phys., 28*(2), 231–244. http://dx.doi.org/10.1023/A:1018752803368. Retrieved from http://arxiv.org/abs/hep-th/9704144.

Fedele, R., Jovanovi, D., De Nicola, S., Mannan, A., & Tanjia, F. (2014). Self-modulation of a relativistic charged-particle beam as thermal matter wave envelope. *J. Phys. Conf. Ser., 482,* 012014. http://dx.doi.org/10.1088/1742-6596/482/1/012014.

Fedele, R., & Man'ko, V. I. (1999). The role of semiclassical description in the quantum-like theory of light rays. *Phys. Rev. E, 60,* 6042–6050. http://dx.doi.org/10.1103/PhysRevE.60.6042.

Fedele, R., Man'ko, M. A., & Man'ko, V. I. (2000). Wave-optics applications in charged-particle-beam transport. *J. Russ. Laser Res., 21*(1), 1–33. http://dx.doi.org/10.1007/BF02539473.

Fedele, R., Miele, G., Palumbo, L., & Vaccaro, V. G. (1993). Thermal wave model for non-linear longitudinal dynamics in particle accelerators. *Phys. Lett. A, 179*(6), 407–413. http://dx.doi.org/10.1016/0375-9601(93)90099-L.

Fedele, R., Tanjia, F., Jovanović, D., De Nicola, S., & Ronsivalle, C. (2014). Wave theories of non-laminar charged particle beams: From quantum to thermal regime. *J. Plasma Phys., 80*(02), 133–145. http://dx.doi.org/10.1017/S0022377813000913.

Ferwerda, H. A., Hoenders, B. J., & Slump, C. H. (1986a). Fully relativistic treatment of electron-optical image formation based on the Dirac equation. *Opt. Acta, 33*(2), 145–157. http://dx.doi.org/10.1080/713821923.

Ferwerda, H. A., Hoenders, B. J., & Slump, C. H. (1986b). The fully relativistic foundation of linear transfer theory in electron optics based on the Dirac equation. *Opt. Acta, 33*(2), 159–183. http://dx.doi.org/10.1080/713821925.

Feshbach, H., & Villars, F. M. H. (1958). Elementary relativistic wave mechanics of spin 0 and spin 1/2 particles. *Rev. Mod. Phys., 30*, 24–45. http://dx.doi.org/10.1103/RevModPhys.30.24.

Fishman, L. (1992). Exact and operator rational approximate solutions of the Helmholtz, Weyl composition equation in underwater acoustics – the quadratic profile. *J. Math. Phys., 33*(5), 1887–1914. http://dx.doi.org/10.1063/1.529666.

Fishman, L. (2004). One-way wave equation modeling in two-way wave propagation problems. In B. Nilsson, & L. Fishman (Eds.), *Mathematical modelling in physics, engineering and cognitive sciences: Vol. 7. Mathematical modelling of wave phenomena 2002* (pp. 91–111). Växjö, Sweden: Växjö University Press.

Fishman, L., De Hoop, M. V., & Van Stralen, M. J. N. (2000). Exact constructions of square root Helmholtz operator symbols: The focusing quadrature profile. *J. Math. Phys., 41*(7), 4881–4938. http://dx.doi.org/10.1063/1.533384.

Fishman, L., & McCoy, J. J. (1984). Derivation and application of extended parabolic wave theories. Part I. The factored Helmholtz equation. *J. Math. Phys., 25*, 285–296. http://dx.doi.org/10.1063/1.526149.

Foldy, L. L. (1952). The electromagnetic properties of the Dirac particles. *Phys. Rev., 87*(5), 682–693. http://dx.doi.org/10.1103/PhysRev.87.688.

Foldy, L. L. (2006). Origins of the FW transformation: A memoir, appendix G. In W. Fickinger (Ed.), *Physics at a research university, Case Western Reserve University 1830–1990* (pp. 347–351). Retrieved from http://www.phys.cwru.edu/history.

Foldy, L. L., & Wouthuysen, S. A. (1950). On the Dirac theory of spin 1/2 particles and its non-relativistic limit. *Phys. Rev., 78*, 29–36. http://dx.doi.org/10.1103/PhysRev.78.29.

Forbes, G. W. (2001). Hamilton's optics: Characterizing ray mapping and opening a link to waves. *Opt. Photonics News, 12*(11), 34–38. http://dx.doi.org/10.1364/OPN.12.11.000034.

Forest, E., Berz, M., & Irwin, J. (1989). Normal form methods for complicated periodic systems. *Part. Accel., 24*, 91–97. Retrieved from http://cds.cern.ch/record/1053511/files/p91.pdf.

Forest, E., & Hirata, K. (1992). *A contemporary guide to beam dynamics.* KEK report 92-12. Tsukuba, Japan: National Laboratory for High Energy Physics. Retrieved from http://ccdb5fs.kek.jp/cgi-bin/img_index?199224012.

Gerrard, A., & Burch, J. M. (1994). *Introduction to matrix methods in optics.* New York, USA: Dover Publications.

Giannetto, E. (1985). A Majorana–Oppenheimer formulation of quantum electrodynamics. *Lett. Nuovo Cimento, 44*(3), 140–144. http://dx.doi.org/10.1007/BF02746912.

Gill, T. L., & Zachary, W. W. (2005). Analytic representation of the square-root operator. *J. Phys. A, Math. Gen., 38*, 2479–2496. http://dx.doi.org/10.1088/0305-4470/38/11/010.

Glaser, W. (1956). Elektronen und Ionenoptik. In S. Flügge (Ed.), *Handbuch der Physik: Vol. 33* (pp. 123–395). Berlin: Springer-Verlag. http://dx.doi.org/10.1007/978-3-642-45852-1_2.

Gloge, D., & Marcuse, D. (1969). Formal quantum theory of light rays. *J. Opt. Soc. Am., 59*(12), 1629–1631. http://dx.doi.org/10.1364/JOSA.59.001629.

Good Jr., R. H. (1957). Particle aspect of the electromagnetic field equations. *Phys. Rev., 105*, 1914–1919. http://dx.doi.org/10.1103/PhysRev.105.1914.

Goodman, J. W. (1996). *Introduction to Fourier optics* (2nd ed.). New York: McGraw-Hill.

Hawkes, P. W., & Kasper, E. (1989a). *Principles of electron optics: Vol. I*. London: Academic Press.

Hawkes, P. W., & Kasper, E. (1989b). *Principles of electron optics: Vol. II*. London: Academic Press.

Hawkes, P. W., & Kasper, E. (1994). *Principles of electron optics: Vol. 3. Wave optics*. London/San Diego: Academic Press.

Heinemann, K., & Barber, D. P. (1999). The semiclassical Foldy–Wouthuysen transformation and the derivation of the Bloch equation for spin-1/2 polarised beams using Wigner functions. In P. Chen (Ed.), *Proceedings of the 15th advanced ICFA beam dynamics workshop on quantum aspects of beam physics*. Singapore: World Scientific. Retrieved from http://arXiv.org/abs/physics/9901044/.

Hill, C. T. (2000). The diffractive quantum limits of particle colliders. Retrieved from http://arxiv.org/abs/hep-ph/0002230.

Inskeep, W. H. (1988). On electromagnetic spinors and quantum theory. *Z. Naturforsch., 43A*, 695–696. http://dx.doi.org/10.1515/zna-1988-0715.

Ivezć, T. (2006). Lorentz invariant Majorana formulation of the field equations and Dirac-like equation for the free photon. *Electron. J. Theor. Phys., 3*(10), 131–142. Retrieved from http://www.ejtp.com/articles/ejtpv3i10p131.pdf, http://arxiv.org/abs/physics/0605030.

Jackson, J. D. (1998). *Classical electrodynamics* (3rd ed.). John Wiley & Sons.

Jagannathan, R. (1990). Quantum theory of electron lenses based on the Dirac equation. *Phys. Rev. A, 42*, 6674–6689. http://dx.doi.org/10.1103/PhysRevA.42.6674.

Jagannathan, R. (1993). Dirac equation and electron optics. In R. Dutt, & A. K. Ray (Eds.), *Dirac and Feynman: Pioneers in quantum mechanics* (pp. 75–82). New Delhi, India: Wiley Eastern. Retrieved from http://isbn.nu/9788122404937.

Jagannathan, R. (1999). The Dirac equation approach to spin-1/2 particle beam optics. In P. Chen (Ed.), *Proceedings of the 15th advanced ICFA beam dynamics workshop on quantum aspects of beam physics* (pp. 670–681). Singapore: World Scientific. Retrieved from http://arxiv.org/abs/physics/9803042.

Jagannathan, R. (2002). Quantum mechanics of Dirac particle beam optics: Single-particle theory. In P. Chen (Ed.), *Proceedings of the 18th advanced ICFA beam dynamics workshop on quantum aspects of beam physics* (pp. 568–577). Singapore: World Scientific. Retrieved from http://arxiv.org/abs/physics/0101060.

Jagannathan, R. (2003). Quantum mechanics of Dirac particle beam transport through optical elements with straight and curved axes. In P. Chen, & K. Reil (Eds.), *Proceedings of the 28th advanced ICFA beam dynamics workshop on quantum aspects of beam physics* (pp. 13–21). Singapore: World Scientific. http://dx.doi.org/10.1142/9789812702333_0002. Retrieved from http://arXiv.org/abs/physics/0304099/.

Jagannathan, R., & Khan, S. A. (1995). Wigner functions in charged particle optics. In R. Sridhar, K. Srinivasa Rao, & V. Lakshminarayanan (Eds.), *Selected topics in mathematical physics – Professor R. Vasudevan memorial volume* (pp. 308–321). Delhi, India: Allied Publishers. Retrieved from http://isbn.nu/9788170234883.

Jagannathan, R., & Khan, S. A. (1996). Quantum theory of the optics of charged particles. In P. W. Hawkes (Ed.), *Advances in imaging and electron physics: Vol. 97* (pp. 257–358). San Diego: Academic Press. http://dx.doi.org/10.1016/S1076-5670(08)70096-X.

Jagannathan, R., & Khan, S. A. (1997). Quantum mechanics of accelerator optics. *ICFA Beam Dyn. Newslett.*, *13*, 21–27. Retrieved from http://icfa-usa.jlab.org/archive/newsletter/icfa_bd_nl_13.pdf.

Jagannathan, R., & Khan, S. A. (2018). *Quantum mechanics of charged particle beam optics.* CRC Press, Taylor & Francis. Retrieved from http://isbn.nu/9781138035928.

Jagannathan, R., Simon, R., Sudarshan, E. C. G., & Mukunda, N. (1989). Quantum theory of magnetic electron lenses based on the Dirac equation. *Phys. Lett. A*, *134*, 457–464. http://dx.doi.org/10.1016/0375-9601(89)90685-3.

Jayaraman, J. (1975). A note on the recent Foldy–Wouthuysen transformations for particles of arbitrary spin. *J. Phys. A, Math. Gen.*, *8*, L1–L4. http://dx.doi.org/10.1088/0305-4470/8/1/001.

Khan, S. A. (1997). *Quantum theory of charged-particle beam optics* (Ph.D. thesis). Chennai, India: University of Madras. Complete thesis available from Dspace of IMSc Library, The Institute of Mathematical Sciences, Chennai, India. Retrieved from http://www.imsc.res.in/xmlui/handle/123456789/75?show=full, http://www.imsc.res.in/xmlui/.

Khan, S. A. (1999a). Quantum theory of magnetic quadrupole lenses for spin-1/2 particles. In P. Chen (Ed.), *Proceedings of the 15th advanced ICFA beam dynamics workshop on quantum aspects of beam physics* (pp. 682–694). Singapore: World Scientific. Retrieved from http://arxiv.org/abs/physics/9809032.

Khan, S. A. (1999b). Quantum aspects of accelerator optics. In A. Luccio, & W. MacKay (Eds.), *Proceedings of the 1999 particle accelerator conference (PAC99)* (pp. 2817–2819). Retrieved from http://arxiv.org/abs/physics/9904063.

Khan, S. A. (2001). The world of synchrotrons. *Resonance J. Sci. Educ.*, *6*(11), 77–84. http://dx.doi.org/10.1007/BF02868247. Retrieved from http://arXiv.org/abs/physics/0112086/.

Khan, S. A. (2002a). Introduction to synchrotron radiation. *Bull. IAPT*, *19*(5), 149–153. Retrieved from http://indapt.org/.

Khan, S. A. (2002b). Quantum formalism of beam optics. In P. Chen (Ed.), *Proceedings of the 18th advanced ICFA beam dynamics workshop on quantum aspects of beam physics* (pp. 517–526). Singapore: World Scientific. Retrieved from http://arXiv.org/abs/physics/0112085/.

Khan, S. A. (2002c). Analogies between light optics and charged-particle optics. *ICFA Beam Dyn. Newslett.*, *27*, 42–48. Retrieved from http://arXiv.org/abs/physics/0210028/.

Khan, S. A. (2002d). Electron beams for radiation. *Kiran*, *13*(3), 40–42. Retrieved from http://www.ila.org.in/kiran/.

Khan, S. A. (2005a). Wavelength-dependent modifications in Helmholtz optics. *Int. J. Theor. Phys.*, *44*(1), 95–125. http://dx.doi.org/10.1007/s10773-005-1488-0.

Khan, S. A. (2005b). An exact matrix representation of Maxwell's equations. *Phys. Scr.*, *71*(5), 440–442. http://dx.doi.org/10.1238/Physica.Regular.071a00440.

Khan, S. A. (2006a). Medieval Arab understanding of the rainbow formation. *Europhys. News*, *37*(3), 10. Retrieved from http://www.europhysicsnews.org/articles/epn/pdf/2006/03/epn2006-37-3.pdf.

Khan, S. A. (2006b). The Foldy–Wouthuysen transformation technique in optics. *Optik*, *117*(10), 481–488. http://dx.doi.org/10.1016/j.ijleo.2005.11.010.

Khan, S. A. (2006c). Wavelength-dependent effects in light optics. In V. Krasnoholovets, & F. Columbus (Eds.), *New topics in quantum physics research* (pp. 163–204). New York: Nova Science Publishers. Retrieved from http://isbn.nu/9781600210280, https://arxiv.org/abs/physics/0210027.

Khan, S. A. (2007). Arab origins of the discovery of the refraction of light: Roshdi Hifni Rashed awarded the 2007 King Faisal International Prize. *Opt. Photonics News*, *18*(10), 22–23. Retrieved from http://www.osa-opn.org/Content/ViewFile.aspx?id=10890. Science historian Roshdi Hifni Rashed from CNRS, France was bestowed the 2007 *King Faisal International Prize*, in recognition of his insightful studies, authentication, commentaries and translations of Muslims' contributions to pure science, in particular their achievements in the fields of mathematics and optics (see Rashed, 1990, 1993). The prizes are awarded every year by the King Faisal Foundation, Saudi Arabia. Further details about Roshdi Rashed, http://chspam.vjf.cnrs.fr/Personnel/Rashed.htm.

Khan, S. A. (2008). The Foldy–Wouthuysen transformation technique in optics. In P. W. Hawkes (Ed.), *Advances in imaging and electron physics: Vol. 152* (pp. 49–78). Amsterdam: Elsevier. http://dx.doi.org/10.1016/S1076-5670(08)00602-2.

Khan, S. A. (2010). Maxwell optics of quasiparaxial beams. *Optik*, *121*(5), 408–416. http://dx.doi.org/10.1016/j.ijleo.2008.07.027. Retrieved from http://arxiv.org/abs/physics/0205084.

Khan, S. A. (2014a). Aberrations in Maxwell optics. *Optik*, *125*(3), 968–978. http://dx.doi.org/10.1016/j.ijleo.2013.07.097. Retrieved from http://arxiv.org/abs/physics/0205085.

Khan, S. A. (2014b). 2015 declared the international year of light and light-based technologies. *Curr. Sci.*, *106*(4), 501. Retrieved from http://www.currentscience.ac.in/Volumes/106/04/0501.pdf.

Khan, S. A. (2014c). X-rays to synchrotrons and the international year of light. *IRPS Bull.*, *28*(1), 9–13. Retrieved from http://www.canberra.edu.au/irps, http://radiationphysics.org/.

Khan, S. A. (2014d). Particle accelerators and the international year of light. *ICFA Beam Dyn. Newslett.*, *63*, 9–15. Retrieved from http://www-bd.fnal.gov/icfabd/Newsletter63.pdf.

Khan, S. A. (2015a). Medieval Islamic achievements in optics. *Nuovo Saggiatore*, *31*(1–2), 36–45. Retrieved from http://prometeo.sif.it/papers/online/sag/031/01-02/pdf/06-percorsi.pdf.

Khan, S. A. (2015b). *International year of light and light-based technologies*. Germany: Lambert Academic Publishing. Retrieved from http://isbn.nu/9783659764820/.

Khan, S. A. (2016a). The international year of light and light-based technologies. Report of the conference *The Islamic golden age of science for today's knowledge-based society: The Ibn Al-Haytham example*. *Am. J. Islam. Soc. Sci.*, *33*(1), 160–163.

Khan, S. A. (2016b). Medieval Arab contributions to optics. *Digest Middle East Stud.*, *25*(1), 19–35. http://dx.doi.org/10.1111/dome.12065.

Khan, S. A. (2016c). International year of light and history of optics. In T. Scott (Ed.), *Advances in photonics engineering, nanophotonics and biophotonics* (pp. 1–56). New York: Nova Science Publishers. Retrieved from http://isbn.nu/978-1-63484-498-7, http://www.novapublishers.com/.

Khan, S. A. (2016d). Quantum aspects of charged-particle beam optics. In A. Al-Kamli, N. Can, G. O. Souadi, M. Fadhali, A. Mahdy, & M. Maoub (Eds.), *AIP conference proceedings: Vol. 1742. Proceedings of the fifth Saudi international meeting on frontiers of physics 2016 (SIMFP 2016)*, 030008. American Institute of Physics. http://dx.doi.org/ 10.1063/1.4953129.

Khan, S. A. (2016e). Reflecting on the international year of light and light-based technologies. *Curr. Sci.*, *111*(4), 627–631. http://dx.doi.org/10.18520/cs/v111/i4/627-631. Retrieved from http://www.currentscience.ac.in/Volumes/111/04/0627.pdf.

Khan, S. A. (2016f). Passage from scalar to vector optics and the Mukunda–Simon–Sudarshan theory for paraxial systems. *J. Mod. Opt.*, *63*(17), 1652–1660. http://dx.doi.org/10.1080/09500340.2016.1164257.

Khan, S. A. (2016g). Quantum methodologies in Helmholtz optics. *Optik*, *127*(20), 9798–9809. http://dx.doi.org/10.1016/j.ijleo.2016.07.071.

Khan, S. A. (2016h). Quantum methods in light beam optics. *Opt. Photonics News*, *27*(12), 47. Retrieved from http://www.osa-opn.org/home/articles/volume_27/ december_2016/features/optics_in_2016/. One of the thirty summaries selected under the theme, *Optics in 2016*, highlighting the most exciting peer-reviewed optics research to have emerged over the past 12 months. The summary of the two selected papers (Khan, 2016f, 2016g) is described in this publication.

Khan, S. A. (2017a). *Hamilton's optical-mechanical analogy in the wavelength-dependent regime. Optik*, *130*, 714–722. http://dx.doi.org/10.1016/j.ijleo.2016.07.071.

Khan, S. A. (2017b). Linearization of wave equations. *Optik*, *131*(C), 350–363. http:// dx.doi.org/10.1016/j.ijleo.2016.11.073.

Khan, S. A. (2017c). Polarization in Maxwell optics. *Optik*, *131*, 733–748. http:// dx.doi.org/10.1016/j.ijleo.2016.11.134.

Khan, S. A. (2017d). Medieval Arab achievements in optics. In B. Azzedine, R. Rashed, & V. Lakshminarayanan (Eds.), *Light-based science: Technology and sustainable development.* CRC Press, Taylor & Francis (Proceedings of *The Islamic golden age of science for today's knowledge-based society: The Ibn Al-Haytham example*). Retrieved from http://isbn.nu/9781498779388.

Khan, S. A., & Jagannathan, R. (1993). Theory of relativistic electron beam transport based on the Dirac equation. In S. N. Chintalapudi (Ed.), *Proceedings of the 3rd national seminar on physics and technology of particle accelerators and their applications, PATPAA-93* (pp. 102–107). Kolkata (Calcutta), India: IUC-DAEF. Retrieved from http://cds.cern.ch/record/263573/files/P00023243.pdf.

Khan, S. A., & Jagannathan, R. (1994). Quantum mechanics of charged-particle beam optics: An operator approach. Presented at the JSPS-KEK international spring school on high energy ion beams – novel beam techniques and their applications, March 1994, Japan. Retrieved from http://cds.cern.ch/record/ 263576/files/P00023244.pdf.

Khan, S. A., & Jagannathan, R. (1995). Quantum mechanics of charged particle beam transport through magnetic lenses. *Phys. Rev. E*, *51*, 2510–2515. http:// dx.doi.org/10.1103/PhysRevE.51.2510.

Khan, S. A., Jagannathan, R., & Simon, R. (2002). Foldy–Wouthuysen transformation and a quasiparaxial approximation scheme for the scalar wave theory of light beams. Retrieved from http://arXiv.org/abs/physics/0209082/.

Khan, S. A., & Pusterla, M. (1999). Quantum mechanical aspects of the halo puzzle. In A. Luccio, & W. MacKay (Eds.), *Proceedings of the 1999 particle accelerator conference: Vol. 5* (pp. 3280–3281). http://dx.doi.org/10.1109/PAC.1999.792276. Retrieved from http://arxiv.org/abs/physics/9904064.

Khan, S. A., & Pusterla, M. (2000a). Quantum-like approaches to the beam halo problem. In D. Han, Y. S. Kim, & S. Solimeno (Eds.), *NASA conference publication series. Proceedings of the 6th international conference on squeezed states and uncertainty relations* (pp. 438–441). Retrieved from http://arxiv.org/abs/physics/9905034.

Khan, S. A., & Pusterla, M. (2000b). Quantum-like approach to the transversal and longitudinal beam dynamics. The halo problem. *Eur. Phys. J. A, 7*(4), 583–587. http://dx.doi.org/10.1007/s100500050430. Retrieved from http://arxiv.org/abs/physics/9910026.

Khan, S. A., & Pusterla, M. (2001). Quantum approach to the halo formation in high current beams. *Nucl. Instrum. Methods Phys. Res. Sect. A, 464*(1–3), 461–464. http://dx.doi.org/10.1016/S0168-9002(01)00108-5. Retrieved from http://arxiv.org/abs/physics/0112082.

Khan, S. A., & Wolf, K. B. (2002). Hamiltonian orbit structure of the set of paraxial optical systems. *J. Opt. Soc. Am. A, 19*(12), 2436–2444. http://dx.doi.org/10.1364/JOSAA.19.002436.

Kogelnik, H. (1965). On the propagation of Gaussian beams of light through lenslike media including those with a loss or gain variation. *Appl. Opt., 4*(12), 1562–1569. http://dx.doi.org/10.1364/AO.4.001562.

Kulyabov, D. S. (2016). In Gh. Adam, J. Buša, & M. Hnatič (Eds.), *EPJ web of conferences: Vol. 108. Spinor-like Hamiltonian for Maxwellian optics mathematical modeling and computational physics (MMCP 2015)*, 02034. http://dx.doi.org/10.1051/epjconf/201610802034.

Lakshminarayanan, L., Ghatak, A., & Thyagarajan, K. (2002). *Lagrangian optics.* Springer. http://dx.doi.org/10.1007/978-1-4615-1711-5.

Lakshminarayanan, V., Sridhar, R., & Jagannathan, R. (1998). Lie algebraic treatment of dioptric power and optical aberrations. *J. Opt. Soc. Am. A, 15*(9), 2497–2503. http://dx.doi.org/10.1364/JOSAA.15.002497.

Laporte, O., & Uhlenbeck, G. E. (1931). Applications of spinor analysis to the Maxwell and Dirac equations. *Phys. Rev., 37,* 1380–1397. http://dx.doi.org/10.1103/PhysRev.37.1380.

Leopold, H. (1997). Obituary of Siegfried A. Wouthuysen. *Phys. Today, 50*(11), 89. http://dx.doi.org/10.1063/1.882018.

Leviandier, L. (2009). The one-way wave equation and its invariance properties. *J. Phys. A, Math. Theor., 42*(26), 265402. http://dx.doi.org/10.1088/1751-8113/42/26/265402.

Lippert, M., Brückel, Th., Köhler, Th., & Schneider, J. R. (1994). High-resolution bulk magnetic scattering of high-energy synchrotron radiation. *Europhys. Lett., 27*(7), 537–541. http://dx.doi.org/10.1209/0295-5075/27/7/008.

Lomont, J. S. (1958). Dirac-like wave equations for particles of zero rest mass and their quantization. *Phys. Rev., 111,* 1710–1716. http://dx.doi.org/10.1103/PhysRev.111.1710.

Magnus, W. (1954). On the exponential solution of differential equations for a linear operator. *Commun. Pure Appl. Math.*, 7, 649–673. http://dx.doi.org/10.1002/cpa.3160070404.

Majorana, E. (1974). Unpublished notes, quoted after Mignani, R., Recami, E., & Baldo, M. About a Diraclike equation for the photon, according to Ettore Majorana. *Lett. Nuovo Cimento*, 11, 568–572. Retreived from http://dx.doi.org/10.1007/BF02812391.

Mananga, E. S., & Charpentier, T. (2016). On the Floquet–Magnus expansion: Applications in solid-state nuclear magnetic resonance and physics. *Phys. Rep.*, 609, 1–49. http://dx.doi.org/10.1016/j.physrep.2015.10.005.

Masters, B. R. (2014). Erwin Schrödinger's path to wave mechanics. *Opt. Photonics News*, 25(2), 33–39. http://dx.doi.org/10.1364/OPN.25.2.000032.

Mehrafarin, M., & Balajany, H. (2010). Paraxial spin transport using the Dirac-like paraxial wave equation. *Phys. Lett. A*, 374(15–16), 1608–1610. http://dx.doi.org/10.1016/j.physleta.2010.01.067.

Mohr, P. J. (2010). Solutions of the Maxwell equations and photon wave functions. *Ann. Phys.*, 325(3), 607–663. http://dx.doi.org/10.1016/j.aop.2009.11.007.

Moses, E. (1959). Solutions of Maxwell's equations in terms of a spinor notation: The direct and inverse problems. *Phys. Rev.*, 113(6), 1670–1679. http://dx.doi.org/10.1103/PhysRev.113.1670.

Mukunda, N., Simon, R., & Sudarshan, E. C. G. (1983a). Paraxial-wave optics and relativistic front description I: The scalar theory. *Phys. Rev. A*, 28, 2921–2932. http://dx.doi.org/10.1103/PhysRevA.28.2921.

Mukunda, N., Simon, R., & Sudarshan, E. C. G. (1983b). Paraxial-wave optics and relativistic front description II: The vector theory. *Phys. Rev. A*, 28, 2933–2942. http://dx.doi.org/10.1103/PhysRevA.28.2933.

Mukunda, N., Simon, R., & Sudarshan, E. C. G. (1985a). Fourier optics for the Maxwell field: Formalism and applications. *J. Opt. Soc. Am. A*, 2(3), 416–426. http://dx.doi.org/10.1364/JOSAA.2.000416.

Mukunda, N., Simon, R., & Sudarshan, E. C. G. (1985b). Paraxial Maxwell beams: Transformation by general linear optical systems. *J. Opt. Soc. Am. A*, 2(8), 1291–1296. http://dx.doi.org/10.1364/JOSAA.2.001291.

Nazarathy, M., & Shamir, J. (1980). Fourier optics described by operator algebra. *J. Opt. Soc. Am.*, 70(2), 150–159. http://dx.doi.org/10.1364/JOSA.70.000150.

Nazarathy, M., & Shamir, J. (1982). First-order optics – a canonical operator representation: Lossless systems. *J. Opt. Soc. Am.*, 72(3), 356–364. http://dx.doi.org/10.1364/JOSA.72.000356.

Ohmura, T. (1956). A new formulation on the electromagnetic field. *Prog. Theor. Phys.*, 16(6), 684–685. http://dx.doi.org/10.1143/PTP.16.684.

Oppenheimer, J. R. (1931). Note on light quanta and the electromagnetic field. *Phys. Rev.*, 38, 725–746. http://dx.doi.org/10.1103/PhysRev.38.725.

Orris, G. J., & Wurmser, D. (1995). Applications of the Foldy–Wouthuysen transformation to acoustic modeling using the parabolic equation method. *J. Acoust. Soc. Am.*, 98(5), 2870. http://dx.doi.org/10.1121/1.413215.

Osche, G. R. (1977). Dirac and Dirac–Pauli equation in the Foldy–Wouthuysen representation. *Phys. Rev. D*, 15(8), 2181–2185. http://dx.doi.org/10.1103/PhysRevD.15.2181.

Ovsiyuk, E. M., Kisel, V. V., & Red'kov, V. M. (2013). Maxwell electrodynamics and boson fields in spaces of constant curvature. V. V. Dvoeglazov (Ed.). New York: Nova Science Publishers. Retrieved from https://www.novapublishers.com/catalog/product_info.php?products_id=42135.

Pachucki, K. (2005). Higher-order effective Hamiltonian for light atomic systems. *Phys. Rev. A, 71*, 012503. http://dx.doi.org/10.1103/PhysRevA.71.012503. Retrieved from http://arxiv.org/abs/physics/0411168.

Panofsky, W. K. H., & Phillips, M. (1962). *Classical electricity and magnetics.* Reading, Massachusetts, USA: Addison-Wesley Publishing Company.

Patton, R. S. (1986). In M. Gutzwiller, A. Inomata, J. R. Klauder, & L. Streit (Eds.), *Path integrals from meV to MeV* (pp. 98–115). Singapore: World Scientific. Retrieved from http://isbn.nu/9789971500665.

Petroni, N. C., De Martino, S., De Siena, S., & Illuminati, F. (2000). Stochastic collective dynamics of charged-particle beams in the stability regime. *Phys. Rev. E, 63*, 016501. http://dx.doi.org/10.1103/PhysRevE.63.016501.

Phan-Van-Loc (1953). *C. R. Acad. Sci. Paris, 237*, 649–651.

Phan-Van-Loc (1954). *C. R. Acad. Sci. Paris, 238*, 2494–2496.

Phan-Van-Loc (1955). *Ann. Fac. Sci. Univ. Toulouse, 18*, 178–192.

Phan-Van-Loc (1958a). Interprétation physique de l'expression mathématique du principe de Huygens en théorie de l'électron de Dirac. *Cah. Phys., 97*(12), 327–340.

Phan-Van-Loc (1958b). *C. R. Acad. Sci. Paris, 246*, 388–390.

Phan-Van-Loc (1960). *Principes de Huygens en théorie de l'electron de Dirac* (Thesis). Toulouse.

Pradhan, T. (1987). Maxwell's equations from geometrical optics. *Phys. Lett. A, 122*(8), 397–398. http://dx.doi.org/10.1016/0375-9601(87)90735-3.

Pryce, M. H. L. (1948). The mass-centre in the restricted theory of relativity and its connexion with the quantum theory of elementary particles. *Proc. R. Soc. Lond. Ser. A, Math. Phys. Sci., 195*, 62–81. http://dx.doi.org/10.1098/rspa.1948.0103.

Radlička, T. (2008). Lie algebraic methods in charged particle optics. In P. W. Hawkes (Ed.), *Advances in imaging and electron physics: Vol. 151* (pp. 241–362). http://dx.doi.org/10.1016/S1076-5670(07)00404-1.

Rangarajan, G., Dragt, A. J., & Neri, F. (1990). Solvable map representation of a nonlinear symplectic map. *Part. Accel., 28*, 119–124. Retrieved from http://cds.cern.ch/record/1108134/files/p119.pdf.

Rangarajan, G., & Sachidanand, M. (1997). Spherical aberrations and its correction using Lie algebraic methods. *Pramana, 49*(6), 635–643. http://dx.doi.org/10.1007/BF02848337.

Rangarajan, G., & Sridharan, S. (2010). Invariant norm quantifying nonlinear content of Hamiltonian systems. *Appl. Math. Comput., 217*(6), 2495–2500. http://dx.doi.org/10.1016/j.amc.2010.07.060.

Rashed, R. (1990). A pioneer in anaclastics – Ibn Sahl on burning mirrors and lenses. *ISIS, 81*, 464–491. http://dx.doi.org/10.1086/355456.

Rashed, R. (1993). *Collection sciences et philosophie Arabes, textes et études. Géométrie et dioptrique au Xe siècle: Ibn Sahl, al-Quhî et Ibn al-Haytham.* Paris, France: Les Belles Lettres.

Red'kov, V. M., Tokarevskaya, N. G., & Spix, G. J. (2012). Majorana–Oppenheimer approach to Maxwell electrodynamics. Part I. Minkowski space. *Adv. Appl. Clifford Algebras, 22*(4), 1129–1149. http://dx.doi.org/10.1007/s00006-012-0320-1.

Rubinowicz, A. (1934). *Acta Phys. Pol., 3*, 143–163.

Rubinowicz, A. (1957). *Die Beugungswelle in der Kirchhoffschen Theorie der Beugung*. Warsaw: Panstwowe Wydawnictwo Naukowe. 2nd ed. (1966), Warsaw/Berlin: PWN/Springer.

Rubinowicz, A. (1963). *Acta Phys. Pol.*, *23*, 727–744.

Rubinowicz, A. (1965). The Miyamoto–Wolf diffraction wave. *Prog. Opt.*, *4*, 199–240. http://dx.doi.org/10.1016/S0079-6638(08)70492-X. See Part II, Section 3: Diffraction wave in the Kirchhoff theory of Dirac electron waves.

Ryne, R. D., & Dragt, A. J. (1991). Magnetic optics calculations for cylindrically symmetric beams. *Part. Accel.*, *35*, 129–165. Retrieved from http://cds.cern.ch/record/1108242/files/p129.pdf.

Sachs, M., & Schwebel, S. L. (1962). On covariant formulations of the Maxwell–Lorentz theory of electromagnetism. *J. Math. Phys.*, *3*(5), 843–848. http://dx.doi.org/10.1063/1.1724297.

Silberstein, L. (1907a). Elektromagnetische Grundgleichungen in bivektorieller Behandlung. *Ann. Phys. (Leipz.)*, *22*, 579–586. http://dx.doi.org/10.1002/andp.19073270313. For a detailed discussion of the Riemann–Silberstein complex vector, see Bialynicki-Birula (1994, 1996a, 1996b).

Silberstein, L. (1907b). Nachtrag zur Abhandlung über Elektromagnetische Grundgleichungen in bivektorieller Behandlung. *Ann. Phys. (Leipz.)*, *24*, 783–784. http://dx.doi.org/10.1002/andp.19073291409. For a detailed discussion of the Riemann–Silberstein complex vector, see Bialynicki-Birula (1994, 1996a, 1996b).

Silenko, A. J. (2016). Exact form of the exponential Foldy–Wouthuysen transformation operator for an arbitrary-spin particle. *Phys. Rev. A*, *94*, 032104. http://dx.doi.org/10.1103/PhysRevA.94.032104. Retrieved from http://arxiv.org/abs/1605.09755.

Simon, R., Sudarshan, E. C. G., & Mukunda, N. (1986). Gaussian–Maxwell beams. *J. Opt. Soc. Am. A*, *3*(4), 536–540. http://dx.doi.org/10.1364/JOSAA.3.000536.

Simon, R., Sudarshan, E. C. G., & Mukunda, N. (1987). Cross polarization in laser beams. *Appl. Opt.*, *26*(9), 1589–1593. http://dx.doi.org/10.1364/AO.26.001589.

Sunilkumar, V., Bambah, B. A., Jagannathan, R., Panigrahi, P. K., & Srinivasan, V. (2000). Coherent states of nonlinear algebras: Applications to quantum optics. *J. Opt. B, Quantum Semiclass. Opt.*, *2*(2), 126–132. http://dx.doi.org/10.1088/1464-4266/2/2/311.

Tani, S. (1951). Connection between particle models and field theories. I. The case spin 1/2. *Prog. Theor. Phys.*, *6*, 267–285. http://dx.doi.org/10.1143/ptp/6.3.267.

Todesco, E. (1999). Overview of single-particle nonlinear dynamics. CERN-LHC-99-1-MMS, 16 pp. Talk given at 16th ICFA beam dynamics workshop on nonlinear and collective phenomena in beam physics, Arcidosso, Italy, 1–5 September 1998. In *AIP conference proceedings: Vol. 468* (pp. 157–172). Retrieved from https://cds.cern.ch/record/379703/files/lhc-99-001.pdf.

Turchetti, G., Bazzani, A., Giovannozzi, M., Servizi, G., & Todesco, E. (1989). Normal forms for symplectic maps and stability of beams in particle accelerators. In *Proceedings of the dynamical symmetries and chaotic behaviour in physical systems* (pp. 203–231).

Veko, O. V., Vlasii, N. D., Ovsiyuk, E. M., Red'kov, V. M., & Sitenko, Yu. A. (2014). Electromagnetic field on de Sitter expanding universe: Majorana–Oppenheimer formalism, exact solutions in non-static coordinates. *Nonlinear Phenom. Complex Syst.*, *17*(1), 17–39. Retrieved from http://www.j-npcs.org/abstracts/vol2014/v17no1/v17no1p17.html.

Wang, Z.-Y., Qiu, Q., Wang, Y.-X., & Shi, S.-J. (2015). The $(1, 0) \oplus (0, 1)$ spinor description of the photon field and its applications. Retrieved from http://arxiv.org/abs/1508.02321.

Wilcox, R. M. (1967). Exponential operators and parameter differentiation in quantum physics. *J. Math. Phys.*, *8*(4), 962–982. http://dx.doi.org/10.1063/1.1705306.

Wolf, K. B. (2004). *Geometric optics on phase space* (pp. 57–58). Berlin, Germany: Springer. Retrieved from http://www.springer.com/us/book/9783540220398.

Wolf, K. B., & Krötzsch, G. (1995). Geometry and dynamics in refracting systems. *Eur. J. Phys.*, *16*(1), 14–20. http://dx.doi.org/10.1088/0143-0807/16/1/003.

Wurmser, D. (2001). A new strategy for applying the parabolic equation to a penetrable rough surface. *J. Acoust. Soc. Am.*, *109*(5), 2300. http://dx.doi.org/10.1121/1.4744070.

Wurmser, D. (2004). A parabolic equation for penetrable rough surfaces: Using the Foldy–Wouthuysen transformation to buffer density jumps. *Ann. Phys.*, *311*, 53–80. http://dx.doi.org/10.1016/j.aop.2003.11.006.

CHAPTER THREE

Information Measures, Mean Differences, and Inequalities

Inder Jeet Taneja
Formerly, Professor of Mathematics, Departamento de Matemática, Universidade Federal de Santa Catarina, Florianópolis, SC, Brazil
e-mail address: ijtaneja@gmail.com

Contents

Advances in Imaging and Electron Physics, Volume 201
ISSN 1076-5670
http://dx.doi.org/10.1016/bs.aiep.2017.02.001

1. INTRODUCTION

Kullback and Leibler (1951) introduced a *measure of information* associated with two probability distributions of a discrete random variable. At the same time, they also developed the idea of Jeffrey's (1946) invariant, famous as *J-divergence*. Sibson (1969) studied the idea of information radius generally referred to as *Jensen difference divergence measure* (Burbea & Rao, 1982). Taneja (1995) presented a divergence measure referring *arithmetic and geometric mean divergence measure*. These three measures bear an interesting inequality. On the other side there are measures like *Hellinger's distance* (Hellinger, 1909) and *triangular discrimination* are also famous in the literature. During past years, researchers worked toward one and two scalar parametric generalizations of these *classical measures of information*. Taneja (2005a) presented unified generalizations of these measures. These measures have found deep applications toward statistics and information theory. Also there is a classical inequality among *arithmetic, geometric* and *harmonic means*. In this work, two kinds of inequalities are studied, one is based on *classical information and divergence measures* and another is based among *classical means*. More precisely, studies are made on inequalities arising due to *difference of divergence measures* and *classical means*. Some previous study toward this direction can be seen in Taneja (2006a, 2006b, 2012, 2013a, 2013b). Finally, the connections with differences arising due to *difference of means* and *classical divergences* are made.

1.1 Information Measures

In this subsection we shall give some information measures in two different situations, one with logarithmic expressions and another with non-logarithmic expressions. For simplicity, let

$$\Gamma_n = \left\{ P = (p_1, p_2, ..., p_n) \middle| p_i > 0, \sum_{i=1}^{n} p_i = 1 \right\}, \ n \geq 2,$$

be the set of all complete finite discrete probability distributions. For all $P, Q \in \Gamma_n$, let us consider the two groups of measures.

Logarithmic Measures

Let us consider the following three measures well-known in the literature:

$$I(P \parallel Q) = \frac{1}{2}\left[\sum_{i=1}^{n} p_i \ln\left(\frac{2p_i}{p_i + q_i}\right) + \sum_{i=1}^{n} q_i \ln\left(\frac{2q_i}{p_i + q_i}\right)\right], \tag{1}$$

$$J(P \parallel Q) = \sum_{i=1}^{n}(p_i - q_i) \ln\left(\frac{p_i}{q_i}\right) \tag{2}$$

and

$$T(P \parallel Q) = \sum_{i=1}^{n}\left(\frac{p_i + q_i}{2}\right) \ln\left(\frac{p_i + q_i}{2\sqrt{p_i q_i}}\right). \tag{3}$$

The above three measures are classical divergence measures in the literature on information theory and statistics known as *Jensen–Shannon divergence* (Burbea & Rao, 1982), *J-divergence* (Jeffreys, 1946) and *Arithmetic–Geometric mean divergence* (Taneja, 1995) respectively. These three measures bear the following two relations:

$$I(P \parallel Q) \leq \frac{1}{8}J(P \parallel Q) \leq T(P \parallel Q) \tag{4}$$

and

$$I(P \parallel Q) + T(P \parallel Q) = \frac{1}{4}J(P \parallel Q). \tag{5}$$

Non-logarithmic Measures

Let us consider the following three non-logarithmic measures:

$$\Delta(P \parallel Q) = \sum_{i=1}^{n} \frac{(p_i - q_i)^2}{p_i + q_i}, \tag{6}$$

$$h(P \parallel Q) = \frac{1}{2}\sum_{i=1}^{n}(\sqrt{p_i} - \sqrt{q_i})^2 \tag{7}$$

and

$$\Psi(P \parallel Q) = \sum_{i=1}^{n} \frac{(p_i - q_i)^2(p_i + q_i)}{p_i q_i}. \tag{8}$$

The above three measures $\Delta(P \parallel Q)$, $h(P \parallel Q)$ and $\Psi(P \parallel Q)$ are respectively known as *triangular discrimination* (LeCam, 1986), *Hellinger's divergence* (Hellinger, 1909) and *symmetric chi-square divergence* (Dragomir, Sunde, & Buse, 2000; Taneja, 2005a). These measures allow the following inequalities among the measures.

$$\frac{1}{4}\Delta(P \parallel Q) \le h(P \parallel Q) \le \frac{1}{16}\Psi(P \parallel Q). \tag{9}$$

Inequalities (4) and (9) can be combined in a single inequality (Taneja, 2005b) given by

$$\frac{1}{4}\Delta(P \parallel Q) \le I(P \parallel Q) \le h(P \parallel Q) \le \frac{1}{8}J(P \parallel Q) \le T(P \parallel Q) \le$$
$$\le \frac{1}{16}\Psi(P \parallel Q). \tag{10}$$

The above inequalities (10) admit many nonnegative differences among the divergence measures. Based on these non-negative differences, the author (Taneja, 2005b) proved the following result:

$$D_{I\Delta} \le \frac{2}{3}D_{h\Delta} \le \left\{ \begin{array}{c} 2D_{hI} \\ \frac{1}{2}D_{J\Delta} \le \frac{1}{3}D_{T\Delta} \end{array} \right\} \le D_{TJ} \le \frac{2}{3}D_{Th} \le$$
$$\le 2D_{Jh} \le \frac{1}{6}D_{\Psi\Delta} \le \frac{1}{5}D_{\Psi I} \le \frac{2}{9}D_{\Psi h} \le \frac{1}{4}D_{\Psi J} \le \frac{1}{3}D_{\Psi T}, \tag{11}$$

where, for example, $D_{I\Delta} := I(P \parallel Q) - \frac{1}{4}\Delta(P \parallel Q)$, $D_{TJ} := T(P \parallel Q) - \frac{1}{8}J(P \parallel Q)$, etc. are the nonnegative differences arising due to inequalities (11). More detailed study on measures and inequalities appearing above can be seen in Taneja (2005b, 2013a). Improvements over the inequalities (11) are given in Section 2.

1.2 Generalized Symmetric Divergence Measures

For all $P, Q \in \Gamma_n$, let us consider the unified generalized measure

$$\zeta_s(P \parallel Q) = \begin{cases} J_s(P \parallel Q) = [s(s-1)]^{-1} \left[\sum_{i=1}^{n} (p_i^s q_i^{1-s} + p_i^{1-s} q_i^s) - 2 \right], & s \ne 0, 1 \\ J(P \parallel Q) = \sum_{i=1}^{n} (p_i - q_i) \ln\left(\frac{p_i}{q_i}\right), & s = 0, 1 \end{cases}$$
$$\tag{12}$$

The measure $\zeta_s(P \parallel Q)$ is *generalized J-divergence* extensively studied in Taneja (2005a, 2005b, 2013a). It admits the following particular cases:

1. $\zeta_{-1}(P \parallel Q) = \zeta_2(P \parallel Q) = \frac{1}{2}\Psi(P \parallel Q)$.
2. $\zeta_0(P \parallel Q) = \zeta_1(P \parallel Q) = J(P \parallel Q)$.
3. $\zeta_{1/2}(P \parallel Q) = 8h(P \parallel Q)$.

Again, for all $P, Q \in \Gamma_n$, let us consider another unified generalized measure

$$
\xi_s(P \parallel Q) =
\begin{cases}
IT_s(P \parallel Q) = [s(s-1)]^{-1}\left[\sum_{i=1}^{n}\left(\dfrac{p_i^{1-s} + q_i^{1-s}}{2}\right)\left(\dfrac{p_i + q_i}{2}\right)^s - 1\right], \\[4pt]
\quad s \neq 0, 1 \\[10pt]
I(P \parallel Q) = \dfrac{1}{2}\left[\sum_{i=1}^{n} p_i \ln\left(\dfrac{2p_i}{p_i + q_i}\right) + \sum_{i=1}^{n} q_i \ln\left(\dfrac{2q_i}{p_i + q_i}\right)\right], \\[4pt]
\quad s = 0 \\[10pt]
T(P \parallel Q) = \sum_{i=1}^{n}\left(\dfrac{p_i + q_i}{2}\right)\ln\left(\dfrac{p_i + q_i}{2\sqrt{p_i q_i}}\right), \\[4pt]
\quad s = 1
\end{cases}
\tag{13}
$$

The measure $\xi_s(P \parallel Q)$ known as *generalized arithmetic and geometric mean divergence measure*. More details on these divergence measures can be seen in Taneja (1995, 2005a). It also admits the following particular cases:

1. $\xi_{-1}(P \parallel Q) = \frac{1}{4}\Delta(P \parallel Q)$.
2. $\xi_0(P \parallel Q) = I(P \parallel Q)$.
3. $\xi_1(P \parallel Q) = T(P \parallel Q)$.
4. $\xi_2(P \parallel Q) = \frac{1}{16}\Psi(P \parallel Q)$.

The unified measure (12) admits as particular case three known measures and the measure (13) as four known measures. The measure $\Psi(P \parallel Q)$ appears as a particular case in both the measures. Thus, the six measures (1)–(3) and (6)–(8) written above are particular cases of generalized unified measures (12) and (13) respectively. The measures $I(P \parallel Q)$ and $T(P \parallel Q)$ can easily be written in terms of arithmetic and geometric means:

$$
I(P \parallel Q) = \sum_{i=1}^{n}\left[A(p_i \ln p_i, q_i \ln q_i) - A(p_i, q_i)\ln A(p_i, q_i)\right]
$$

and

$$
T(P \parallel Q) = \sum_{i=1}^{n} A(p_i, q_i)\ln\left(\frac{A(p_i, q_i)}{G(p_i, q_i)}\right),
$$

respectively, where $A(a, b)$ and $G(a, b)$ are well-known arithmetic and geometric means. In view of (5), J-divergence can also be written in terms of arithmetic and geometric means.

1.3 Mean Divergence Measures

The author (Taneja, 2006b) studied the following inequality

$$G(P \parallel Q) \leq N_1(P \parallel Q) \leq N_2(P \parallel Q) \leq A(P \parallel Q), \tag{14}$$

where

$$G(P \parallel Q) = \sum_{i=1}^{n} \sqrt{p_i q_i}, \tag{15}$$

$$N_1(P \parallel Q) = \sum_{i=1}^{n} \left(\frac{p_i + q_i}{2} \right)^2, \tag{16}$$

$$N_2(P \parallel Q) = \sum_{i=1}^{n} \sqrt{\frac{p_i + q_i}{2}} \left(\frac{\sqrt{p_i} + \sqrt{q_i}}{2} \right) \tag{17}$$

and

$$A(P \parallel Q) = \sum_{i=1}^{n} \frac{p_i + q_i}{2} = 1. \tag{18}$$

The inequalities (14) admit non-negative differences given by

$$M_1(P \parallel Q) = D_{N_2 N_1}(P \parallel Q)$$
$$= \sum_{i=1}^{n} \left(\sqrt{\frac{p_i + q_i}{2}} \left(\frac{\sqrt{p_i} + \sqrt{q_i}}{2} \right) - \left(\frac{\sqrt{p_i} + \sqrt{q_i}}{2} \right)^2 \right), \tag{19}$$

$$M_2(P \parallel Q) = D_{N_2 G}(P \parallel Q) = \sum_{i=1}^{n} \left[\left(\frac{\sqrt{p_i} + \sqrt{q_i}}{2} \right) \sqrt{\frac{p_i + q_i}{2}} - \sqrt{p_i q_i} \right], \tag{20}$$

$$M_3(P \parallel Q) = D_{A N_2}(P \parallel Q)$$
$$= \sum_{i=1}^{n} \left[\left(\frac{p_i + q_i}{2} \right) - \left(\frac{\sqrt{p_i} + \sqrt{q_i}}{2} \right) \left(\sqrt{\frac{p_i + q_i}{2}} \right) \right] \tag{21}$$

and

$$h(P \parallel Q) = D_{AN_1}(P \parallel Q) = D_{AG}(P \parallel Q) = D_{N_1G}(P \parallel Q). \qquad (22)$$

The nonnegative differences are understood as, $D_{AN_1} := A(P \parallel Q) - N_1(P \parallel Q)$, $D_{AG}(P \parallel Q) := A(P \parallel Q) - G(P \parallel Q)$, etc. Also,

$$\xi_{1/2}(P \parallel Q) = 4M_3(P \parallel Q),$$

where $\xi_s(P \parallel Q)$ is as given in (13).

1.4 Unified Parametric Generalizations

For all $P, Q \in \Delta_n$, let us consider

$$L_t(P \parallel Q) = \sum_{i=1}^{n} \frac{(p_i - q_i)^2 (p_i + q_i)^t}{2^t (\sqrt{p_i q_i})^{t+1}}, \quad t \in \mathbb{Z}. \qquad (23)$$

In particular, $L_{-1}(P \parallel Q) = 2\Delta(P \parallel Q)$, and $L_1(P \parallel Q) = \frac{1}{2}\Psi(P \parallel Q)$, where Δ and Ψ are given by (6) and (8) respectively. Still, we have

$$L_0(P \parallel Q) = K(P \parallel Q) = \sum_{i=1}^{n} \frac{(p_i - q_i)^2}{\sqrt{p_i q_i}}, \qquad (24)$$

$$L_2(P \parallel Q) = \frac{1}{2}F(P \parallel Q) = \sum_{i=1}^{n} \frac{(p_i^2 - q_i^2)^2}{4(p_i q_i)^{3/2}} \qquad (25)$$

and

$$L_3(P \parallel Q) = \frac{1}{8}L(P \parallel Q) = \sum_{i=1}^{n} \frac{(p_i - q_i)^2 (p_i + q_i)^3}{8(p_i q_i)^2} \qquad (26)$$

The measures K and F given by (24) and (25) are studied respectively by Jain and Srivastava (2007), and Kumar and Johnson (2005). The measure L given by (26) is new. Some studies on it can be seen in Taneja (2013c). For $t = -1, 0, 1, 2$ and 3, the measure (23) is convex in a pair of probability distributions $(P, Q) \in \Delta_n \times \Delta_n$. Also for any $t > -1$, the measure (23) is monotonically increasing in t. This gives

$$\frac{1}{4}\Delta(P \parallel Q) \le \frac{1}{8}K(P \parallel Q) \le \frac{1}{16}\Psi(P \parallel Q) \le \frac{1}{16}F(P \parallel Q) \le \frac{1}{64}L(P \parallel Q). \qquad (27)$$

2. INEQUALITIES AMONG DIFFERENCES OF DIVERGENCE MEASURES

Following the author's work (Taneja, 2013a, 2013b), the 12 measures given by (1)–(3), (6)–(8), (19)–(21), (24)–(26) satisfy the following inequality.

$$\frac{1}{4}\Delta \leq I \leq 4M_1 \leq \frac{4}{3}M_2 \leq h \leq 4M_3 \leq \frac{1}{8}J \leq T \leq \frac{1}{8}K \leq \frac{1}{16}\Psi \leq \frac{1}{16}F \leq \frac{1}{64}L. \tag{28}$$

The 12 measures appearing in (28) admit 66 non-negative differences. These 66 non-negative differences satisfy some obvious inequalities given below in the form of *pyramid-type*:

$$D^1_{I\Delta}.$$
$$D^2_{M_1 I} \leq D^3_{M_1 \Delta}.$$
$$D^4_{M_2 M_1} \leq D^5_{M_2 I} \leq D^6_{M_2 \Delta}.$$
$$D^7_{hM_2} \leq D^8_{hM_1} \leq D^9_{hI} \leq D^{10}_{h\Delta}.$$
$$D^{11}_{M_3 h} \leq D^{12}_{M_3 M_2} \leq D^{13}_{M_3 M_1} \leq D^{14}_{M_3 I} \leq D^{15}_{M_3 \Delta}.$$
$$D^{16}_{JM_3} \leq D^{17}_{Jh} \leq D^{18}_{JM_2} \leq D^{19}_{JM_1} \leq D^{20}_{JI} \leq D^{21}_{J\Delta}.$$
$$D^{22}_{TJ} \leq D^{23}_{TM_3} \leq D^{24}_{Th} \leq D^{25}_{TM_2} \leq D^{26}_{TM_1} \leq D^{27}_{TI} \leq D^{28}_{T\Delta}.$$
$$D^{29}_{KT} \leq D^{30}_{KJ} \leq D^{31}_{KM_3} \leq D^{32}_{Kh} \leq D^{33}_{KM_2} \leq D^{34}_{KM_1} \leq D^{35}_{KI} \leq D^{36}_{K\Delta}.$$
$$D^{37}_{\Psi K} \leq D^{38}_{\Psi T} \leq D^{39}_{\Psi J} \leq D^{40}_{\Psi M_3} \leq D^{41}_{\Psi h} \leq D^{42}_{\Psi M_2} \leq D^{43}_{\Psi M_1} \leq D^{44}_{\Psi I} \leq D^{45}_{\Psi \Delta}.$$
$$D^{46}_{F\Psi} \leq D^{47}_{FK} \leq D^{48}_{FT} \leq D^{49}_{FJ} \leq D^{50}_{FM_3} \leq D^{51}_{Fh} \leq D^{52}_{FM_2} \leq D^{53}_{FM_1} \leq D^{54}_{FI} \leq D^{55}_{F\Delta}.$$
$$D^{56}_{LF} \leq D^{57}_{L\Psi} \leq D^{58}_{LK} \leq D^{59}_{LT} \leq D^{60}_{LJ} \leq D^{61}_{LM_3} \leq D^{62}_{Lh} \leq D^{63}_{LM_2} \leq D^{64}_{LM_1} \leq D^{65}_{LI} \leq D^{66}_{L\Delta}$$

$$\dots \ (29)$$

The *difference measures* are understood, for example, $D^{55}_{F\Delta} = \frac{1}{16}F - \frac{1}{16}\Psi$, $D^{25}_{TM_2} = T - \frac{4}{3}M_2$, etc. Some of these differences become equal by multiplicative constants. These are given by

$$D^4_{M_2 M_1} = 2D^7_{hM_2} = \frac{2}{3}D^8_{hM_1} = \frac{1}{3}D^{11}_{M_3 h} = \frac{1}{2}D^{12}_{M_3 M_2} = \frac{1}{3}D^{13}_{M_3 M_1} \tag{30}$$

and

$$D^{22}_{TJ} = \frac{1}{2}D^{27}_{TI} = D^{20}_{JI}. \tag{31}$$

The 66 measures appearing in the *pyramid-type inequalities* given by (29) can be written as

$$D_{AB} := \sum_{i=1}^{n} q_i f_{AB}\left(\frac{p_i}{q_i}\right), \tag{32}$$

where, $f_{AB}(x) = f_A(x) - f_B(x)$, $A \geq B$, for all $x > 0$, $x \neq 1$.

The aim of this section is to put 66 measures in *nested or continuous* form extending considerably the inequalities (11). Before writing the main Theorem, below are two lemmas. These shall be used frequently to prove important results.

Lemma 1. *If the function $f : [0, \infty) \to \mathbb{R}$ is convex and normalized, i.e., $f(1) = 0$, then the f-divergence, $C_f(P \parallel Q)$ given by*

$$C_f(P \parallel Q) = \sum_{i=1}^{n} q_i f\left(\frac{p_i}{q_i}\right), \tag{33}$$

is nonnegative and convex in a pair of probability distribution $(P, Q) \in \Gamma_n \times \Gamma_n$.

In view of Lemma 1, one can prove that the 12 measures appearing in (28) and 66 measures appearing in (29) are convex in a pair of probability distributions $(P, Q) \in \Gamma_n \times \Gamma_n$.

Lemma 2. *Let $f_1, f_2 : I \subset \mathbb{R}_+ \to \mathbb{R}$ be two functions satisfying the assumptions:*
1. *$f_1(1) = f_2(1) = 0$, $f_1'(1) = f_2'(1) = 0$;*
2. *f_1 and f_2 are twice differentiable on (a, b);*
3. *there exist the real constants α, β such that $\alpha < \beta$ and*

$$\alpha \leq \frac{f_1''(x)}{f_2''(x)} \leq \beta, \quad f_2''(x) \neq 0, \; \forall x > 0,$$

then we have the inequalities

$$\alpha C_{f_2}(P \parallel Q) \leq C_{f_1}(P \parallel Q) \leq \beta C_{f_2}(P \parallel Q).$$

Lemma 1 is due to Csiszár (1967), and Lemma 2 is due to Taneja (2004, 2005a). For applications on divergence measures refer to Taneja and Kumar (2004).

Remarks 1. Lemma 2 is proved in Taneja (2004, 2005a) with the conditions that the functions are convex. From proofs, we observe that convexity of the functions is not necessary, existence of second derivatives is sufficient, but the division $f_1''(x)/f_2''(x)$ should be nonnegative $\forall x > 0$, particularly at $x = 1$.

The following two notes are important, and shall be used frequently throughout the work.

Note 1. Let us write, $\eta(x) = \beta f_2(x) - f_1(x)$, $\forall x > 0$. This gives $\eta(1) = \eta'(1) = 0$. In order to obtain β, let us consider $\eta''(1) = 0$. This gives, $\beta = f_1''(1)/f_2''(1)$.

Note 2. Let a and b be two positive numbers, i.e., $a > 0$ and $b > 0$. If $a^2 - b^2 \geq 0$, then we can conclude that $a \geq b$ because $a - b = (a^2 - b^2)/(a+b)$.

In (29) we have 66 nonnegative divergence measures arising due to difference of measures. Due to equalities among some of the measures we are left with 59 measures. These 59 measures are connected through inequalities among each other given in theorem below:

Theorem 1. *The following inequalities hold:*

$$D_{I\Delta}^1 \leq \frac{8}{9} D_{M_1\Delta}^3 \leq \frac{8}{11} D_{M_2\Delta}^6 \leq \frac{2}{3} D_{h\Delta}^{10} \leq \frac{8}{15} D_{M_3\Delta}^{15} \leq$$

$$\leq \left\{ \begin{array}{c} \frac{1}{2} D_{J\Delta}^{21} \\ \frac{8}{3} D_{hM_1}^8 \end{array} \right\} \leq \frac{1}{3} D_{T\Delta}^{28} \left. \begin{array}{c} \\ \\ \\ \end{array} \right\} \leq \frac{8}{3} D_{M_2I}^5 \leq$$

$$\frac{8}{3} D_{hM_1}^8 \leq \frac{8}{7} D_{M_3I}^{14} \leq 2 D_{hI}^9$$

$$\leq \left\{ \begin{array}{c} \frac{1}{3} D_{K\Delta}^{36} \\ D_{TJ}^{22} \\ \frac{8}{15} D_{TM_1}^{26} \\ \frac{8}{7} D_{JM_1}^{19} \\ 8 D_{M_1I}^2 \end{array} \right\} \leq \frac{8}{13} D_{TM_2}^{25} \leq \frac{2}{3} D_{Th}^{24} \leq \left\{ \begin{array}{c} \leq \frac{8}{5} D_{JM_2}^{18} \leq 2 D_{Jh}^{17} \\ \frac{8}{9} D_{TM_3}^{23} \end{array} \right\} \leq$$

$$\leq \frac{1}{2} D_{KI}^{35} \leq \frac{8}{15} D_{KM_1}^{34} \leq$$

$$\leq \frac{8}{13} D_{KM_2}^{33} \leq \frac{2}{3} D_{Kh}^{32} \leq \left\{ \begin{array}{c} D_{KJ}^{30} \\ \frac{1}{6} D_{\Psi\Delta}^{45} \\ \frac{8}{9} D_{KM_3}^{31} \end{array} \right\} \leq \frac{1}{5} D_{\Psi I}^{44} \right\} \leq$$

$$\leq \frac{8}{39} D_{\Psi M_1}^{43} \leq \frac{8}{37} D_{\Psi M_2}^{42} \leq \frac{2}{9} D_{\Psi h}^{41} \leq$$

$$\leq \left\{ \begin{array}{c} \frac{1}{4} D_{\Psi J}^{39} \\ \frac{8}{33} D_{\Psi M_3}^{40} \end{array} \right\} \leq \frac{1}{3} D_{\Psi K}^{37} \leq \left\{ \begin{array}{c} \frac{1}{3} D_{\Psi T}^{38} \\ \frac{1}{9} D_{FA}^{55} \end{array} \right\} \leq \frac{1}{8} D_{FI}^{54} \leq$$

$$\leq \frac{8}{63} D_{FM_1}^{53} \leq \frac{8}{61} D_{FM_2}^{52} \leq$$

$$\leq \frac{2}{15} D_{Fh}^{51} \leq \begin{Bmatrix} \frac{1}{7} D_{FJ}^{49} \\ \frac{8}{57} D_{FM_3}^{50} \end{Bmatrix} \leq \frac{1}{6} D_{FK}^{47} \leq \frac{1}{6} D_{FT}^{48} \leq \frac{1}{3} D_{F\Psi}^{46}, \qquad (34)$$

$$\begin{Bmatrix} \frac{1}{4} D_{Jh}^{17} \\ D_{M_1 I}^{2} \\ \frac{1}{9} D_{TM_3}^{23} \end{Bmatrix} \leq \begin{Bmatrix} D_{JM_3}^{16} \\ \frac{1}{6} D_{FT}^{48} \end{Bmatrix} \leq \frac{1}{24} D_{F\Psi}^{46} \qquad (35)$$

and

$$\frac{1}{6} D_{FT}^{48} \leq \begin{Bmatrix} \frac{1}{11} D_{LI}^{65} \leq \frac{8}{87} D_{LM_1}^{64} \leq \frac{8}{85} D_{LM_2}^{63} \leq \frac{2}{21} D_{Lh}^{62} \leq \begin{Bmatrix} \frac{8}{81} D_{LM_3}^{61} \\ \frac{1}{10} D_{LJ}^{60} \end{Bmatrix} \\ \frac{1}{3} D_{F\Psi}^{46} \end{Bmatrix} \leq$$

$$\leq \frac{1}{9} D_{LK}^{58} \leq \frac{1}{9} D_{LT}^{59} \leq \begin{Bmatrix} \frac{1}{6} D_{L\Psi}^{57} \leq \frac{1}{3} D_{LF}^{56} \\ \frac{1}{9} D_{L\Delta}^{66} \end{Bmatrix}. \qquad (36)$$

Proof. We observe that the inequalities (36) starts, where (34) and (35) ends, i.e., at D_{FT}^{48} and $D_{F\Psi}^{46}$. The proofs of inequalities (34) and (35) are given in Taneja (2013a). Moreover, the proofs are without the measure L. We shall prove the inequalities appearing in (36). These are based on individual results:

1. $D_{FT}^{48} \leq \frac{6}{11} D_{LI}^{65}$: Let us consider, $g_{FT\_LI}(x) = \frac{f_{FT}''(x)}{f_{LI}''(x)}$. This gives

$$g_{FT\_LI}(x) = \frac{\sqrt{x}\left(\begin{array}{c} 15x^4 + 30x^{7/2} + 60x^3 + 58x^{5/2} + \\ +58x^2 + 58x^{3/2} + 60x + 30\sqrt{x} + 15 \end{array}\right)}{4(\sqrt{x}+1)^2(3x^4 + 10x^3 + 18x^2 + 10x + 3)}, \qquad x \neq 1$$

and

$$\beta_{FT\_LI} = \lim_{x \to 1} g_{FT\_LI}(x) = \frac{6}{11}.$$

Calculating the first order derivative of the function $g_{FT\_LI}(x)$, we get

$$g_{FT\_LI}'(x) = -\frac{k_1(x)}{8x^{3/2}(x-1)^3(3x^4 + 10x^3 + 18x^2 + 10x + 3)^2}, \qquad x \neq 1$$

where

$$k_1(x) = x(\sqrt{x} - 1)^4\big(45x^8 + 180x^{15/2} + 570x^7 + 996x^{13/2} + 1404x^6 +$$
$$+ 1100x^{11/2} + 806x^5 - 164x^{9/2} - 18x^4 - 164x^{7/2} + 806x^3 +$$
$$+ 1100x^{5/2} + 1404x^2 + 996x^{3/2} + 570x + 180\sqrt{x} + 45\big).$$

In order to get rid negative expressions appearing in $k_1(x)$, let us simplify it again

$$k_1(x) = x(\sqrt{x} - 1)^4\big[45x^8 + 180x^{15/2} + 570x^7 + 996x^{13/2} +$$
$$+ 1404x^6 + 1100x^{11/2} + 82x^3(x+1)(\sqrt{x} - 1)^2 +$$
$$+ 633x^5 + 91x^3(x-1)^2 + 633x^3 + 1100x^{5/2} +$$
$$+ 1404x^2 + 996x^{3/2} + 570x + 180\sqrt{x} + 45\big] \geq 0, \quad \forall x > 0$$

This gives

$$g'_{FT\_LI}(x) \begin{cases} > 0 & x < 1 \\ < 0 & x > 1 \end{cases}.$$

By the application of Lemma 2, we get the required result.

2. $D_{LI}^{65} \leq \frac{88}{87}D_{LM_1}^{64}$: Let us consider the function, $g_{LI\_LM_1}(x) = \frac{f''_{LI}(x)}{f''_{LM_1}(x)}$. This gives

$$g_{LI\_LM_1}(x) = \frac{(x-1)^2(2x+2)^{3/2}x^{3/2}(3x^4 + 10x^3 + 18x^2 + 10x + 3)}{x^{3/2}(x+1)\left(\begin{array}{l} 32x^{5/2}(\sqrt{x}+1)(x-\sqrt{x}+1) + \\ +(2x+2)^{3/2}(3x^5 + x^4 - 16x^{5/2} + x + 3)\end{array}\right)},$$

$$x \neq 1$$

and

$$\beta_{LI\_LM_1} = \lim_{x \to 1} g_{LI\_LM_1}(x) = \frac{88}{87}.$$

Calculating the first order derivative of the function $g_{LI\_LM_1}(x)$, we get

$$g'_{LI\_LM_1}(x) = -\frac{8x^{5/2}\sqrt{2x+2}(\sqrt{x}-1)k_2(x)}{x^3(x+1)^2\left(\begin{array}{l} 32x^{5/2}(\sqrt{x}+1)(x-\sqrt{x}+1) + \\ +(2x+2)^{3/2}(3x^5 + x^4 - 16x^{5/2} + x + 3)\end{array}\right)^2},$$

$$x \neq 1,$$

where

$$k_2(x) = 4x^2(x+1)(\sqrt{x}+1)^2 \times$$

$$\times \left(\begin{array}{l} 15x^7 - 15x^{13/2} + 54x^6 - 30x^{11/2} + 78x^5 - \\ -9x^{9/2} + 73x^4 + 20x^{7/2} + 73x^3 - 9x^{5/2} \\ +78x^2 - 30x^{3/2} + 54x - 15\sqrt{x} + 15 \end{array} \right) -$$

$$- x^2(2x+2)^{5/2}(\sqrt{x}+1)$$

$$\times \left(\begin{array}{l} 15x^6 - 18x^{11/2} + 48x^5 - 34x^{9/2} + \\ 69x^4 - 36x^{7/2} + 88x^3 - 36x^{5/2} + \\ +69x^2 - 34x^{3/2} + 48x - 18\sqrt{x} + 15 \end{array} \right).$$

Since, there are negative in between the expressions, let us simplify again:

$$k_2(x) = 4x^2(x+1)(\sqrt{x}+1)^2 \times$$

$$\times \left(\begin{array}{l} (15x^6 + 15x^5 + 9x^4 + 9x^2 + 15x + 15)(\sqrt{x}-1)^2 + \\ \sqrt{x}(15x^6 + 24x^{11/2} + 54x^{9/2} + 9x^4 + 64x^{7/2} + \\ +20x^3 + 64x^{5/2} + 9x^2 + 54x^{3/2} + 24\sqrt{x} + 15) \end{array} \right) -$$

$$- x^2(2x+2)^{5/2}(\sqrt{x}+1) \times$$

$$\times \left(\begin{array}{l} (\sqrt{x}-1)^2 \times (9x^5 + 17x^4 + 18x^3 + \\ +19x^{5/2} + 18x^2 + 17x + 9) + 6x^6 + 22x^5 + \\ +16x^4 + 53x^{7/2} + 17x^{5/2} + 34x^2 + 22x + 6 \end{array} \right).$$

In order to show $k_2(x) \geq 0$, $\forall x > 0$, we shall apply Note 2. Let us consider

$$v_2(x) = \left[4x^2(x+1)(\sqrt{x}+1)^2 \times \right.$$

$$\left. \times \left(\begin{array}{l} (15x^6 + 15x^5 + 9x^4 + 9x^2 + 15x + 15)(\sqrt{x}-1)^2 + \\ \sqrt{x}(15x^6 + 24x^{11/2} + 54x^{9/2} + 9x^4 + 64x^{7/2} + \\ +20x^3 + 64x^{5/2} + 9x^2 + 54x^{3/2} + 24\sqrt{x} + 15) \end{array} \right) \right]^2 -$$

$$- \left[x^2(2x+2)^{5/2}(\sqrt{x}+1) \times \right.$$

$$\left. \times \left(\begin{array}{l} (\sqrt{x}-1)^2 \times (9x^5 + 17x^4 + 18x^3 + \\ +19x^{5/2} + 18x^2 + 17x + 9) + 6x^6 + 22x^5 + \\ +16x^4 + 53x^{7/2} + 17x^{5/2} + 34x^2 + 22x + 6 \end{array} \right) \right]^2.$$

After simplifications, we have

$$v_2(x) = 16x^4(\sqrt{x}+1)^2(\sqrt{x}-1)^4(x+1)^2 \times$$

$$\times \left(\begin{array}{l} 225 + 1638x + 34953x^5 + 23840x^4 - 924x^{3/2} + 13280x^3 - \\ -4492x^{13/2} - 4320x^{11/2} - 180\sqrt{x} + 5700x^2 + 42332x^7 + \\ +42332x^6 + 225x^{13} - 180x^{25/2} - 3460x^{7/2} + 34953x^8 + \\ +23840x^9 + 5700x^{11} - 4320x^{15/2} - 3460x^{19/2} + 13280x^{10} - \\ -924x^{23/2} + 1638x^{12} - 2250x^{21/2} - 4044x^{17/2} - 4044x^{9/2} - \\ -2250x^{5/2} \end{array} \right).$$

Again, we have negative expression. In order to apply Note 2, we need the positive expression. After simplifying again, we get

$$v_2(x) = 16x^4(\sqrt{x}+1)^2(\sqrt{x}-1)^4(x+1)^2 \times$$

$$\times \left(\begin{array}{l} (\sqrt{x}-1)^2(90x^{12} + 462x^{11} + 1125x^{10} + 1730x^9 + \\ +2022x^8 + 2160x^7 + 2246x^6 + 2160x^5 + \\ +2022x^4 + 1730x^3 + 1125x^2 + 462x + 90) + 135x^{13} + \\ +1086x^{12} + 4113x^{11} + 10400x^{10} + 20088x^9 + \\ +30771x^8 + 37926x^7 + 37926x^6 + 30771x^5 + \\ +20088x^4 + 10400x^3 + 4113x^2 + 1086x + 135 \end{array} \right).$$

Since $v_2(x) \geq 0$, $\forall x > 0$, proving $k_2(x) \geq 0$, $\forall x > 0$. This allows us to conclude

$$g'_{LI\_LM_1}(x) \begin{cases} > 0 & x < 1 \\ < 0 & x > 1 \end{cases}.$$

By the application of Lemma 2, we get required result.

3. $D^{64}_{LM_1} \leq \frac{87}{85} D^{63}_{LM_2}$: Let us consider the function, $g_{LM_1\_LM_2}(x) = \frac{f''_{LM_1}(x)}{f''_{LM_2}(x)}$. This gives

$$g_{LM_1\_LM_2}(x) = \frac{3\left(\begin{array}{l} 32x^4(\sqrt{x}+1)(x-\sqrt{x}+1) + \\ +x^{3/2}(2x+2)^{3/2}(3x^5 + x^4 - 16x^{5/2} + x + 3) \end{array} \right)}{\left(\begin{array}{l} 32x^4(\sqrt{x}+1)(x-\sqrt{x}+1) + \\ +x^{3/2}(2x+2)^{3/2}(9x^5 + 3x^4 - 32x^{5/2} + 3x + 9) \end{array} \right)},$$

$$x \neq 1$$

and

$$\beta_{LM_1\_LM_2} = \lim_{x \to 1} g_{LM_1\_LM_2}(x) = \frac{87}{85}.$$

Calculating the first order derivative of the function $g_{LI\_LM_1}(x)$, we get

$$g'_{LM_1\_LM_2}(x) = -\frac{72x^{9/2}\sqrt{2x+2}(\sqrt{x}-1)k_3(x)}{\left(\begin{array}{c} 32x^4(\sqrt{x}+1)(x-\sqrt{x}+1)+ \\ +x^{3/2}(2x+2)^{3/2}(9x^5+3x^4-32x^{5/2}+3x+9) \end{array}\right)^2},$$

$$x \neq 1,$$

where

$$k_3(x) = 8\left(\begin{array}{c} 5x^7 + 5x^{13/2} + 8x^6 + 16x^{11/2} + 16x^5 + 23x^{9/2} + 23x^4 + \\ +32x^{7/2} + 23x^3 + 23x^{5/2} + 16x^2 + 16x^{3/2} + 8x + 5\sqrt{x} + 5 \end{array}\right)$$
$$- (2x+2)^{5/2}(\sqrt{x}+1)(5x^4 + 6x^3 + 6x^2 + 6x + 5)$$

In order to show $k_3(x) \geq 0$, $\forall x > 0$, we shall apply Note 2. Let us consider

$$k_3(x) = 64\left(\begin{array}{c} 5x^7 + 5x^{13/2} + 8x^6 + 16x^{11/2} + 16x^5 + 23x^{9/2} + 23x^4 + \\ +32x^{7/2} + 23x^3 + 23x^{5/2} + 16x^2 + 16x^{3/2} + 8x + 5\sqrt{x} + 5 \end{array}\right)^2$$
$$- \left[(2x+2)^{5/2}(\sqrt{x}+1)(5x^4 + 6x^3 + 6x^2 + 6x + 5)\right]^2$$

After simplifications, we have

$$v_3(x) = 32(\sqrt{x}-1)^4 \times$$
$$\times \left(\begin{array}{c} 25 + 450x + 17180x^5 + 12039x^4 + 12039x^8 + 14844x^{15/2} + \\ +25x^{12} + 450x^{11} + 150x^{23/2} + 1110x^{21/2} + 9070x^{17/2} + \\ +3998x^{19/2} + 2252x^{10} + 6306x^9 + 18700x^{13/2} + \\ +150\sqrt{x} + 2252x^2 + 1110x^{3/2} + 9070x^{7/2} + 6306x^3 + \\ +19272x^6 + 14844x^{9/2} + 17180x^7 + 18700x^{11/2} + 3998x^{5/2} \end{array}\right).$$

This gives

$$g'_{LI\_LM_1}(x) \begin{cases} > 0 & x < 1 \\ < 0 & x > 1 \end{cases}.$$

By the application of Lemma 2, we get required result.

4. $D_{LM_2}^{63} \leq \frac{85}{84}D_{Lh}^{62}$: Let us consider the function, $g_{LM_2\_Lh}(x) = \frac{f''_{LM_2}(x)}{f''_{Lh}(x)}$.
This gives

$$g_{LM_2\_Lh}(x) = \frac{\left(\begin{array}{c} 32x^4(\sqrt{x}+1)(x-\sqrt{x}+1) + \\ +x^{3/2}(2x+2)^{3/2}(9x^5 + 3x^4 - 32x^{5/2} + 3x + 9) \end{array}\right)}{3(2x+2)^{3/2}(3x^{13/2} + x^{11/2} - 8x^4 + x^{5/2} + 3x^{3/2})}, \quad x \neq 1$$

and

$$\beta_{LM_2\_Lh} = \lim_{x \to 1} g_{LM_2\_Lh}(x) = \frac{85}{84}.$$

Calculating the first order derivative of the function $g_{LI\_LM_1}(x)$, we get

$$g'_{LM_2\_Lh}(x) = -\frac{x^{5/2}(\sqrt{x}-1)k_4(x)}{(2x+2)^{5/2}(3x^{13/2} + x^{11/2} - 8x^4 + 3x^{3/2} + x^{5/2})^2}, \quad x \neq 1,$$

where

$$k_4(x) = k_3(x) \geq 0, \quad \forall x > 0$$

This gives

$$g'_{LM_2\_Lh}(x) \begin{cases} > 0 & x < 1 \\ < 0 & x > 1 \end{cases}.$$

By the application of Lemma 2, we get required result.

5. $D_{Lh}^{62} \leq \frac{28}{27}D_{LM_3}^{61}$: Let us consider the function, $g_{Lh\_LM_3}(x) = \frac{f''_{Lh}(x)}{f''_{LM_3}(x)}$. This gives

$$g_{Lh\_LM_3}(x) =$$

$$= \frac{(2x+2)^{3/2}x^{3/2}(\sqrt{x}-1)^2\left(\begin{array}{c} 3x^4 + 6x^{7/2} + 10x^3 + 14x^{5/2} + \\ +18x^2 + 14x^{3/2} + 10x + 6\sqrt{x} + 3 \end{array}\right)}{\left(\begin{array}{c} (2x+2)^{3/2}x^{3/2}(x+1)(3x^4 - 2x^3 + 2x^2 - 2x + 3) - \\ -32x^4(\sqrt{x}+1)(x-\sqrt{x}+1) \end{array}\right)}$$

and

$$\beta_{Lh\_LM_3} = \lim_{x \to 1} g_{Lh\_LM_3}(x) = \frac{28}{27}.$$

Calculating the first order derivative of the function $g_{Lh\_LM_3}(x)$, we get

$$g'_{Lh\_LM_3}(x) = -\frac{12x^{9/2}\sqrt{2x+2}(\sqrt{x}-1)k_5(x)}{\left(\begin{array}{c} x^{3/2}(x+1)(2x+2)^{3/2}(3x^4 - 2x^3 + 2x^2 - 2x + 3) - \\ -32x^4(\sqrt{x}+1)(x-\sqrt{x}+1) \end{array}\right)^2},$$

$$x \neq 1$$

where

$$k_5(x) = k_3(x) \geq 0, \quad \forall x > 0$$

This gives

$$g'_{Lh\_LM_3}(x) \begin{cases} > 0 & x < 1 \\ < 0 & x > 1 \end{cases}.$$

By the application of Lemma 2, we get required result.

6. $D^{62}_{Lh} \leq \frac{21}{20} D^{60}_{LJ}$: Let us consider the function, $g_{Lh\_LJ}(x) = \frac{f''_{Lh}(x)}{f''_{LM_3}(x)}$. This gives

$$g_{Lh\_LJ}(x) = \frac{\left(\begin{array}{c} 3x^4 + 6x^{7/2} + 10x^3 + 14x^{5/2} + \\ +18x^2 + 14x^{3/2} + 10x + 6\sqrt{x} + 3 \end{array} \right)}{(x+1)(\sqrt{x}+1)^2(3x^2+4x+3)}, \quad x \neq 1$$

and

$$\beta_{Lh\_LJ} = \lim_{x\to 1} g_{Lh\_LJ}(x) = \frac{21}{20}.$$

Calculating the first order derivative of the function $g_{Lh\_LM_3}(x)$, we get

$$g'_{Lh\_LJ}(x) = -\frac{4k_6(x)}{x^{3/2}(x-1)^3(x+1)^2(3x^2+4x+3)^2}, \quad x \neq 1,$$

where

$$k_6(x) = x^{5/2}(\sqrt{x}-1)^4\left[(\sqrt{x}-1)^2 + x + \sqrt{x} + 1\right] \times \\ \times \left(3x^2 + 6x^{3/2} + 8x + 6\sqrt{x} + 3\right) \geq 0, \quad x > 0$$

This gives

$$g'_{Lh\_LJ}(x) \begin{cases} > 0 & x < 1 \\ < 0 & x > 1 \end{cases}.$$

By the application of Lemma 2, we get required result.

7. $D^{61}_{LM_3} \leq \frac{9}{8} D^{58}_{LK}$: Let us consider the function, $g_{LM_3\_LK}(x) = \frac{f''_{LM_3}(x)}{f''_{LK}(x)}$. This gives

$$\beta_{LM_3\_LK} = \lim_{x\to 1} g_{LM_3\_LK}(x) = \frac{9}{8}.$$

In order to show $\frac{9}{8}D_{LK}^{58} - D_{LM_3}^{61} \geq 0$, let us consider

$$\frac{9}{8}D_{LK}^{58} - D_{LM_3}^{61} = \sum_{i=1}^{n} q_i f_{LK\_KM_3}\left(\frac{p_i}{q_i}\right),$$

where

$$f_{LK\_KM_3}(x) = \frac{k_7(x)}{512x^{5/2}}$$

with

$$k_7(x) = \sqrt{x}\left(x^5 + x^4 - 72x^{7/2} + 1022x^3 + 144x^{5/2} + 1022x^2 - \right.$$
$$\left. - 72x^{3/2} + x + 1\right) - 512x^{5/2}(\sqrt{x} + 1)\sqrt{2x + 2}$$
$$= \sqrt{x}\left[x^2(x^{3/2} - 5)^2 + x^3(\sqrt{x} - 31)^2 + 36x^3 + 144x^{5/2} + 36x^2 + \right.$$
$$\left. + x(31\sqrt{x} - 1)^2 + (5x^{3/2} - 1)^2\right] - 512x^{5/2}(\sqrt{x} + 1)\sqrt{2x + 2}.$$

Let us consider

$$v_7(x) = x\left[x^2(x^{3/2} - 5)^2 + x^3(\sqrt{x} - 31)^2 + 36x^3 + 144x^{5/2} + 36x^2 + \right.$$
$$\left. + x(31\sqrt{x} - 1)^2 + (5x^{3/2} - 1)^2\right]^2 - \left[512x^{5/2}(\sqrt{x} + 1)\sqrt{2x + 2}\right]^2.$$

After simplifications, we get

$$v_7(x) = x(\sqrt{x} - 1)^6(x^7 + 6x^{13/2} + 23x^6 - 76x^{11/2} + 1349x^5 + \right.$$
$$+ 9754x^{9/2} + 45731x^4 + 9304x^{7/2} + 45731x^3 + $$
$$+ 9754x^{5/2} + 1349x^2 - 76x^{3/2} + 23x + 6\sqrt{x} + 1).$$

Simplifying again, we get

$$v_7(x) = x(\sqrt{x} - 1)^6[x^7 + 6x^{13/2} + 21x^6 + x^5(\sqrt{x} - 2)^2 + $$
$$+ x^5(\sqrt{x} - 36)^2 + 49x^5 + 9754x^{9/2} + 45731x^4 + $$
$$+ 9304x^{7/2} + 45731x^3 + 9754x^{5/2} + 49x^2 + $$
$$+ x(36\sqrt{x} - 1)^2 + x(2\sqrt{x} - 1)^2 + 21x + 6\sqrt{x} + 1] \geq 0, \quad x > 0.$$

By applications of Note 2 and Lemma 2, we get the required result.

8. $D_{LJ}^{60} \le \frac{10}{9} D_{LK}^{58}$: Let us consider the function, $g_{LJ\_LK}(x) = \frac{f_{LJ}''(x)}{f_{LK}''(x)}$. This gives

$$g_{LJ\_LK}(x) = \frac{(x+1)(\sqrt{x}+1)^2(3x^2+4x+3)}{(x+\sqrt{x}+1)\left(\begin{array}{l} 3x^4+6x^{7/2}+10x^3+14x^{5/2}+ \\ +18x^2+14x^{3/2}+10x+6\sqrt{x}+3 \end{array}\right)}, \quad x \ne 1$$

and

$$\beta_{LJ\_LK} = \lim_{x\to 1} g_{LJ\_LK}(x) = \frac{10}{9}.$$

Calculating the first order derivative of the function $g_{LJ\_LK}(x)$, we get

$$g_{LJ\_LK}'(x) = -\frac{x(x-1)k_8(x)}{3\sqrt{x}(x+\sqrt{x}+1)^2\left(\begin{array}{l} 3x^4+6x^{7/2}+10x^3+14x^{5/2}+ \\ +18x^2+14x^{3/2}+10x+6\sqrt{x}+3 \end{array}\right)^2},$$

$$x \ne 1,$$

where

$$k_8(x) = \left(27x^4+60x^{7/2}+138x^3+172x^{5/2}+ \right.$$
$$\left. +14x^2+172x^{3/2}+138x+60\sqrt{x}+27\right) \ge 0, \quad x > 0$$

This gives

$$g_{Lh\_LJ}'(x) \begin{cases} > 0 & x < 1 \\ < 0 & x > 1 \end{cases}.$$

By the application of Lemma 2, we get required result.

9. $D_{LK}^{58} \le D_{LT}^{59}$: After simplifying the difference, we get $D_{LT}^{59} - D_{LK}^{58} = \frac{1}{8}K - T \ge 0$. In view of (28) we have the required result.

10. $D_{LT}^{59} \le D_{LA}^{66}$: After simplifying the difference, we get $D_{LA}^{66} - D_{LT}^{59} = T - \frac{1}{4}\Delta \ge 0$. In view of (28), we have required result.

11. $D_{LT}^{59} \le \frac{3}{2} D_{L\Psi}^{57}$: Let us consider the function, $g_{LT\_L\Psi}(x) = \frac{f_{LT}''(x)}{f_{L\Psi}''(x)}$. This gives

$$g_{LT\_L\Psi}(x) = \frac{3x^4+10x^3+10x^2+10x+3}{3(x+1)^2(x^2+1)}, \quad x \ne 1$$

and

$$\beta_{LT\_L\Psi} = \lim_{x\to 1} g_{LT\_L\Psi}(x) = \frac{3}{2}.$$

Calculating the first order derivative of the function $g_{LJ\_LK}(x)$, we get

$$g'_{LT\_L\Psi}(x) = -\frac{4(x-1)(x^4 + 2x^3 + 4x^2 + 2x + 1)}{3(x+1)^3(x^2+1)^2}, \quad x \neq 1.$$

This gives

$$g'_{LT\_L\Psi}(x) \begin{cases} > 0 & x < 1 \\ < 0 & x > 1 \end{cases}.$$

By the application of Lemma 2, we get required result.

12. $D_{L\Psi}^{57} \leq 2D_{LF}^{56}$: In this case, $\beta_{L\Psi\_LF} = g_{L\Psi\_LF}(1) = \frac{f''_{LT}(1)}{f''_{L\Psi}(1)} = 2$. It is sufficient to show, $2D_{LF}^{56} - D_{L\Psi}^{57} \geq 0$. Let us write

$$2D_{LF}^{56} - D_{L\Psi}^{57} = \sum_{i=1}^{n} q_i f_{LF\_L\Psi}\left(\frac{p_i}{q_i}\right),$$

where

$$f_{LF\_L\Psi}(x) = \frac{x^5 - 4x^{9/2} + 5x^4 - 6x^3 + 8x^{5/2} - 6x^2 + 5x - 4\sqrt{x} + 1}{64x^2}$$

$$= \frac{(x+1)(\sqrt{x}+1)^2(\sqrt{x}-1)^6}{64x^2} \geq 0, \quad \forall x > 0.$$

13. $D_{F\Psi}^{46} \leq \frac{1}{3}D_{LK}^{58}$: In this case, $\beta_{F\Psi\_LK} = g_{F\Psi\_LK}(1) = \frac{f''_{F\Psi}(1)}{f''_{LK}(1)} = \frac{1}{3}$. It is sufficient to show, $\frac{1}{3}D_{LK}^{58} - D_{F\Psi}^{46} \geq 0$. Let us write

$$\frac{1}{3}D_{LK}^{58} - D_{F\Psi}^{46} = \sum_{i=1}^{n} q_i f_{F\Psi\_LK}\left(\frac{p_i}{q_i}\right),$$

where

$$f_{F\Psi\_LK}(x) = \frac{\left(\begin{array}{c} x^{13/2} - 6x^6 + 13x^{11/2} - 8x^5 - 14x^{9/2} + \\ +28x^4 - 14x^{7/2} - 8x^3 + 13x^{5/2} - 6x^2 + x^{3/2} \end{array}\right)}{192x^{7/2}}$$

$$= \frac{(\sqrt{x}+1)^2(\sqrt{x}-1)^8}{192x^2} \geq 0, \quad \forall x > 0.$$

Results 1–13 complete the proof of (35).   □

2.1 Reverse Inequalities

The *pyramid-type inequalities* appearing in (29) are obvious. Below are inequalities written in reverse order of (29). We will observe that not all the measures can be written in reverse order, but majority is possible.

- $D^3_{M_1\Delta} \leq 9D^2_{M_1I}.$

- $D^6_{M_2\Delta} \leq \frac{11}{2}D^4_{M_2M_1} \leq \frac{11}{3}D^5_{M_2I}.$

- $D^{10}_{h\Delta} \leq 12D^7_{hM_2} = 4D^8_{hM_1} \leq 3D^9_{hI}.$

- $D^{15}_{M_3\Delta} \leq 5D^{11}_{M_3h} = \frac{15}{4}D^{12}_{M_3M_2} = \frac{5}{2}D^{13}_{M_3M_1} \leq \frac{15}{7}D^{14}_{M_3I}.$

- $D^{21}_{J\Delta} \leq \left\{ \begin{array}{c} \frac{16}{7}D^{19}_{JM_1} \\ 2D^{20}_{JI} \end{array} \right\} \leq \frac{16}{5}D^{18}_{JM_2} \leq 2D^{17}_{Jh} \leq 8D^{16}_{JM_3}.$

- $D^{28}_{T\Delta} \leq \left\{ \begin{array}{c} D^{26}_{TM_1} \\ \frac{3}{2}D^{27}_{TI} = 3D^{22}_{TJ} \end{array} \right\} \leq D^{25}_{TM_2} \leq 2D^{24}_{Th} \leq \frac{8}{3}D^{23}_{TM_3}.$

- $D^{36}_{K\Delta} \leq \frac{3}{2}D^{35}_{KI} \leq \frac{8}{5}D^{34}_{KM_1} \leq \frac{24}{13}D^{33}_{KM_2} \leq 2D^{32}_{Kh} \leq \left\{ \begin{array}{c} 3D^{30}_{KJ} \\ \frac{8}{3}D^{31}_{KM_3} \end{array} \right\}.$

- $D^{45}_{\Psi\Delta} \leq \frac{6}{5}D^{44}_{\Psi I} \leq \frac{16}{15}D^{43}_{\Psi M_1} \leq \frac{48}{37}D^{42}_{\Psi M_2} \leq \frac{4}{3}D^{41}_{\Psi h} \leq \left\{ \begin{array}{c} \frac{3}{2}D^{39}_{\Psi J} \\ \frac{16}{11}D^{40}_{\Psi M_3} \end{array} \right\} \leq 2D^{37}_{\Psi K} \leq 2D^{38}_{\Psi T}.$

- $D^{55}_{F\Delta} \leq \frac{9}{8}D^{54}_{FI} \leq \frac{8}{7}D^{53}_{FM_1} \leq \frac{72}{61}D^{52}_{FM_2} \leq \frac{6}{5}D^{51}_{Fh} \leq \left\{ \begin{array}{c} \frac{9}{7}D^{49}_{FJ} \\ \frac{24}{19}D^{50}_{FM_3} \end{array} \right\} \leq \frac{3}{2}D^{47}_{FK} \leq \frac{3}{2}D^{48}_{FT} \leq 3D^{46}_{F\Psi}.$

- $D^{65}_{LI} \leq \frac{88}{87}D^{64}_{LM_1} \leq \frac{88}{85}D^{63}_{LM_2} \leq \frac{22}{21}D^{62}_{Lh} \leq \left\{ \begin{array}{c} \frac{88}{81}D^{61}_{LM_3} \\ \frac{11}{10}D^{60}_{LJ} \end{array} \right\} \leq \frac{11}{9}D^{58}_{LK} \leq \frac{11}{9}D^{59}_{LT} \leq \left\{ \begin{array}{c} \frac{11}{6}D^{57}_{L\Psi} \leq \frac{11}{3}D^{56}_{LF} \\ \frac{11}{9}D^{66}_{L\Delta} \end{array} \right\}.$

Some of the above inequalities can be seen in (34), (35) and (36). Moreover, the last line is exactly (36).

Remarks 2. (i) From the inequalities (34), (35) and (36), it is interesting to observe that all the measures are in between $D^1_{I\Delta}$ and D^{56}_{LF} or $D^{66}_{L\Delta}$, i.e., in between the first members $D^1_{I\Delta}$ of first row and first member $D^{56}_{F\Psi}$ or the last member $D^{66}_{L\Delta}$ of last row of (29).

(ii) The last members of each row (corner members) of (29) are connected in an increasing order, i.e.,

$$D_{I\Delta}^1 \leq \frac{8}{9}D_{M_1\Delta}^3 \leq \frac{8}{11}D_{M_2\Delta}^6 \leq \frac{2}{3}D_{h\Delta}^{10} \leq \frac{8}{15}D_{M_3\Delta}^{15} \leq$$

$$\leq \frac{1}{2}D_{J\Delta}^{21} \leq \frac{1}{3}D_{T\Delta}^{28} \leq \frac{1}{3}D_{K\Delta}^{36} \leq \frac{1}{6}D_{\Psi\Delta}^{45} \leq \frac{1}{9}D_{F\Delta}^{55} \leq \frac{1}{9}D_{L\Delta}^{66}. \qquad (37)$$

(iii) The measures $M_1(P \parallel Q)$, $M_2(P \parallel Q)$ and $M_3(P \parallel Q)$ are with square-roots. In order to solve inequalities (34), (35) and (36) with square-root it requires little more efforts, i.e., we always use Note 2. Without use of these three measures, the inequalities (34), (35) and (36) stands as single expression:

$$D_{I\Delta}^1 \leq \frac{2}{3}D_{h\Delta}^{10} \leq \left\{\begin{matrix} \frac{1}{2}D_{J\Delta}^{21} \leq \frac{1}{3}D_{T\Delta}^{28} \\ 2D_{hI}^9 \end{matrix}\right\} \leq \left\{\begin{matrix} \frac{1}{3}D_{K\Delta}^{36} \\ D_{TJ}^{22} \leq \frac{2}{3}D_{Th}^{24} \end{matrix}\right\} \leq 2D_{Jh}^{17} \leq \frac{1}{2}D_{KI}^{35} \leq$$

$$\leq \frac{2}{3}D_{Kh}^{32} \leq \left\{\begin{matrix} D_{KJ}^{30} \\ \frac{1}{6}D_{\Psi\Delta}^{45} \end{matrix}\right\} \leq \frac{1}{5}D_{\Psi I}^{44} \leq \frac{2}{9}D_{\Psi h}^{41} \leq \frac{1}{4}D_{\Psi J}^{39} \leq \frac{1}{3}D_{\Psi K_0}^{37} \leq$$

$$\leq \left\{\begin{matrix} \frac{1}{3}D_{\Psi T}^{38} \\ \frac{1}{9}D_{F\Delta}^{55} \end{matrix}\right\} \leq \frac{1}{8}D_{FI}^{54} \leq \frac{2}{15}D_{Fh}^{51} \leq \frac{1}{7}D_{FJ}^{49} \leq \frac{1}{6}D_{FK}^{47} \leq$$

$$\leq \frac{1}{6}D_{FT}^{48} \leq \left\{\begin{matrix} \frac{1}{11}D_{LI}^{65} \leq \frac{8}{87}D_{LM_1}^{64} \leq \frac{8}{85}D_{LM_2}^{63} \leq \frac{2}{21}D_{Lh}^{62} \leq \left\{\begin{matrix} \frac{8}{81}D_{LM_3}^{61} \\ \frac{1}{10}D_{LJ}^{60} \end{matrix}\right\} \\ \frac{1}{3}D_{F\Psi}^{46} \end{matrix}\right\} \leq$$

$$\leq \frac{1}{9}D_{LK}^{58} \leq \frac{1}{9}D_{LT}^{59} \leq \left\{\begin{matrix} \frac{1}{6}D_{L\Psi}^{57} \leq \frac{1}{3}D_{LF}^{56} \\ \frac{1}{9}D_{L\Delta}^{66} \end{matrix}\right\}. \qquad (38)$$

2.2 Equivalent Inequalities

We used 72 results to prove the inequalities (34), (35) and (36). These 72 results are as follows:

1. $D_{I\Delta}^1 \leq \frac{8}{9}D_{M_1\Delta}^3$.
2. $D_{M_1\Delta}^3 \leq \frac{9}{11}D_{M_2\Delta}^6$.
3. $D_{M_2\Delta}^6 \leq \frac{11}{12}D_{h\Delta}^{10}$.
4. $D_{h\Delta}^{10} \leq \frac{4}{5}D_{M_3\Delta}^{15}$.
5. $D_{M_3\Delta}^{15} \leq 5D_{hM_1}^8$.
6. $D_{M_3\Delta}^{15} \leq \frac{15}{16}D_{J\Delta}^{21}$.
7. $D_{J\Delta}^{21} \leq \frac{2}{3}D_{T\Delta}^{28}$.
8. $D_{hM_1}^8 \leq \frac{1}{8}D_{T\Delta}^{28}$.
9. $D_{M_3h}^{11} \leq \frac{3}{7}D_{M_3I}^{14}$.
10. $D_{M_3I}^{14} \leq \frac{7}{4}D_{hI}^9$.

11. $D_{hI}^9 \leq \frac{4}{3}D_{M_2I}^5$.
12. $D_{M_2I}^5 \leq \frac{1}{8}D_{K_0\Delta}^{36}$.
13. $D_{M_2I}^5 \leq \frac{3}{8}D_{TJ}^{22}$.
14. $D_{M_2I}^5 \leq \frac{1}{5}D_{TM_1}^{26}$.
15. $D_{M_2I}^5 \leq \frac{3}{7}D_{JM_1}^{19}$.
16. $D_{M_2I}^5 \leq 3D_{M_1I}^2$.
17. $D_{TJ}^{22} \leq \frac{8}{13}D_{TM_2}^{25}$.
18. $D_{TM_1}^{26} \leq \frac{15}{13}D_{TM_2}^{25}$.
19. $D_{TM_2}^{25} \leq \frac{13}{12}D_{Th}^{24}$.
20. $D_{Th}^{24} \leq \frac{4}{3}D_{TM_3}^{23}$.

21. $D_{Th}^{24} \leq \frac{12}{5}D_{JM_2}^{18}$.
22. $D_{JM_2}^{18} \leq \frac{5}{4}D_{Jh}^{17}$.
23. $D_{K\Delta}^{36} \leq \frac{3}{2}D_{KI}^{35}$.
24. $D_{TM_3}^{23} \leq \frac{9}{16}D_{KI}^{35}$.
25. $D_{M_1I}^2 \leq \frac{1}{16}D_{KI}^{35}$.
26. $D_{Jh}^{17} \leq \frac{1}{4}D_{KI}^{35}$.
27. $D_{KI}^{35} \leq \frac{16}{15}D_{KM_1}^{34}$.
28. $D_{KM_1}^{34} \leq \frac{15}{13}D_{KM_2}^{33}$.
29. $D_{KM_2}^{33} \leq \frac{13}{12}D_{Kh}^{32}$.
30. $D_{Kh}^{32} \leq \frac{4}{3}D_{KM_3}^{31}$.

31. $D_{Kh}^{32} \le \frac{3}{2} D_{KJ}^{30}$.

32. $D_{Kh}^{32} \le \frac{1}{4} D_{\Psi\Delta}^{45}$.

33. $D_{KJ}^{30} \le \frac{1}{5} D_{\Psi I}^{44}$.

34. $D_{\Psi\Delta}^{45} \le \frac{6}{5} D_{\Psi I}^{44}$.

35. $D_{KM_3}^{31} \le \frac{3}{13} D_{\Psi M_1}^{43}$.

36. $D_{\Psi I}^{44} \le \frac{40}{39} D_{\Psi M_1}^{43}$.

37. $D_{\Psi M_1}^{43} \le \frac{39}{37} D_{\Psi M_2}^{42}$.

38. $D_{\Psi M_2}^{42} \le \frac{37}{36} D_{\Psi h}^{41}$.

39. $D_{\Psi h}^{41} \le \frac{9}{8} D_{\Psi J}^{39}$.

40. $D_{\Psi h}^{41} \le \frac{12}{11} D_{\Psi M_3}^{40}$.

41. $D_{\Psi M_3}^{40} \le \frac{11}{8} D_{\Psi K}^{37}$.

42. $D_{\Psi J}^{39} \le \frac{4}{3} D_{\Psi K}^{37}$.

43. $D_{\Psi K}^{37} \le D_{\Psi T}^{38}$.

44. $D_{\Psi K}^{37} \le \frac{1}{3} D_{F\Delta}^{55}$.

45. $D_{\Psi T}^{38} \le \frac{3}{8} D_{FI}^{54}$.

46. $D_{F\Delta}^{55} \le \frac{9}{8} D_{FI}^{54}$.

47. $D_{FI}^{54} \le \frac{64}{63} D_{FM_1}^{53}$.

48. $D_{FM_1}^{53} \le \frac{63}{61} D_{FM_2}^{52}$.

49. $D_{FM_2}^{52} \le \frac{61}{60} D_{Fh}^{51}$.

50. $D_{Fh}^{51} \le \frac{15}{14} D_{FJ}^{49}$.

51. $D_{Fh}^{51} \le \frac{20}{19} D_{FM_3}^{50}$.

52. $D_{FM_3}^{50} \le \frac{19}{16} D_{FK}^{47}$.

53. $D_{FJ}^{49} \le \frac{7}{6} D_{FK}^{47}$.

54. $D_{FK}^{47} \le D_{FT}^{48}$.

55. $D_{FT}^{48} \le 2 D_{F\Psi}^{46}$.

56. $D_{Jh}^{17} \le 4 D_{JM_3}^{16}$.

57. $D_{M_1 I}^{2} \le D_{JM_3}^{16}$.

58. $D_{TM_3}^{23} \le 9 D_{JM_3}^{16}$.

59. $D_{JM_3}^{16} \le \frac{1}{24} D_{F\Psi}^{46}$.

60. $D_{FT}^{48} \le \frac{6}{11} D_{LI}^{65}$.

61. $D_{LI}^{65} \le \frac{88}{87} D_{LM_1}^{64}$.

62. $D_{LM_1}^{64} \le \frac{87}{85} D_{LM_2}^{63}$.

63. $D_{LM_2}^{63} \le \frac{85}{84} D_{Lh}^{62}$.

64. $D_{Lh}^{62} \le \frac{28}{27} D_{LM_3}^{61}$.

65. $D_{Lh}^{62} \le \frac{21}{20} D_{LJ}^{60}$.

66. $D_{LM_3}^{61} \le \frac{9}{8} D_{LK}^{58}$.

67. $D_{LJ}^{60} \le \frac{10}{9} D_{LK}^{58}$.

68. $D_{LK}^{58} \le D_{LT}^{57}$.

69. $D_{LT}^{59} \le D_{L\Delta}^{66}$.

70. $D_{LT}^{59} \le \frac{3}{2} D_{L\Psi}^{57}$.

71. $D_{L\Psi}^{57} \le 2 D_{LF}^{56}$.

72. $D_{F\Psi}^{46} \le \frac{1}{3} D_{LK}^{58}$.

Simplifying each part separately, we get some interesting inequalities among the measures. These are divided in two groups. One group is with single measure on left side, and another group is with two measures on both sides. See below:

Group 1

1. $I \le \frac{\Delta + 128 M_1}{36}$.

2. $I \le \frac{4\Delta + K}{24}$.

3. $I \le \frac{20\Delta + \Psi}{96}$.

4. $I \le \frac{32\Delta + F}{144}$.

5. $4M_1 \le \frac{\Delta + 24 M_2}{22}$.

6. $4M_1 \le \frac{120I + K}{128}$.

7. $4M_1 \le \frac{624I + \Psi}{640}$.

8. $4M_1 \le \frac{1008I + F}{1024}$.

9. $4M_1 \ge \frac{2(3I + 2M_2)}{9}$.

10. $4M_1 \le \frac{L + 5568 I}{5632}$.

11. $\frac{4}{3} M_2 \le \frac{\Delta + 44h}{48}$.

12. $\frac{4}{3} M_2 \le \frac{2(T + 26 M_1)}{15}$.

13. $\frac{4}{3} M_2 \le \frac{K + 208 M_1}{60}$.

14. $\frac{4}{3} M_2 \le \frac{\Psi + 1184 M_1}{312}$.

15. $\frac{4}{3} M_2 \le \frac{F + 1952 M_1}{504}$.

16. $\frac{4}{3} M_2 \ge \frac{3I + 9h}{12}$.

17. $\frac{4}{3} M_2 \le \frac{L + 10880 M_1}{2784}$.

18. $h \le \frac{\Delta + 64 M_3}{20}$.

19. $h \le \frac{3J + 128 M_2}{120}$.

20. $h \le \frac{T + 16 M_2}{13}$.

21. $h \le \frac{K + 128 M_2}{104}$.

22. $h \le \frac{\Psi + 768 M_2}{592}$.

23. $h \le \frac{F + 1280 M_2}{976}$.

24. $h \ge \frac{3I + 16 M_3}{7}$.

25. $h \le \frac{L + 7186 M_2}{5440}$.

26. $4M_3 \le \frac{2\Delta + 15 J}{128}$.

27. $4M_3 \le \frac{3J + 8h}{32}$.

28. $4M_3 \le \frac{T + 3h}{4}$.

29. $4M_3 \le \frac{K + 24h}{32}$.

30. $4M_3 \le \frac{F + 304h}{320}$.

31. $4M_3 \leq \frac{\Psi+176h}{192}$.

32. $4M_3 \leq \frac{L+1728h}{1792}$.

33. $\frac{1}{8}J \leq \frac{K+16h}{12}$.

34. $\frac{1}{8}J \leq \frac{\Delta+8T}{12}$.

35. $\frac{1}{8}J \leq \frac{\Psi+128h}{144}$.

36. $\frac{1}{8}J \leq \frac{F+224h}{240}$.

37. $\frac{1}{8}J \geq \frac{120T+256M_2}{312}$.

38. $\frac{1}{8}J \geq \frac{8T+256M_3}{72}$.

39. $\frac{1}{8}J \leq \frac{L+1280h}{1344}$.

40. $\frac{1}{8}K \leq \frac{6J+\Psi}{64}$.

41. $\frac{1}{8}K \leq \frac{12J+F}{112}$.

42. $\frac{1}{8}K \leq \frac{3\Psi+512M_3}{176}$.

43. $\frac{1}{8}K \leq \frac{3F+1024M_3}{304}$.

44. $\frac{1}{8}K \leq \frac{L+2048M_3}{576}$.

45. $\frac{1}{8}K \leq \frac{L+72J}{640}$.

46. $\frac{1}{16}\Psi \leq \frac{F+16T}{32}$.

47. $\frac{1}{16}\Psi \leq \frac{L+128T}{192}$.

48. $\frac{1}{16}F \leq \frac{L+4\Psi}{128}$.

Group 2

1. $80M_1 + 16M_3 \leq \Delta + 20h$.

2. $\Delta + 32h \leq 4T + 128M_1$.

3. $6\Delta + 256M_2 \leq 192I + 3K$.

4. $288M_1 + 224M_2 \leq 168I + 9J$.

5. $12M_1 + 20M_2 \leq 15I + 3T$.

6. $9J + 256M_2 \leq 192I + 72T$.

7. $10T + 32M_2 \leq 3J + 10h$.

8. $72I + 128T \leq 9K + 512M_3$.

9. $8I + 4J \leq K + 32h$.

10. $4\Delta + 8K \leq \Psi + 64h$.

11. $16I + 10K \leq \Psi + 10J$.

12. $26K + 192M_1 \leq 3\Psi + 832M_3$.

13. $32M_1 + 32M_3 \leq J + 8T$.

14. $4\Delta + 3\Psi \leq F + 6K$.

15. $48I + 8\Psi \leq 3F + 128T$.

16. $48J + \Psi \leq 2F + 1536M_3$.

17. $192I + 22F \leq 3L + 352T$.

18. $8K + 12F \leq L + 12\Psi$.

2.3 Refinement Inequalities

Based on the inequalities given in Groups 1 and 2 above, we can improve the inequalities (28). This improvement is given in the theorem below:

Theorem 2. *The following inequalities hold:*

$$\frac{1}{4}\Delta \leq I \leq \begin{cases} \frac{1}{36}\Delta + \frac{32}{9}M_1 \leq \frac{1}{20}\Delta + \frac{4}{3}M_2 - \frac{4}{5}M_3 \\ \frac{2}{3}I + \frac{4}{9}M_2 \end{cases} \leq 4M_1 \leq$$

$$\leq \frac{1}{22}\Delta + \frac{12}{11}M_2 \leq \frac{4}{3}M_2 \leq \frac{1}{48}\Delta + \frac{11}{12}h \leq h \leq \frac{1}{20}\Delta + \frac{16}{5}M_3 \leq$$

$$\leq 4M_3 \leq \frac{1}{4}T + \frac{3}{4}h \leq \frac{5}{13}T + \frac{32}{39}M_2 \leq \frac{1}{8}J \leq$$

$$\leq \begin{cases} \frac{1}{12}\Delta + \frac{2}{3}T \\ \frac{1}{8}K + h - \frac{1}{4}I \\ \frac{8}{3}I + T - \frac{32}{9}M_2 \end{cases} \leq T \leq$$

$$\leq \frac{9}{128}K + 4M_3 - \frac{9}{16}I \leq \frac{1}{8}K \leq \frac{1}{64}\Psi + h - \frac{1}{16}\Delta \leq$$

$$\leq \begin{cases} \frac{1}{64}\Psi + \frac{3}{32}J \\ \frac{3}{176}\Psi + \frac{32}{11}M_3 \end{cases} \leq$$

$$\leq \frac{1}{16}\Psi \leq \frac{3}{128}F + T - \frac{3}{8}I \leq \frac{1}{32}F + \frac{1}{2}T \leq \frac{1}{16}F. \qquad (39)$$

Proof. It is based on the following individual results.

1. $I \leq \frac{2}{3}I + \frac{4}{9}M_2$: It is true in view of (28).

2. $I \leq \frac{1}{36}\Delta + \frac{32}{9}M_1$: It follows in view of G1-1.

3. $\frac{1}{36}\Delta + \frac{32}{9}M_1 \leq \frac{1}{20}\Delta + \frac{4}{3}M_2 - \frac{4}{5}M_3$: In order to show $\Delta + 60M_1 - 36M_3 - 160M_1 \geq 0$, let us consider $\Delta + 60M_1 - 36M_3 - 160M_1 = bf_1(a/b)$, where $f_1(x) = \frac{1}{45}u_1(x)/(x+1)$, with

$$u_1(x) = 23x^2 + 20x^{3/2} + 42x + 20\sqrt{x} + 23 - $$
$$- 16\sqrt{2x+2}(\sqrt{x}+1)(x+1).$$

In order to apply Note 2, let us consider,

$$v_1(x) = \left(23x^2 + 20x^{3/2} + 42x + 20\sqrt{x} + 23\right)^2 - $$
$$- \left(16\sqrt{2x+2}(\sqrt{x}+1)(x+1)\right)^2.$$

After simplifications, we get

$$v_1(x) = (\sqrt{x} - 1)^6(17x - 2\sqrt{x} + 17)$$
$$= (\sqrt{x} - 1)^6\left[16(x+1) + (\sqrt{x} - 1)^2\right] \geq 0, \quad \forall x > 0.$$

4. $\frac{1}{20}\Delta + \frac{4}{3}M_2 - \frac{4}{5}M_3 \leq 4M_1$: In order to prove $240M_1 + 48M_3 - 3\Delta - 80M_2 \geq 0$, let us consider $240M_1 + 48M_3 - 3\Delta - 80M_2 = bf_2(a/b)$, where $f_2(x) = \frac{1}{60}u_2(x)/(x+1)$, with

$$u_1(x) = 28\sqrt{2x+2}\left(x^{3/2} + \sqrt{x} + x + 1\right)$$
$$- \left(39x^2 + 40x^{3/2} + 66x + 40\sqrt{x} + 39\right).$$

In order to apply Note 2, let us consider,

$$lv_1(x) = \left[28\sqrt{2x+2}\left(x^{3/2} + \sqrt{x} + x + 1\right)\right]^2 - $$
$$- \left(39x^2 + 40x^{3/2} + 66x + 40\sqrt{x} + 39\right)^2.$$

After simplifications, we get

$$v_1(x) = (\sqrt{x} - 1)^4 \left(47x^2 + 204x^{3/2} + 58x + 204\sqrt{x} + 47\right) \geq 0, \quad \forall x > 0.$$

5. $\frac{2}{3}I + \frac{4}{9}M_2 \leq 4M_1$: It is same as G1-9.

6. $4M_1 \leq \frac{1}{22}D + \frac{12}{11}M_2$: It is same as G1-6.

7. $\frac{1}{22}\Delta + \frac{12}{11}M_2 \leq \frac{4}{3}M_2$: True in view of (28).

8. $\frac{1}{4}I + \frac{3}{4}h \leq \frac{4}{3}M_2$: It is same as G1-15.

9. $\frac{4}{3}M_2 \leq \frac{1}{48}D + \frac{11}{12}h$: It is same as G1-10.

10. $\frac{1}{48}\Delta + \frac{11}{12}h \leq h$: It is true in view of (28).

11. $\frac{3}{7}I + \frac{16}{7}M_3 \leq h$: It is same as G1-22.

12. $h \leq \frac{1}{13}T + \frac{16}{13}M_2$: It is same as G1-18.

13. $h \leq \frac{1}{20}\Delta + \frac{16}{5}M_3$: It is same as G1-16.

14. $\frac{1}{20}\Delta + \frac{16}{5}M_3 \leq 4M_3$: It is true in view of (28).

15. $4M_3 \leq \frac{1}{4}T + \frac{3}{4}h$: It is same as G1-25.

16. $4M_3 \leq \frac{1}{9}T + \frac{32}{9}M_3$: It is true in view of (28).

17. $\frac{1}{4}T + \frac{3}{4}h \leq \frac{3}{32}J + \frac{1}{4}h$: After simplifications, we get $D_{Th}^{24} \leq 3D_{Jh}^{17}$. It is true in view of (34).

18. $\frac{5}{13}T + \frac{32}{39}M_2 \leq \frac{1}{8}J$: It is same as G1-33.

19. $\frac{1}{8}J \leq \frac{1}{12}\Delta + \frac{2}{3}T$: It is same as G1-30.

20. $\frac{1}{8}J \leq \frac{1}{8}K + h - \frac{1}{4}I$: It is same as G2-9.

21. $\frac{1}{8}J \leq \frac{8}{3}I + T - \frac{32}{9}M_2$: It is same as G2-6.

22. $\frac{1}{12}\Delta + \frac{2}{3}T \leq T$: It is true in view of (28).

23. $\frac{8}{3}I + T - \frac{32}{9}M_2 \leq T$: It is true in view of (28).

24. $T \leq \frac{3}{10}J + h - \frac{16}{5}M_2$: It is same as G2-7.

25. $T \leq \frac{9}{128}K + 4M_3 - \frac{9}{16}I$: It is same as G2-8.

26. $\frac{9}{128}K + 4M_3 - \frac{9}{16}I \leq \frac{1}{8}K$: After simplifications, we get $D_{M_3I}^{14} \leq \frac{7}{16}D_{KI}^{35}$. It is true in view of (34).

27. $\frac{1}{8}K \leq \frac{1}{80}\Psi + \frac{1}{8}J - \frac{1}{5}I$: It is same as G2-11.

28. $\frac{1}{8}K \leq \frac{1}{64}\Psi + h - \frac{1}{16}\Delta$: It is same as G2-10.

29. $\frac{1}{80}\Psi + \frac{1}{8}J - \frac{1}{5}I \leq \frac{1}{64}\Psi + \frac{3}{32}J$: After simplifications, we get $D_{JI}^{20} = D_{TJ}^{22} \leq \frac{1}{5}D_{\Psi I}^{44}$. It is true in view of (35).

30. $\frac{1}{64}\Psi + h - \frac{1}{16}\Delta \leq \frac{1}{64}\Psi + \frac{3}{32}J$: After simplifications, we get $D_{h\Delta}^{10} \leq 3D_{Jh}^{17}$. It is true in view of (34).

31. $\frac{1}{64}\Psi + \frac{3}{32}J \leq \frac{1}{16}\Psi$: It is true in view of (28).

32. $\frac{3}{176}\Psi + \frac{32}{11}M_3 \leq \frac{1}{16}\Psi$: It is true in view of (28).

33. $\frac{1}{16}\Psi \leq \frac{1}{48}F + \frac{1}{8}K - \frac{1}{12}D$: After simplifications, we get $D_{\Psi K}^{37} \leq 3D_{FA}^{55}$. It is true in view of (34).

34. $\frac{3}{128}F + T - \frac{3}{8}I \leq \frac{1}{32}F + \frac{1}{2}T$: After simplifications, we get $\frac{1}{2}D_{TI} = D_{TJ}^{22} \leq \frac{1}{6}D_{FT}^{48}$. It is true in view of (34).

35. $\frac{1}{32}F + \frac{1}{2}T \leq \frac{1}{16}F$: It is true in view of (28).

Combining 35 parts proved above, we get the inequalities (39). This completes the proof of Theorem 2. $\qquad\square$

Remarks 3. Theorem 2 is based on the results given by Groups 1 and 2. We observe that not all the results in these two groups are used. Also, there are no expression with measure L. Theorem 2 can be improved considerably. This shall be dealt elsewhere.

3. MEANS AND INEQUALITIES

This section deals with different kinds of means written in an inequality. Initially, we consider *seven means* already known in the literature (Eves, 2003), and establish inequalities arising due to *positive differences*.

3.1 Seven Means

Let $a, b > 0$ be two positive numbers. Eves (2003) studied the geometrical interpretations of the following seven means:

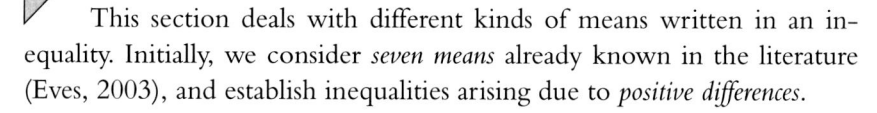

- **Arithmetic mean:** $A(a, b) = \frac{a+b}{2}$.
- **Geometric mean:** $G(a, b) = \sqrt{ab}$.
- **Harmonic mean:** $H(a, b) = \frac{2ab}{a+b}$.
- **Heronian mean:** $N(a, b) = \frac{a+\sqrt{ab}+b}{3}$.
- **Contra-harmonic mean:** $C(a, b) = \frac{a^2+b^2}{a+b}$.
- **Root-mean-square:** $S(a, b) = \sqrt{\frac{a^2+b^2}{2}}$.
- **Centroidal mean:** $R(a, b) = \frac{2(a^2+ab+b^2)}{3(a+b)}$.

Eves (2003) gave geometrically interpreted above *seven means* and proved the following inequalities among them:

$$H \leq G \leq N \leq A \leq R \leq S \leq C. \qquad (40)$$

The inequalities appearing in (40) admits 21 nonnegative differences. These differences satisfies some obvious *pyramid-type inequalities* given by

$$D_{GH}.$$
$$D_{NG} \leq D_{NH}.$$
$$D_{AN} \leq D_{AG} \leq D_{AH}.$$
$$D_{RA} \leq D_{RN} \leq D_{RG} \leq D_{RH}.$$
$$D_{SR} \leq D_{SA} \leq D_{SN} \leq D_{SG} \leq D_{SH}.$$
$$D_{CS} \leq D_{CR} \leq D_{CA} \leq D_{CN} \leq D_{CG} \leq D_{CH}.$$

The differences are understood as, $D_{GH} = G - H$, $D_{NG} = N - G$, $D_{CS} = C - S$, etc. Some of these differences are equal by multiplicative constants. See below:

1. $3D_{CR} = 2D_{AH} = 2D_{CA} = D_{CH} = 6D_{RA} = \frac{3}{2}D_{RH} = \frac{(a-b)^2}{(a+b)} := \Delta.$
2. $3D_{AN} = D_{AG} = 3D_{NG} = \frac{1}{2}(\sqrt{a} - \sqrt{b})^2 := h.$
3. $D_{CG} = D_{RN}.$

The measures Δ and h appearing in 1. and 2. are the same as given in (6) and (7) respectively for the probability distributions, known as *triangular discrimination* and *Hellinger's distance* respectively. From now onward, we shall work with *positive real numbers, instead of probability distributions*.

Lemmas 1 and 2 are rewritten below in terms of positive real numbers, and shall be used frequently.

Lemma 3. *Let $f : I \subset \mathbb{R}_+ \to \mathbb{R}$ be a convex and differentiable function satisfying $f(1) = 0$. Consider a function*

$$\phi_f(a, b) = af\left(\frac{b}{a}\right), \quad a, b > 0, \tag{41}$$

then the function $\phi_f(a, b)$ is convex in $\mathbb{R}_+^2$. Additionally, if $f'(1) = 0$, then the following inequality holds:

$$0 \leq \phi_f(a, b) \leq \left(\frac{b - a}{a}\right)\phi_{f'}(a, b). \tag{42}$$

Lemma 4. *Let $f_1, f_2 : I \subset \mathbb{R}_+ \to \mathbb{R}$ be two functions satisfying the assumptions:*
1. *$f_1(1) = f_1'(1) = 0$, $f_2(1) = f_2'(1) = 0$;*
2. *f_1 and f_2 are twice differentiable in $\mathbb{R}_+$;*
3. *there exists the real constants α, β such that $0 \leq \alpha < \beta$ and*

$$\alpha \leq \frac{f_1''(x)}{f_2''(x)} \leq \beta, \quad f_2''(x) > 0, \quad \forall x > 0, \tag{43}$$

then we have the inequalities:

$$\alpha \phi_{f_2}(a, b) \leq \phi_{f_1}(a, b) \leq \beta \phi_{f_2}(a, b), \tag{44}$$

for all $a, b \in (0, \infty)$, *where the function* $\phi_{(.)}(a, b)$ *is as defined in Lemma 3.*

Remarks 4. (i) Lemmas 3 and 4 are extensions of Lemmas 1 and 2 respectively. Here we use positive real numbers instead of probability distributions. For proof refer to Taneja (2004).

(ii) Applying Lemma 3, we can easily prove that mean differences are convex in pair of real numbers a and b, except the mean differences D_{SR}, D_{NH} and D_{GH}. In order to apply Lemma 4, convexity is not necessary, existence of second derivative is sufficient. The author (Taneja, 2013c) worked without these three difference of means. Here these are under study.

(iii) The idea of Note 1 is to apply Lemma 4 to find the values of β. Note 2 helps to prove the nonnegativity of the expressions with squareroots. These two notes shall be used frequently without any specifications.

For simplicity, let us consider, $\Phi_{(.)}(a, b) = b f_{(.)}(b/a)$, where $\Phi :=$ H, G, N, A, R, S and C, then (40) can be rewritten as

$$f_H(x) \leq f_G(x) \leq f_N(x) \leq f_A(x) \leq f_R(x) \leq f_S(x) \leq f_C(x), \tag{45}$$

i.e.,

$$\frac{2x}{1+x} \leq \sqrt{x} \leq \frac{x + \sqrt{x} + 1}{3} \leq \frac{x+1}{2} \leq \frac{2(x^2 + x + 1)}{3(x+1)} \leq \sqrt{\frac{x^2 + 1}{2}} \leq$$
$$\leq \frac{x^2 + 1}{x + 1}, \quad \forall x > 0.$$

The first and second order derivatives of the functions given in (45) are as follows:

1. $f'_H(x) = \frac{2}{(x+1)^2}, f''_H(x) = -\frac{4}{(x+1)^3}$ with $f''_H(1) = -\frac{1}{2}$.
2. $f'_G(x) = \frac{1}{2\sqrt{x}}, f''_G(x) = -\frac{1}{4x^{3/2}}$ with $f''_G(1) = -\frac{1}{4}$.
3. $f'_N(x) = \frac{2\sqrt{x}+1}{6\sqrt{x}}, f''_N(x) = -\frac{1}{12x^{3/2}}$ with $f''_N(x) = -\frac{1}{12}$.
4. $f'_A(x) = \frac{1}{2}, f''_A(x) = 0$ with $f''_A(x) = 0$.
5. $f'_R(x) = \frac{2}{3}\frac{x(x+2)}{(x+1)^2}, f''_R(x) = \frac{4}{3(x+1)^3}$ with $f''_R(1) = \frac{1}{6}$.
6. $f'_S(x) = \frac{x}{\sqrt{2x^2+2}}, f''_S(x) = \frac{4}{(2x^2+2)^{3/2}}$ with $f''_S(1) = \frac{1}{4}$.
7. $f'_C(x) = \frac{x^2+2x-1}{(x+1)^2}, f''_C(x) = \frac{4}{(x+1)^3}$ and $f''_C(1) = \frac{1}{2}$.

Remarks 5. From the values of second order derivatives we observe that not all the measures are convex. According to Remarks 1, the nonnegativity of division of second order derivatives at $x = 1$ is sufficient to apply Lemma 4.

The inequalities (40) admit 21 nonnegative differences. In view of equalities given in 1., 2. and 3., we are left with only 11 measures. These 11 nonnegative differences are related with each other by the inequalities given in theorem below.

Theorem 3. *The following inequalities hold:*

$$
\left\{ \begin{array}{c} 3D_{SR} \\ D_{GH} \end{array} \right\} \leq \left\{ \begin{array}{c} D_{SA} \leq \left\{ \begin{array}{c} \frac{3}{4}D_{SN} \\ \frac{1}{3}D_{SH} \leq \frac{1}{2}D_{AH} \end{array} \right. \\ \frac{3}{5}D_{NH} \leq \frac{1}{2}D_{AH} \end{array} \right\} \leq
$$

$$
\leq \left\{ \begin{array}{c} \frac{3}{7}D_{CN} \leq \left\{ \begin{array}{c} D_{CS} \\ \frac{1}{3}D_{CG} \leq \frac{3}{5}D_{RG} \end{array} \right. \\ \frac{1}{2}D_{SG} \leq \frac{3}{5}D_{RG} \end{array} \right\} \leq 3D_{AN}. \qquad (46)
$$

Proof. Based on the values of second order derivatives of the means given above, we can easily calculate the values of second order derivatives of 11 mean differences appearing in (46). These values are helpful in applying Lemma 4 and Note 1. Each part of (46) is proved separately. See the results below:

1. For $D_{SR} \leq \frac{1}{5}D_{NH}$: In this case, we have

$$
\beta_1 = \frac{f_{SR}''(1)}{f_{NH}''(1)} = \frac{f_S''(1) - f_R''(1)}{f_N''(1) - f_H''(1)} = \frac{\frac{1}{4} - \frac{1}{6}}{-\frac{1}{12} + \frac{1}{2}} = \frac{1}{5}.
$$

In order to prove, $\frac{1}{5}D_{NH} - D_{SR} \geq 0$, let us write, $\frac{1}{5}D_{NH} - D_{SR} = bg_1(a/b)$, where

$$
g_1(x) = \frac{1}{5}f_{NH}(x) - f_{SR}(x) = \frac{u_1(x)}{30(x+1)}
$$

with

$$
u_1(x) = 22x^2 + 12x + 2x^{3/2} + 2\sqrt{x} + 22 - 15(x+1)\sqrt{2x^2 + 2}.
$$

Let us consider

$$
v_1(x) = \left(22x^2 + 12x + 2x^{3/2} + 2\sqrt{x} + 22\right)^2 - \left[15(x+1)\sqrt{2x^2 + 2}\right]^2
$$
$$
= (\sqrt{x} - 1)^4\left(34x^2 + 224x^{3/2} + 324x + 224\sqrt{x} + 34\right).
$$

In view of Note 2, $v_1(x) \geq 0$ implies $u_1(x) \geq 0$, $\forall x > 0$, hence proving the required result.

2. For $D_{GH} \leq \frac{3}{5}D_{NH}$: In this case, $\beta_2 = f''_{GH}(1)/f''_{NH}(1) = \frac{3}{5}$. For proving $\frac{3}{5}D_{NH} - D_{GH} \geq 0$, let us consider $\frac{3}{5}D_{NH} - D_{GH} = bg_2(a/b)$, where

$$g_2(x) = \frac{3}{5}f_{NH}(x) - f_{GH}(x)$$
$$= \frac{x^2 + 6x - 4x^{3/2} - 4\sqrt{x} + 1}{5(x+1)} = \frac{(\sqrt{x} - 1)^4}{5(x+1)} \geq 0, \quad \forall x > 0.$$

Since $g_2(x) \geq 0$, $\forall x > 0$, hence proving the required result.

3. For $D_{NH} \leq \frac{5}{6}D_{AH}$: In this case, $\beta_3 = f''_{NH}(1)/f''_{AH}(1) = \frac{5}{6}$. For proving $\frac{5}{6}D_{AH} - D_{NH} \geq 0$, let us consider $\frac{5}{6}D_{AH} - D_{NH} = bg_3(a/b)$, where

$$g_3(x) = \frac{5}{6}f_{AH}(x) - f_{NH}(x)$$
$$= \frac{x^2 + 6x - 4x^{3/2} - 4\sqrt{x} + 1}{12(x+1)} = \frac{(\sqrt{x} - 1)^4}{12(x+1)} \geq 0, \quad \forall x > 0.$$

Since $g_3(x) \geq 0$, $\forall x > 0$, hence proving the required result.

4. For $D_{GH} \leq D_{SA}$: In this case, $\beta_4 = f''_{GH}(1)/f''_{SA}(1) = 1$. For proving $D_{SA} - D_{GH} \geq 0$, let us consider $D_{SA} - D_{GH} = bg_4(a/b)$, where

$$g_4(x) = f_{SA}(x) - f_{GH}(x) = \frac{u_4(x)}{2(x+1)}$$

with

$$u_4(x) = (x+1)\sqrt{2x^2 + 2} - \left(x^2 + 2x^{3/2} - 2x + 2\sqrt{x} + 1\right).$$

Let us consider

$$v_4(x) = \left[(x+1)\sqrt{2x^2 + 2}\right]^2 - \left(x^2 + 2x^{3/2} - 2x + 2\sqrt{x} + 1\right)^2$$
$$= (\sqrt{x} + 1)^2(\sqrt{x} - 1)^6 \geq 0, \quad \forall x > 0.$$

In view of Note 2, $v_4(x) \geq 0$ implies $u_4(x) \geq 0$, $\forall x > 0$, hence proving the required result.

5. For $D_{SR} \leq \frac{1}{3}D_{SA}$: In this case, $\beta_5 = f''_{SR}(1)/f''_{SA}(1) = \frac{1}{3}$. For proving $\frac{1}{3}D_{SA} - D_{SR} \geq 0$, let us consider $\frac{1}{3}D_{SA} - D_{SR} = bg_5(a/b)$, where

$$g_5(x) = \frac{1}{3}f_{SA}(x) - f_{SR}(x) = \frac{u_5(x)}{6(x+1)}$$

with

$$u_5(x) = 2(x+1)\sqrt{2x^2+2} - (3x^2+2x+3).$$

Let us consider

$$v_5(x) = \left[2(x+1)\sqrt{2x^2+2}\right]^2 - \left(3x^2+2x+3\right)^2 = (x-1)^4 \geq 0, \quad \forall x > 0.$$

In view of Note 2, $v_5(x) \geq 0$ implies $u_5(x) \geq 0$, $\forall x > 0$, hence proving the required result.

6. For $D_{SA} \leq \frac{3}{4}D_{SN}$: In this case, $\beta_6 = f''_{SA}(1)/f''_{SN}(1) = \frac{3}{4}$. For proving $\frac{3}{4}D_{SN} - D_{SA} \geq 0$, let us consider $\frac{3}{4}D_{SN} - D_{SA} = bg_6(a/b)$, where

$$g_6(x) = \frac{3}{4}f_{SN}(x) - f_{SA}(x) = \frac{-\sqrt{2x^2+2} + 2x - 2\sqrt{x} + 2}{8}.$$

Let us consider

$$v_6(x) = (2x - 2\sqrt{x} + 2)^2 - \left(\sqrt{2x^2+2}\right)^2 = 2(\sqrt{x} - 1)^2 \geq 0, \quad \forall x > 0.$$

In view of Note 2, $v_6(x) \geq 0$ implies $u_6(x) \geq 0$, $\forall x > 0$, hence proving the required result.

7. For $D_{SA} \leq \frac{1}{3}D_{SH}$: In this case, $\beta_7 = f''_{SA}(1)/f''_{SH}(1) = \frac{1}{3}$. To show $\frac{1}{3}D_{SH} - D_{SA} \geq 0$, let us consider, $\frac{1}{3}D_{SH} - D_{SA} = bg_7(a/b)$, where

$$g_7(x) = \frac{1}{3}f^{11}_{SH}(x) - f^8_{SA}(x) = \frac{u_7(x)}{6(x+1)}$$

with

$$u_7(x) = 3x^2 + 2x + 3 - 2(x+1)\sqrt{2x^2+2}$$

Let us consider

$$v_7(x) = \left(3x^2+2x+3\right)^2 - \left[2(x+1)\sqrt{2x^2+2}\right]^2$$
$$= (x-1)^4 \geq 0, \quad \forall x > 0.$$

In view of Note 2, $v_7(x) \geq 0$ implies $u_7(x) \geq 0$, $\forall x > 0$, hence proving the required result.

8. For $D_{SH} \leq \frac{3}{2}D_{AH}$: In this case, $\beta_8 = f''_{SH}(1)/f''_{AH}(1) = \frac{3}{2}$. To show $\frac{3}{2}D_{AH} - D_{SH} \geq 0$, let us consider $\frac{3}{2}D_{AH} - D_{SH} = bg_8(a/b)$, where

$$g_8(x) = \frac{3}{2}f_{AH}(x) - f_{SH}(x) = \frac{u_8(x)}{4(x+1)}$$

with

$$u_8(x) = 3x^2 + 2x + 3 - 2(x+1)\sqrt{2x^2+2}$$

Let us consider

$$v_8(x) = \left(3x^2 + 2x + 3\right)^2 - \left[2(x+1)\sqrt{2x^2+2}\right]^2$$
$$= (x-1)^4 \geq 0, \quad \forall x > 0.$$

In view of Note 2, $v_8(x) \geq 0$ implies $u_8(x) \geq 0$, $\forall x > 0$, hence proving the required result.

9. For $D_{SN} \leq \frac{4}{7}D_{CN}$: In this case, $\beta_9 = f''_{SN}(1)/f''_{CN}(1) = \frac{4}{7}$. To show $\frac{4}{7}D_{CN} - D_{SN} \geq 0$, let us consider $\frac{4}{7}D_{CN} - D_{SN} = bg_9(a/b)$, where

$$g_9(x) = \frac{4}{7}f_{CN}(x) - f_{SN}(x) = \frac{u_9(x)}{14(x+1)}$$

with

$$u_9(x) = 10x^2 + 2x^{3/2} + 4x + 2\sqrt{x} + 10 - 7(x+1)\sqrt{2x^2+2}.$$

Let us consider

$$v_9(x) = \left(10x^2 + 2x^{3/2} + 4x + 2\sqrt{x} + 10\right)^2 - \left[7(x+1)\sqrt{2x^2+2}\right]^2$$
$$= (\sqrt{x}-1)^4\left(2x^2 + 48x^{3/2} + 68x + 48\sqrt{x} + 2\right) \geq 0, \quad \forall x > 0.$$

In view of Note 2, $v_9(x) \geq 0$ implies $u_9(x) \geq 0$, $\forall x > 0$, hence proving the required result.

10. For $D_{SN} \leq \frac{2}{3}D_{SG}$: In this case, $\beta_{10} = f''_{SN}(1)/f''_{SG}(1) = \frac{2}{3}$. To show $\frac{2}{3}D_{SG} - D_{SN} \geq 0$, let us consider $\frac{2}{3}D_{SG} - D_{SN} = bg_{10}(a/b)$, where

$$g_{10}(x) = \frac{2}{3}f_{SG}(x) - f_{SN}(x) = 2x - 2\sqrt{x} + 2 - \sqrt{2x^2+2}.$$

Let us consider

$$v_{10}(x) = (2x - 2\sqrt{x} + 2)^2 - \left(\sqrt{2x^2+2}\right)^2 = 2(\sqrt{x}-1)^4 \geq 0, \quad \forall x > 0.$$

In view of Note 2, $v_{10}(x) \geq 0$ implies $u_{10}(x) \geq 0$, $\forall x > 0$, hence proving the required result.

11. For $D_{AH} \leq \frac{6}{7}D_{CN}$: In this case, $\beta_{11} = f''_{AH}(1)/f''_{CN}(1) = \frac{6}{7}$. To show $\frac{6}{7}D_{CN} - D_{AH} \geq 0$, let us consider $\frac{6}{7}D_{CN} - D_{AH} = bg_{11}(a/b)$, where

$$g_{11}(x) = \frac{6}{7}f_{CN}(x) - f_{AH}(x)$$
$$= \frac{x^2 + 6x - 4x^{3/2} - 4\sqrt{x} + 1}{14(x+1)} = \frac{(\sqrt{x}-1)^4}{14(x+1)} \geq 0, \quad \forall x > 0.$$

In view of Note 2, $v_{11}(x) \geq 0$ implies $u_{11}(x) \geq 0$, $\forall x > 0$, hence proving the required result.

12. For $D_{AH} \leq D_{SG}$: In this case, $\beta_{12} = f''_{AH}(1)/f''_{SG}(1) = 1$. For proving $D_{SG} - D_{AH} \geq 0$, let us consider $D_{SG} - D_{AH} = bg_{12}(a/b)$, where

$$g_{12}(x) = f_{SG}(x) - f_{AH}(x) = \frac{u_{12}(x)}{3(x+1)}$$

with

$$u_{12}(x) = (x+1)\sqrt{2x^2 + 2} - \left(x^2 + 2x^{3/2} - 2x + 2\sqrt{x} + 1\right).$$

Let us consider

$$v_{12}(x) = \left[(x+1)\sqrt{2x^2 + 2}\right]^2 - \left(x^2 + 2x^{3/2} - 2x + 2\sqrt{x} + 1\right)^2$$
$$= (\sqrt{x}-1)^6(\sqrt{x}+1)^4 \geq 0, \quad \forall x > 0.$$

In view of Note 2, $v_{12}(x) \geq 0$ implies $u_{12}(x) \geq 0$, $\forall x > 0$, hence proving the required result.

13. For $D_{SG} \leq \frac{6}{5}D_{RG}$: We have $\beta_{13} = f''_{SG}(1)/f''_{RG}(1) = \frac{6}{5}$. For proving $\frac{6}{5}D_{RG} - D_{SG} \geq 0$, let us consider $\frac{6}{5}D_{RG} - D_{SG} = bg_{13}(a/b)$, where

$$g_{13}(x) = \frac{6}{5}f_{RG}(x) - f_{SG}(x) = \frac{u_{13}(x)}{15(x+1)}$$

with

$$u_{13}(x) = 9(x+1)\sqrt{2x^2 + 2} - \left(10x^2 + 3x^{3/2} + 10x + 3\sqrt{x} + 10\right).$$

Let us consider

$$v_{13}(x) = \left[9(x+1)\sqrt{2x^2 + 2}\right]^2 - \left(10x^2 + 3x^{3/2} + 10x + 3\sqrt{x} + 10\right)^2$$
$$= (\sqrt{x}-1)^2\left(62x^3 + 64x^{5/2} + 181x^2 + \right.$$
$$\left. + 178x^{3/2} + 181x + 64\sqrt{x} + 62\right) \geq 0, \quad \forall x > 0.$$

In view of Note 2, $v_{13}(x) \geq 0$ implies $u_{13}(x) \geq 0$, $\forall x > 0$, hence proving the required result.

14. For $D_{CN} \leq \frac{7}{3}D_{CS}$: In this case, $\beta_{14} = f''_{CN}(1)/f''_{CS}(1) = \frac{7}{3}$. For proving $\frac{7}{3}D_{CS} - D_{CN} \geq 0$, let us consider $\frac{7}{3}D_{CS} - D_{CN} = bg_{14}(a/b)$, where

$$g_{14}(x) = \frac{7}{3}f_{CS}(x) - f_{CN}(x) = \frac{u_{14}(x)}{6(x+1)}$$

with

$$u_{14}(x) = 10x^2 + 2x^{3/2} + 4x + 2\sqrt{x} + 10 - 7(x+1)\sqrt{2x^2+2}.$$

Let us consider

$$\begin{aligned}
v_{14}(x) &= \left(10x^2 + 2x^{3/2} + 4x + 2\sqrt{x} + 10\right)^2 - \left[7(x+1)\sqrt{2x^2+2}\right]^2 \\
&= (\sqrt{x}-1)^4\left(2x^2 + 48x^{3/2} + 68x + 48\sqrt{x} + 2\right) \geq 0, \quad \forall x > 0.
\end{aligned}$$

In view of Note 2, $v_{14}(x) \geq 0$ implies $u_{14}(x) \geq 0$, $\forall x > 0$, hence proving the required result.

15. For $D_{CS} \leq 3D_{AN}$: In this case, $\beta_{15} = f''_{CS}(1)/f''_{AN}(1) = 3$. To show $3D_{AN} - D_{CS} \geq 0$, let us consider $3D_{AN} - D_{CS} = bg_{15}(a/b)$, where

$$g_{15}(x) = 3f_{AN}(x) - f_{CS}(x) = \frac{u_{15}(x)}{2(x+1)}$$

with

$$u_{15}(x) = (x+1)\sqrt{2x^2+2} - \left(x^2 + 2x^{3/2} - 2x + 2\sqrt{x} + 1\right).$$

Let us consider

$$\begin{aligned}
v_{15}(x) &= \left[(x+1)\sqrt{2x^2+2}\right]^2 - \left(x^2 + 2x^{3/2} - 2x + 2\sqrt{x} + 1\right)^2 \\
&= (\sqrt{x}-1)^6(\sqrt{x}+1)^2 \geq 0, \quad \forall x > 0.
\end{aligned}$$

In view of Note 2, $v_{15}(x) \geq 0$ implies $u_{15}(x) \geq 0$, $\forall x > 0$, hence proving the required result.

16. For $D_{CN} \leq \frac{7}{9}D_{CG}$: In this case, $\beta_{16} = f''_{CN}(1)/f''_{CG}(1) = \frac{7}{9}$. To show $\frac{7}{9}D_{CG} - D_{CN} \geq 0$, let us consider $\frac{7}{9}D_{CG} - D_{CN} = bg_{16}(a/b)$, where

$$\begin{aligned}
g_{16}(x) &= \frac{7}{9}f_{CG}(x) - f_{CN}(x) \\
&= \frac{x^2 - 4x^{3/2} + 6x - 4\sqrt{x} + 1}{9(x+1)} = \frac{(\sqrt{x}-1)^4}{9(x+1)} \geq 0, \quad \forall x > 0.
\end{aligned}$$

In view of Note 2, $v_{16}(x) \geq 0$ implies $u_{16}(x) \geq 0$, $\forall x > 0$, hence proving the required result.

17. For $D_{CG} \leq \frac{9}{5}D_{RG}$: In this case, $\beta_{17} = f''_{CG}(1)/f''_{RG}(1) = \frac{9}{5}$. To show $\frac{9}{5}D_{RG} - D_{CG} \geq 0$, let us consider $\frac{9}{5}D_{RG} - D_{CG} = bg_{17}(a/b)$, where

$$g_{17}(x) = \frac{9}{5}f_{RG}(x) - f_{CG}(x)$$
$$= \frac{x^2 - 4x^{3/2} + 6x - 4\sqrt{x} + 1}{5(x+1)} = \frac{(\sqrt{x} - 1)^4}{5(x+1)} \geq 0, \quad \forall x > 0.$$

Since $g_{17}(x) \geq 0$, $\forall x > 0$, hence proving the required result.

18. For $D_{RG} \leq 5D_{AN}$: In this case, $\beta_{18} = f''_{RG}(1)/f''_{AN}(1) = 5$. To show $5D_{AN} - D_{RG} \geq 0$, let us consider $5D_{AN} - D_{RG} = bg_{18}(a/b)$, where

$$g_{18}(x) = 5f_{AN}(x) - f_{RG}(x)$$
$$= \frac{x^2 - 4x^{3/2} + 6x - 4\sqrt{x} + 1}{6(x+1)} = \frac{(\sqrt{x} - 1)^4}{6(x+1)} \geq 0, \quad \forall x > 0.$$

Since $g_{17}(x) \geq 0$, $\forall x > 0$, hence proving the required result.

Combining Results 1–18, we get the proof of Theorem 2. □

Remarks 6. Except the measures D_{SR}, D_{NH} and D_{GH}, the other results are studied in Taneja (2013c). Theorem 3 is an improvement over the work of the author (Taneja, 2006b, 2013c).

3.2 Refinement Inequalities

As a consequence of above 18 results, we have the following individual relations.

Group 3

1. $G \leq \frac{3N+2H}{5}$.
2. $\frac{S+2G}{3} \leq N$.
3. $\frac{2C+7G}{9} \leq N$.
4. $N \leq \frac{5A+H}{6}$.
5. $\frac{S+3N}{4} \leq A$.
6. $\frac{2S+H}{3} \leq A$.
7. $\frac{G+5R}{6} \leq S$.
8. $S \leq \frac{4C+3N}{7}$.
9. $\frac{4G+5C}{9} \leq R$.
10. $\frac{2S+A}{3} \leq R$.
11. $H + 5S \leq N + 5R$.
12. $A + G \leq S + H$.
13. $6N + 7A \leq 6C + 7H$.
14. $3N + C \leq 3A + S$.
15. $5N + R \leq 5A + G$.

Instead of 18 relations, we have 15. The reason is that in some cases they are same. The theorem below gives a refinement inequality improving the result (40).

Theorem 4. *The following inequalities hold:*

$$H \leq \left\{ \begin{matrix} G \\ N + 5R - 5S \end{matrix} \right\} \leq \left\{ \begin{matrix} \frac{3}{5}N + \frac{2}{5}H \\ S + II - A \leq \left\{ \begin{matrix} \frac{1}{3}S + \frac{2}{3}G \\ \frac{2}{9}C + \frac{7}{9}G \end{matrix} \right\} \end{matrix} \right\} \leq$$

$$\leq N \leq \left\{ \begin{matrix} A + \frac{1}{3}S - \frac{1}{3}C \\ A + \frac{1}{5}G - \frac{1}{5}R \end{matrix} \right\} \leq$$

$$\leq \frac{5}{6}A + \frac{1}{6}H \leq \left\{ \begin{matrix} \frac{1}{4}S + \frac{3}{4}N \\ \frac{2}{3}S + \frac{1}{3}H \end{matrix} \right\} \leq A \leq$$

$$\leq \left\{ \begin{matrix} \frac{1}{6}G + \frac{5}{6}R \leq \frac{4}{9}G + \frac{5}{9}C \\ \frac{1}{6}G + \frac{5}{6}R \\ \frac{6}{7}C + H - \frac{6}{7}N \\ S + H - G \end{matrix} \right\} \leq \frac{1}{3}A + \frac{2}{3}S \right\} \leq R \right\} \leq$$

$$\leq \left\{ \begin{matrix} S \leq \frac{1}{5}N + R - \frac{1}{5}H \\ 5A + G - 5N \end{matrix} \right\} \leq C \leq \frac{4}{7}C + \frac{3}{7}N \leq 3A + S - 3N. \qquad (47)$$

Proof. In view of 15 inequalities given in Groups 3, we don't require to prove all the parts. Some of them follow immediately. Below are proofs of other parts. Proof of these parts is much more simple than given in Theorem 3.

1. For $H + 5S \leq N + 5R$: Let us write, $N + 5R - H - 5S = bg_1(a/b)$, where $g_1(x) = \frac{1}{6}u_1(x)/(x+1)$ with

$$u_1(x) = 2\left(11x^2 + x^{3/2} + 6x + \sqrt{x} + 11\right) - 15(x+1)\sqrt{2x^2 + 2}.$$

Let us consider

$$v_1(x) = 4\left(11x^2 + x^{3/2} + 6x + \sqrt{x} + 11\right)^2 - \left[15(x+1)\sqrt{2x^2 + 2}\right]^2$$
$$= 2(\sqrt{x} - 1)^4\left(17x^2 + 112x^{3/2} + 162x + 112\sqrt{x} + 17\right) \geq 0, \quad \forall x > 0.$$

In view of Note 2, $v_1(x) \geq 0$ implies $u_1(x) \geq 0$, $\forall x > 0$, hence proving the required result.

2. For $2N + 25R \leq 2H + 25S$: Let us write, $2H + 25S - 2N - 25R = bg_2(a/b)$, where $g_2(x) = \frac{1}{30}u_2(x)/(x+1)$ with

$$u_2(x) = 75(x+1)\sqrt{2x^2 + 2} - 4\left(26x^2 + 21x + x^{3/2} + \sqrt{x} + 26\right)$$

Let us consider

$$v_2(x) = \left[75(x+1)\sqrt{2x^2+2}\right]^2 - 16\left(26x^2 + 21x + x^{3/2} + \sqrt{x} + 26\right)^2$$
$$= 2(\sqrt{x}-1)^4\left(217x^3 + 18x^{5/2} + 2325x^2 + \right.$$
$$\left. + 3880x^{3/2} + 2325x + 18\sqrt{x} + 217\right) \geq 0, \quad \forall x > 0.$$

In view of Note 2, $v_2(x) \geq 0$ implies $u_2(x) \geq 0$, $\forall x > 0$, hence proving the required result.

3. For $A + N + 5R \leq H + 6S$: Let us write, $6S + H - A - N - 5R = bg_3(a/b)$, where $g_3(x) = \frac{1}{6}u_3(x)/(x+1)$ with

$$u_3(x) = 18(x+1)\sqrt{2x^2+2} - \left(25x^2 + 2x^{3/2} + 18x + 2\sqrt{x} + 25\right)$$

Let us consider

$$v_3(x) = \left(18(x+1)\sqrt{2x^2+2}\right)^2 - \left(25x^2 + 2x^{3/2} + 18x + 2\sqrt{x} + 25\right)^2$$
$$= (\sqrt{x}-1)^2\left(23x^3 - 54x^{5/2} + 261x^2 + \right.$$
$$\left. + 404x^{3/2} + 261x - 54\sqrt{x} + 23\right) \geq 0, \quad \forall x > 0.$$

In view of Note 2, $v_3(x) \geq 0$ implies $u_3(x) \geq 0$, $\forall x > 0$, hence proving the required result.

4. For $2S + 3H \leq 2G + 3A$: Let us write, $2G + 3A - 2S - 3H = bg_4(a/b)$, where $g_4(x) = \frac{1}{6}u_4(x)/(x+1)$ with

$$u_4(x) = 3x^2 + 4x^{3/2} - 6x + 4\sqrt{x} + 3 - 2(x+1)\sqrt{2x^2+2}$$

Let us consider

$$v_4(x) = \left(3x^2 + 4x^{3/2} - 6x + 4\sqrt{x} + 3\right)^2 - \left(2(x+1)\sqrt{2x^2+2}\right)^2$$
$$= (x-1)^2\left(x^2 + 24x^{3/2} - 34x + 24\sqrt{x} + 1\right) \geq 0, \quad \forall x > 0.$$

In view of Note 2, $v_4(x) \geq 0$ implies $u_4(x) \geq 0$, $\forall x > 0$, hence proving the required result.

5. For $S + H \leq \frac{2C+7G+9A}{9}$: Let us write, $2C + 7G + 9A - 9S - 9H = bg_5(a/b)$, where $g_5(x) = \frac{1}{18}u_5(x)/(x+1)$ with

$$u_5(x) = 13x^2 + 14x^{3/2} - 18x + 14\sqrt{x} + 13 - 9(x+1)\sqrt{2x^2+2}$$

Let us consider

$$v_5(x) = \left(13x^2 + 14x^{3/2} - 18x + 14\sqrt{x} + 13\right)^2 - \left(9(x+1)\sqrt{2x^2+2}\right)^2$$
$$= (\sqrt{x} - 1)^2\left(7x^3 + 378x^{5/2} + 153x^2 - 212x^{3/2} + \right.$$
$$\left. + 153x + 378\sqrt{x} + 7\right) \geq 0, \quad \forall x > 0.$$

In view of Note 2, $v_5(x) \geq 0$ implies $u_5(x) \geq 0$, $\forall x > 0$, hence proving the required result.

6. For $A + 2S \leq H + 2C$: Let us write, $H + 2C - A - 2S = bg_6(a/b)$, where $g_6(x) = \frac{1}{12}u_6(x)/(x+1)$ with

$$u_6(x) = 3x^2 + 2x + 3 - 2(x+1)\sqrt{2x^2+2}$$

Let us consider

$$v_6(x) = \left(3x^2 + 2x + 3\right)^2 - \left(2(x+1)\sqrt{2x^2+2}\right)^2 = (x-1)^4 \geq 0, \quad \forall x > 0.$$

In view of Note 2, $v_6(x) \geq 0$ implies $u_6(x) \geq 0$, $\forall x > 0$, hence proving the required result.

7. For $5A + 6G \leq 5H + 6R$: Let us write, $5H + 6R - 5A - 6G = bg_7(a/b)$, where

$$g_7(x) = \frac{x^2 - 4x^{3/2} + 6x - 4\sqrt{x} + 1}{20(x+1)} = \frac{(\sqrt{x} - 1)^4}{20(x+1)} \geq 0, \quad \forall x > 0.$$

Since $g_7(x) \geq 0$, $\forall x > 0$, hence proving the required result.

8. For $10A + 2H \leq 9N + 3S$: Let us write, $3S + 9N - 10A - 2H = bg_8(a/b)$, where $g_8(x) = \frac{1}{24}u_8(x)/(x+1)$ with

$$u_8(x) = 3(x+1)\sqrt{2x^2+2} - 2\left(2x^2 - 3x^{3/2} + 8x - 3\sqrt{x} + 2\right)$$

Let us consider

$$v_8(x) = \left(3(x+1)\sqrt{2x^2+2}\right)^2 - 4\left(2x^2 - 3x^{3/2} + 8x - 3\sqrt{x} + 2\right)^2$$
$$= 2(\sqrt{x} - 1)^2\left(x^3 + 26x^{5/2} - 13x^2 + \right.$$
$$\left. + 68x^{3/2} - 13x + 26\sqrt{x} + 1\right) \geq 0, \quad \forall x > 0.$$

In view of Note 2, $v_8(x) \geq 0$ implies $u_8(x) \geq 0$, $\forall x > 0$, hence proving the required result.

9. For $5A \leq 4S + H$: Let us write, $4S + H - 5A = bg_9(a/b)$, where $g_9(x) = \frac{1}{12}u_9(x)/(x+1)$ with

$$u_9(x) = 4(x+1)\sqrt{2x^2 + 2} - \left(5x^2 + 6x + 5\right)$$

Let us consider

$$v_9(x) = \left(4(x+1)\sqrt{2x^2+2}\right)^2 - \left(5x^2 + 6x + 5\right)^2$$
$$= (x-1)^2\left(7x^2 + 18x + 7\right) \geq 0, \quad \forall x > 0.$$

In view of Note 2, $v_9(x) \geq 0$ implies $u_9(x) \geq 0$, $\forall x > 0$, hence proving the required result.

10. For $A \leq \frac{G+5R}{6}$: Let us write, $G + 5R - 6A = bg_{10}(a/b)$, where $g_{10}(x) = \frac{1}{12}u_{10}(x)/(x+1)$ with

$$u_{10}(x) = 4(x+1)\sqrt{2x^2 + 2} - \left(5x^2 + 6x + 5\right)$$

The rest part of proof is same as Part 9.

11. For $6N + 7A \leq 6C + 7H$: Let us write, $6C + 7H - 6N + 7A = bg_{11}(a/b)$, where

$$g_{11}(x) = \frac{x^2 - 4x^{3/2} + 6x - 4\sqrt{x} + 1}{14(x+1)} = \frac{(\sqrt{x} - 1)^4}{14(x+1)} \geq 0, \quad \forall x > 0.$$

Since $g_{11}(x) \geq 0$, $\forall x > 0$, hence proving the required result.

12. For $R \leq \frac{G+2C}{3}$: Let us write, $G + 2C - 3R = bg_{12}(a/b)$, where

$$g_{12}(x) = \frac{5(4x^{3/2} - 2x + \sqrt{x})}{18(x+1)} = \frac{\sqrt{x}(\sqrt{x} - 1)^2}{18(x+1)} \geq 0, \quad \forall x > 0.$$

Since $g_{12}(x) \geq 0$, $\forall x > 0$, hence proving the required result.

13. For $G + 5R \leq 4S + 2A$: Let us write $4S + 2A - G - 5R = bg_{13}(a/b)$, where $g_{13}(x) = \frac{1}{18}u_{13}(x)/(x+1)$ with

$$u_{13}(x) = 6(x+1)\sqrt{2x^2 + 2} - \left(7x^2 + 3x^{3/2} + 4x + 3\sqrt{x} + 7\right).$$

Let us consider

$$v_{13}(x) = \left(6(x+1)\sqrt{2x^2 + 2}\right)^2 - \left(7x^2 + 3x^{3/2} + 4x + 3\sqrt{x} + 7\right)^2$$
$$= (\sqrt{x} - 1)^2\left(23x^3 + 4x^{5/2} + 64x^2 + \right.$$
$$\left. + 58x^{3/2} + 64x + 4\sqrt{x} + 23\right) \geq 0, \quad \forall x > 0.$$

In view of Note 2, $v_{13}(x) \geq 0$ implies $u_{13}(x) \geq 0$, $\forall x > 0$, hence proving the required result.

14. For $18C + 21H \leq 14S + 7A + 18N$: Let us write $14S + 7A + 18N - 18C - 21H = bg_{14}(a/b)$, where $g_{14}(x) = \frac{1}{42}u_{14}(x)/(x+1)$ with

$$u_{14}(x) = 14(x+1)\sqrt{2x^2+2} - \left(17x^2 - 12x^{3/2} + 46x - 12\sqrt{x} + 17\right)$$

Let us consider

$$\begin{aligned} v_{14}(x) &= \left(14(x+1)\sqrt{2x^2+2}\right)^2 - \left(17x^2 - 12x^{3/2} + 46x - 12\sqrt{x} + 17\right)^2 \\ &= (\sqrt{x}-1)^2 \big(103x^3 + 614x^{5/2} + 201x^2 + \\ &\quad + 1300x^{3/2} + 201x + 614\sqrt{x} + 103\big) \geq 0, \quad \forall x > 0. \end{aligned}$$

In view of Note 2, $v_{14}(x) \geq 0$ implies $u_{14}(x) \geq 0$, $\forall x > 0$, hence proving the required result.

15. For $N + 5R \leq H + 5C$: Let us write, $H + 5C - N - 5R = bg_{15}(a/b)$, where

$$\begin{aligned} g_{15}(x) &= \frac{4x^2 - x^{3/2} - 6x - \sqrt{x} + 4}{15(x+1)} \\ &= \frac{(\sqrt{x}-1)^2(4x + 7\sqrt{x} + 4)}{15(x+1)} \geq 0, \quad \forall x > 0. \end{aligned}$$

16. For $G + 5A \leq C + 5N$: Let us write, $C + 5N - G - 5A = bg_{16}(a/b)$, where

$$\begin{aligned} g_{16}(x) &= \frac{x^2 + 4x^{3/2} - 10x + 4\sqrt{x} + 1}{6(x+1)} \\ &= \frac{(\sqrt{x}-1)^2(x + 6\sqrt{x} + 1)}{6(x+1)} \geq 0, \quad \forall x > 0. \end{aligned}$$

Since $g_{16}(x) \geq 0$, $\forall x > 0$, hence proving the required result.

17. For $6N + C \leq \frac{7(3A+S)}{4}$: Let us write, $21A + 7S - 24N - 4C = bg_{17}(a/b)$, where $g_{17}(x) = \frac{1}{14}u_{17}(x)/(x+1)$ with

$$u_{17}(x) = 7(x+1)\sqrt{2x^2+2} - \left(3x^2 + 16x^{3/2} - 10x + 16\sqrt{x} + 3\right)$$

Let us consider

$$\begin{aligned} v_{14}(x) &= \left(7(x+1)\sqrt{2x^2+2}\right)^2 - \left(3x^2 + 16x^{3/2} - 10x + 16\sqrt{x} + 3\right)^2 \\ &= (\sqrt{x}-1)^2 \big(89x^3 + 82x^{5/2} + 75x^2 + \\ &\quad + 292x^{3/2} + 75x + 82\sqrt{x} + 89\big) \geq 0, \quad \forall x > 0. \end{aligned}$$

In view of Note 2, $v_{14}(x) \geq 0$ implies $u_{14}(x) \geq 0$, $\forall x > 0$, hence proving the required result. $\square$

In continuation with 15 inequalities given in Group 3, below are more arising due to Theorem 4.

Group 4

1. $A \leq \frac{G+5R}{6}$.
2. $A \leq \frac{4S+H}{5}$.
3. $R \leq \frac{G+2C}{3}$.
4. $6N + C \leq \frac{7(3A+S)}{4}$.
5. $S + H \leq \frac{2C+7G+9A}{9}$.
6. $6C + 7H \leq \frac{14S+7A+18N}{3}$.
7. $H + 5S \leq N + 5R$.
8. $2N + 25R \leq 2H + 25S$.
9. $2S + 3H \leq 2G + 3A$.

10. $A + 2S \leq H + 2C$.
11. $5A + 6G \leq 5H + 6R$.
12. $10A + 2H \leq 9N + 3S$.
13. $6N + 7A \leq 6C + 7H$.
14. $G + 5R \leq 4S + 2A$.
15. $N + 5R \leq H + 5C$.
16. $G + 5A \leq C + 5N$.
17. $A + N + 5R \leq H + 6S$.

Excluding the expressions with negative signs in (47), we get the following proposition.

Proposition 1. *The following refinement over the inequalities (40) holds:*

$$H \leq G \leq \left\{ \begin{array}{c} \frac{3}{5}N + \frac{2}{5}H \\ \frac{1}{3}S + \frac{2}{3}G \\ \frac{2}{9}C + \frac{7}{9}G \end{array} \right\} \leq N \leq \frac{5}{6}A + \frac{1}{6}H \leq \left\{ \begin{array}{c} \frac{1}{4}S + \frac{3}{4}N \\ \frac{2}{3}S + \frac{1}{3}H \end{array} \right\} \leq A \leq$$

$$\leq \frac{1}{6}G + \frac{5}{6}R \leq \left\{ \begin{array}{c} \frac{4}{9}G + \frac{5}{8}C \\ \frac{2}{3}S + \frac{1}{3}A \end{array} \right\} \leq R \leq S \leq C \leq \frac{4}{7}C + \frac{3}{7}N. \qquad (48)$$

4. INEQUALITIES WITH REPEATED DIFFERENCES

In this section, we will bring inequalities in different stages based on *repeated differences*. Combining (28) and (46), we get the following unified inequalities:

$$\begin{Bmatrix} 3D_{SR}^7 \\ D_{GH}^{21} \end{Bmatrix} \le \begin{Bmatrix} D_{SA}^8 \le \begin{cases} \frac{3}{4}D_{SN}^9 \\ \frac{1}{3}D_{SH}^{11} \le \frac{1}{4}\Delta \\ \frac{3}{5}D_{NH}^{20} \le \frac{1}{4}\Delta \end{cases} \end{Bmatrix} \le$$

$$\le \begin{Bmatrix} \frac{3}{7}D_{CN}^4 \le \begin{cases} D_{CS}^1 \\ \frac{1}{3}D_{CG}^5 \le \frac{3}{5}D_{RG}^{14} \\ \frac{1}{2}D_{SG}^{10} \le \frac{3}{5}D_{RG}^{14} \end{cases} \end{Bmatrix} \le$$

$$\le h \le \frac{1}{8}K \le \frac{1}{16}\Psi \le \frac{1}{16}F \le \frac{1}{64}L. \tag{49}$$

As a part of (49) let us consider the following inequalities

$$2\Delta \le \frac{24}{7}D_{CN}^4 \le \frac{8}{3}D_{CG}^5 \le \frac{24}{5}D_{RG}^{14} \le 8h \le K \le \frac{1}{2}\Psi \le \frac{1}{2}F \le \frac{1}{8}L. \tag{50}$$

For simplicity, let us write (50) as

$$W_1 \le W_2 \le W_3 \le W_4 \le W_5 \le W_6 \le W_7 \le W_8 \le W_9, \tag{51}$$

where, $W_1 = 2\Delta$, $W_2 = \frac{24}{7}D_{CN}$, $W_3 = \frac{8}{3}D_{CG}$, $W_4 = \frac{24}{5}D_{RG}$, $W_5 = 8h$, $W_6 = K$, $W_7 = \frac{1}{2}\Psi$, $W_8 = \frac{1}{2}F$ and $W_9 = \frac{1}{8}L$.

We shall work on differences arises due to (51) in different stages, i.e., successively. Initially, let us consider the nonnegative differences arising due to (51). This we shall do many times until reaching a single measure.

4.1 First Stage

For all $a > 0$, $b > 0$, $a \ne b$, let us consider

$$W_t(a, b) := bf\left(\frac{a}{b}\right), \quad t = 1, 2, ..., 9, \tag{52}$$

where

$$f_{W_1}(x) := 2f_\Delta(x) = \frac{2(x-1)^2}{x+1},$$

$$f_{W_2}(x) := \frac{24}{7}f_{CN}(x) = \frac{8(\sqrt{x}-1)^2(2x+3\sqrt{x}+2)}{7(x+1)},$$

$$f_{W_3}(x) := \frac{8}{3}f_{CG}(x) = \frac{8(\sqrt{x}-1)^2(x+\sqrt{x}+1)}{3(x+1)}, \quad \text{etc.}$$

The inequalities (51) admit 55 nonnegative differences. These differences satisfy some natural *pyramid-type inequalities* given by

$$D^1_{W_2 W_1}.$$
$$D^2_{W_3 W_2} \leq D^3_{W_3 W_1}.$$
$$D^4_{W_4 W_3} \leq D^5_{W_4 W_2} \leq D^6_{W_4 W_1}.$$
$$D^7_{W_5 W_4} \leq D^8_{W_5 W_3} \leq D^9_{W_5 W_2} \leq D^{10}_{W_5 W_1}.$$
$$D^{11}_{W_6 W_5} \leq D^{12}_{W_6 W_4} \leq D^{13}_{W_6 W_3} \leq D^{14}_{W_6 W_2} \leq D^{15}_{W_6 W_1}.$$
$$D^{16}_{W_7 W_6} \leq D^{17}_{W_7 W_5} \leq D^{18}_{W_7 W_4} \leq D^{19}_{W_7 W_3} \leq D^{20}_{W_7 W_2} \leq D^{21}_{W_7 W_1}.$$
$$D^{22}_{W_8 W_7} \leq D^{23}_{W_8 W_6} \leq D^{24}_{W_8 W_5} \leq D^{25}_{W_8 W_4} \leq D^{26}_{W_8 W_3} \leq D^{27}_{W_8 W_2} \leq D^{28}_{W_8 W_1}.$$
$$D^{29}_{W_9 W_8} \leq D^{30}_{W_9 W_7} \leq D^{31}_{W_9 W_6} \leq D^{32}_{W_9 W_5} \leq D^{33}_{W_9 W_4} \leq D^{34}_{W_9 W_3} \leq D^{35}_{W_9 W_2} \leq D^{36}_{W_9 W_1}.$$

$$\dots (53)$$

The differences are understood as, $D^1_{W_2 W_1} := W_2 - W_1$, $D^{16}_{W_7 W_6} := W_7 - W_6$, etc.

After simplifications, the first four lines of (53) give us the following equalities:

$$\frac{7}{2} D^1_{W_4 W_3} = \frac{21}{8} D^2_{W_5 W_4} = \frac{3}{2} D^3_{W_5 W_2} = \frac{15}{8} D^4_{W_6 W_5} = \frac{35}{32} D^5_{W_6 W_3}$$
$$= \frac{5}{6} D^6_{W_6 W_3} = \frac{5}{4} D^7_{W_7 W_6} = \frac{3}{4} D^8_{W_7 W_4} = \frac{7}{12} D^9_{W_7 W_3} = \frac{1}{2} D^{10}_{W_7 W_2}$$
$$= \frac{(\sqrt{a} - \sqrt{b})^4}{a + b}. \tag{54}$$

The expression (54) leads us to the following relations among the means. For simplicity, let us write them as Group 5.

Group 5

1. $4A = 2(C + H) = 3R + H$.
2. $3R = C + 2A = 2C + H$.
3. $3N = 2A + G$.
4. $3C + 2H = 3R + 2A$.
5. $C + 6A = H + 6R$.
6. $C + 3N = G + 3R$.
7. $3N + 2A = 2C + 2H + G$.
8. $27R + 2G = 14A + 9C + 6N$.
9. $3(N + 3R) = 8A + 3C + G$.
10. $3G + 8H + 9C = 3R + 8A + 9N$.

11. $4G + 14H + 17C = 9R + 14A + 12N.$

12. $5G + 24H + 31C = 21R + 24A + 15N.$

In view of above equalities given in (54), there lefts 27 measures instead of 45. These 27 can be put in single inequality given in theorem below.

Theorem 5. *The following continued inequalities hold:*

$$D^1_{W_2 W_1} \leq \frac{1}{14} D^{15}_{W_6 W_1} \leq \frac{1}{13} D^{14}_{W_6 W_2} \leq D^{13}_{W_6 W_3} \leq D^{12}_{W_6 W_4} \leq D^{11}_{W_6 W_5} \leq$$

$$\leq D^{21}_{W_7 W_1} \leq D^{20}_{W_7 W_2} \leq D^{19}_{W_7 W_3} \leq D^{18}_{W_7 W_4} \leq D^{17}_{W_7 W_5} \leq D^{16}_{W_7 W_6} \leq$$

$$\leq D^{28}_{W_8 W_1} \leq D^{27}_{W_8 W_2} \leq D^{26}_{W_8 W_3} \leq D^{25}_{W_8 W_4} \leq D^{24}_{W_8 W_5} \leq D^{23}_{W_8 W_6} \leq$$

$$\leq \left\{ \begin{array}{c} D^{22}_{W_8 W_7} \\ D^{36}_{W_9 W_1} \end{array} \right\} \leq$$

$$\leq D^{35}_{W_9 W_2} \leq D^{34}_{W_9 W_3} \leq D^{33}_{W_9 W_4} \leq D^{32}_{W_9 W_5} \leq D^{31}_{W_9 W_6} \leq$$

$$\leq D^{30}_{W_9 W_7} \leq D^{29}_{W_9 W_8}. \tag{55}$$

The proof of above theorem is based on the following 27 results.

1. $D^1_{W_2 W_1} \leq \frac{1}{14} D^{15}_{W_6 W_1}.$

2. $D^{15}_{W_6 W_1} \leq \frac{14}{13} D^{14}_{W_6 W_2}.$

3. $D^{14}_{W_6 W_2} \leq \frac{39}{35} D^{13}_{W_6 W_3}.$

4. $D^{13}_{W_6 W_3} \leq \frac{25}{21} D^{12}_{W_6 W_4}.$

5. $D^{12}_{W_6 W_4} \leq \frac{7}{5} D^{11}_{W_6 W_5}.$

6. $D^{11}_{W_6 W_5} \leq \frac{1}{4} D^{21}_{W_7 W_1}.$

7. $D^{21}_{W_7 W_1} \leq \frac{28}{27} D^{20}_{W_7 W_2}.$

8. $D^{20}_{W_7 W_2} \leq \frac{81}{77} D^{19}_{W_7 W_3}.$

9. $D^{19}_{W_7 W_3} \leq \frac{55}{51} D^{18}_{W_7 W_4}.$

10. $D^{18}_{W_7 W_4} \leq \frac{17}{15} D^{17}_{W_7 W_5}.$

11. $D^{17}_{W_7 W_5} \leq \frac{3}{2} D^{16}_{W_7 W_6}.$

12. $D^{16}_{W_7 W_6} \leq \frac{1}{3} D^{28}_{W_8 W_1}.$

13. $D^{28}_{W_8 W_1} \leq \frac{42}{41} D^{27}_{W_8 W_2}.$

14. $D^{27}_{W_8 W_2} \leq \frac{123}{119} D^{26}_{W_8 W_3}.$

15. $D^{26}_{W_8 W_3} \leq \frac{85}{81} D^{25}_{W_8 W_4}.$

16. $D^{25}_{W_8 W_4} \leq \frac{27}{25} D^{24}_{W_8 W_5}.$

17. $D^{24}_{W_8 W_5} \leq \frac{5}{4} D^{23}_{W_8 W_6}.$

18. $D^{23}_{W_8 W_6} \leq 2 D^{22}_{W_8 W_7}.$

19. $D^{23}_{W_8 W_6} \leq \frac{1}{2} D^{36}_{W_9 W_1}.$

20. $D^{36}_{W_9 W_1} \leq \frac{56}{55} D^{35}_{W_9 W_2}.$

21. $D^{35}_{W_9 W_2} \leq \frac{165}{161} D^{34}_{W_9 W_3}.$

22. $D^{34}_{W_9 W_3} \leq \frac{115}{111} D^{33}_{W_9 W_4}.$

23. $D^{33}_{W_9 W_4} \leq \frac{37}{35} D^{32}_{W_9 W_5}.$

24. $D^{32}_{W_9 W_5} \leq \frac{7}{6} D^{31}_{W_9 W_6}.$

25. $D^{22}_{W_8 W_7} \leq \frac{1}{3} D^{31}_{W_9 W_6}.$

26. $D^{31}_{W_9 W_6} \leq \frac{3}{2} D^{30}_{W_9 W_7}.$

27. $D^{30}_{W_9 W_7} \leq 2 D^{29}_{W_9 W_8}.$

$$\dots \tag{56}$$

The proof of inequalities (56) is similar to Theorems 3 and 4. For details refer to Taneja (2013a).

4.2 Reverse Inequalities

The first four lines of (53) are equal with some multiplicative constants. The other four lines satisfy reverse inequalities given by

1.

$$D^{11}_{W_6 W_5} \leq D^{12}_{W_6 W_4} \leq D^{13}_{W_6 W_3} \leq D^{14}_{W_6 W_2} \leq D^{15}_{W_6 W_1} \leq$$
$$\leq \frac{14}{13} D^{14}_{W_6 W_2} \leq \frac{6}{5} D^{13}_{W_6 W_3} \leq \frac{10}{7} D^{12}_{W_6 W_4} \leq 2D^{11}_{W_6 W_5}.$$

2.

$$D^{16}_{W_7 W_6} \leq D^{17}_{W_7 W_5} \leq D^{18}_{W_7 W_4} \leq D^{19}_{W_7 W_3} \leq D^{20}_{W_7 W_2} \leq D^{21}_{W_7 W_1} \leq$$
$$\leq \frac{28}{27} D^{20}_{W_7 W_2} \leq \frac{12}{11} D^{19}_{W_7 W_3} \leq \frac{20}{17} D^{18}_{W_7 W_4} \leq \frac{4}{3} D^{17}_{W_7 W_5} \leq 2D^{16}_{W_7 W_6}.$$

3.

$$D^{22}_{W_8 W_7} \leq D^{23}_{W_8 W_6} \leq D^{24}_{W_8 W_5} \leq D^{25}_{W_8 W_4} \leq D^{26}_{W_8 W_3} \leq D^{27}_{W_8 W_2} \leq D^{28}_{W_8 W_1} \leq$$
$$\leq \frac{42}{41} D^{27}_{W_8 W_2} \leq \frac{18}{17} D^{26}_{W_8 W_3} \leq \frac{10}{9} D^{25}_{W_8 W_4} \leq \frac{6}{5} D^{24}_{W_8 W_5} \leq$$
$$\leq \frac{3}{2} D^{23}_{W_8 W_6} \leq 3D^{22}_{W_8 W_7}.$$

4.

$$D^{29}_{W_9 W_8} \leq D^{30}_{W_9 W_7} \leq D^{31}_{W_9 W_6} \leq D^{32}_{W_9 W_5} \leq D^{33}_{W_9 W_4} \leq D^{34}_{W_9 W_3} \leq D^{35}_{W_9 W_2} \leq$$
$$\leq D^{36}_{W_9 W_1} \leq \frac{56}{55} D^{35}_{W_9 W_2} \leq \frac{24}{23} D^{34}_{W_9 W_3} \leq \frac{40}{37} D^{33}_{W_9 W_4} \leq \frac{8}{7} D^{32}_{W_9 W_5} \leq$$
$$\leq \frac{4}{3} D^{31}_{W_9 W_6} \leq 2D^{30}_{W_9 W_7} \leq 4D^{29}_{W_9 W_8}.$$

Proof is based on the similar lines of Theorems 3 and 4.

4.3 Second Stage

In Section 4.1, we studied first order differences arising due (51). The 27 results given in (56) generate 14 new nonnegative differences given by

1. $W_6 + 13W_1 - 14W_2 := V_1.$
2. $W_6 + 13W_1 - 14W_2 := V_1.$
3. $4W_6 + 35W_2 - 39W_2 := 4V_1.$
4. $4W_6 + 21W_3 - 25W_4 := 4V_1.$
5. $2W_6 + 5W_4 - 7W := 2V_1.$
6. $W_7 + 4W_5 - W_1 - 4W_6 := \frac{1}{2}V_2.$
7. $W_7 + 27W_1 - 28W := \frac{1}{2}V_3.$
8. $4W_7 + 77W_2 - 81W_3 := 2V_3.$
9. $4W_7 + 51W_3 - 55W_4 := 2V_3.$
10. $2W_7 + 15W_4 - 17W_5 := V_3.$
11. $W_7 + 2W_5 - 3W_6 := \frac{1}{2}V_4.$
12. $W_8 + 3W_6 - W_1 - 3W_7 := \frac{1}{4}V_5.$
13. $W_8 + 41W_1 - 42W_2 := \frac{1}{4}V_6.$
14. $4W_8 + 119W_2 - 123K_3 := V_6.$
15. $4W_8 + 81W_3 - 85W_4 := V_6.$
16. $2W_8 + 25W_4 - 27K_5 := V_6.$
17. $W_8 + 4W_5 - 5W := V_7.$
18. $W_8 + W_6 - 2W_7 := \frac{1}{4}V_8.$
19. $W_9 + 2W_6 - K_1 - 2W := \frac{1}{8}V_9.$
20. $W_9 + 55W_2 - 56K_2 := \frac{1}{8}V_{10}.$
21. $4W_9 + 161W_2 - 165W := \frac{1}{2}V_{10}.$
22. $4W_9 + 111W_3 - 115W := \frac{1}{2}V_{10}.$
23. $2W_9 + 35W_4 - 37W_5 := \frac{1}{4}V_{10}.$
24. $W_9 + 6W_5 - 7W_6 := \frac{1}{8}V_{11}.$
25. $W_9 + 3W_7 - W_6 - 3W_8 := \frac{1}{4}V_{12}.$
26. $W_9 + 2W_6 - 3W := \frac{1}{8}V_{13}.$
27. $W_9 + W_7 - 2W_8 := \frac{1}{8}V_{14}.$

The new measures $V_{(.)}$ can be written as

$$V_t(P \parallel Q) := bf_{V_t}\left(\frac{b}{a}\right), \quad t = 1, 2, ..., 14, \tag{57}$$

where $f_{V_t}(x)$, $t = 1, 2, ..., 14$, $x > 0$, $x \neq 1$ are given by

$$f_{V_1}(x) = \frac{(\sqrt{x} - 1)^6}{\sqrt{x}(x + 1)},$$

$$f_{V_2}(x) = \frac{(\sqrt{x} - 1)^8}{x(x + 1)},$$

$$f_{V_3}(x) = \frac{(x + 6\sqrt{x} + 1)(\sqrt{x} - 1)^6}{x(x + 1)},$$

$$f_{V_4}(x) = \frac{(\sqrt{x} - 1)^6}{x},$$

$$f_{V_5}(x) = \frac{(\sqrt{x} + 1)^2(\sqrt{x} - 1)^8}{x^{3/2}(x + 1)},$$

$$f_{V_6}(x) = \frac{(x^2 + 6x^{3/2} + 22x + 6\sqrt{x} + 1)(\sqrt{x} - 1)^6}{x^{3/2}(x + 1)},$$

$$f_{V_7}(x) = \frac{(x + 6\sqrt{x} + 1)(\sqrt{x} - 1)^6}{x^{3/2}},$$

$$f_{V_8}(x) = \frac{(\sqrt{x} + 1)^2(\sqrt{x} - 1)^6}{x^{3/2}},$$

$$f_{V_9}(x) = \frac{(\sqrt{x} + 1)^4(\sqrt{x} - 1)^8}{16x^2(x + 1)},$$

$$f_{V_{10}}(x) = \frac{(x^3 + 6x^{5/2} + 23x^2 + 68x^{3/2} + 23x + 6\sqrt{x} + 1)(\sqrt{x} - 1)^6}{x^2(x + 1)},$$

$$f_{V_{11}}(x) = \frac{(x^2 + 6x^{3/2} + 22x + 6\sqrt{x} + 1)(\sqrt{x} - 1)^6}{x^2},$$

$$f_{V_{12}}(x) = \frac{(\sqrt{x} + 1)^2(\sqrt{x} - 1)^8}{x^2},$$

$$f_{V_{13}}(x) = \frac{(x - 1)^2(x + 4\sqrt{x} + 1)(\sqrt{x} - 1)^4}{x^2}$$

and

$$f_{V_{14}}(x) = \frac{(x + 1)(\sqrt{x} + 1)^2(\sqrt{x} - 1)^6}{8x^2}.$$

In all the cases we have $f_{V_t}(1) = 0$, $t = 1, 2, ..., 14$. The theorem below gives inequalities among these 14 new measures.

Theorem 6. *The following inequalities hold:*

$$V_1 \le \frac{1}{8}V_3 \le \left\{ \begin{array}{l} \frac{1}{2}V_4 \le \frac{1}{16}V_7 \le \frac{1}{8}V_8 \\ \frac{1}{36}V_6 \le \frac{1}{128}V_{10} \end{array} \right\} \le \frac{1}{72}V_{11} \le \frac{1}{48}V_{13} \le \frac{1}{16}V_{14} \quad (58)$$

and

$$V_2 \le \frac{1}{4}V_5 \le \frac{1}{16}V_9 \le \frac{1}{8}V_{12}. \quad (59)$$

The proof of inequalities given in (58) and (59) follows on the similar lines of Theorems 4 and 5. The inequalities (58) and (59) again generate new nonnegative differences given by

1. $\frac{1}{8}V_3 - V_1 = \frac{1}{8}U_1.$ 8. $2V_8 - V_7 = U_2.$
2. $\frac{1}{8}V_7 - V_4 = \frac{1}{8}U_2.$ 9. $4V_2 - V_4 = \frac{1}{9}U_7.$
3. $\frac{1}{8}V_7 - V_4 = \frac{1}{8}U_2.$ 10. $\frac{3}{2}V_{13} - V_{11} = \frac{1}{2}U_8.$
4. $\frac{2}{9}V_6 - V_3 = \frac{1}{9}U_3.$ 11. $3V_{14} - V_{13} = 2U_9.$
5. $\frac{9}{32}V_{10} - V_6 = \frac{1}{32}U_4.$ 12. $\frac{1}{4}V_5 - V_2 = \frac{1}{4}U_{10}.$
6. $\frac{16}{9}V_{11} - V_{10} = \frac{7}{9}U_5.$ 13. $4V_2 - V_4 = \frac{1}{4}U_{11}.$
7. $\frac{9}{4}V_7 - V_6 = \frac{5}{4}U_6.$ 14. $2V_{12} - V_9 = U_{11}.$

$$\cdots \qquad (60)$$

Based on the measures U_t, $t = 1, 2, ..., 11$ appearing in (60) we derive more inequalities given in subsection below.

4.4 Third Stage

The difference measures given in (60) generate new measures given by

$$U_t(a, b) := bf_{U_t}\left(\frac{b}{a}\right), \quad t = 1, 2, ..., 11, \qquad (61)$$

where the functions $f_{U_t}(x)$, $t = 1, 2, ..., 11$ are

$$f_{U_1}(x) = \frac{(\sqrt{x} - 1)^8}{x(x + 1)}$$

$$f_{U_2}(x) = \frac{(\sqrt{x} - 1)^8}{x^{3/2}}$$

$$f_{U_3}(x) = \frac{(\sqrt{x} - 1)^8(2x + 7\sqrt{x} + 2)}{x^{3/2}(x + 1)}$$

$$f_{U_4}(x) = \frac{(\sqrt{x} - 1)^8(9x^2 + 40x^{3/2} + 86x + 40\sqrt{x} + 9)}{x^2(x + 1)}$$

$$f_{U_5}(x) = \frac{(\sqrt{x} - 1)^8(x^2 + 8x^{3/2} + 38x + 8\sqrt{x} + 1)}{x^2(x + 1)}$$

$$f_{U_6}(x) = \frac{(\sqrt{x} - 1)^8(x + 8\sqrt{x} + 1)}{x^{3/2}(x + 1)}$$

$$f_{U_7}(x) = \frac{(\sqrt{x}-1)^8(x-\sqrt{x}+1)}{x^2}$$

$$f_{U_8}(x) = \frac{(\sqrt{x}-1)^8(x+8\sqrt{x}+1)}{x^2}$$

$$f_{U_9}(x) = \frac{(\sqrt{x}-1)^8(\sqrt{x}+1)^2}{x^2}$$

$$f_{U_{10}}(x) = \frac{(\sqrt{x}-1)^{10}}{x^{3/2}(x+1)}$$

and

$$f_{U_{11}}(x) = \frac{(\sqrt{x}+1)^2(\sqrt{x}-1)^{10}}{x^2(x+1)}$$

In all the cases, we have $f_{U_t}(1) = 0$, $t = 1, 2, ..., 11$. The 11 measures given in (61) satisfy some inequalities given in theorem below.

Theorem 7. *The following inequalities hold:*

$$U_1 \leq \frac{1}{10}U_6 \leq \frac{1}{11}U_3 \leq \begin{cases} \frac{1}{2}U_2 \\ \frac{1}{56}U_5 \leq \frac{1}{184}U_4 \end{cases} \leq \frac{1}{20}U_8 \leq \frac{1}{8}U_9 \leq \frac{1}{2}U_7. \quad (62)$$

It leads us to following new nonnegative differences:

1. $\frac{1}{10}U_6 - U_1 = \frac{1}{10}U_{10}$.

2. $\frac{10}{11}U_3 - U_6 = \frac{9}{11}U_{10}$.

3. $\frac{11}{2}U_2 - U_3 = \frac{1}{56}U_{12}$.

4. $\frac{1}{10}U_8 - U_2 = \frac{7}{2}U_{10}$.

5. $\frac{11}{56}U_5 - U_3 = \frac{1}{10}U_{13}$.

6. $\frac{7}{23}U_4 - U_5 = \frac{9}{5}U_{14}$.

7. $\frac{46}{5}U_8 - U_4 = \frac{9}{5}U_{15}$.

8. $\frac{5}{2}U_9 - U_8 = \frac{3}{2}U_{13}$.

9. $4U_7 - U_9 = 3U_{13}$.

The measures U_{10} and U_{11} are already defined above. The other measures appearing are given by

$$U_t(a, b) := bf_{U_t}\left(\frac{b}{a}\right), \quad t = 12, 13, 14 \text{ and } 15, \quad (63)$$

where the functions $f_{U_t}(x)$, $t = 12, 13, 14$ and 15 are given by

$$f_{U_{12}}(x) = \frac{(\sqrt{x}-1)^{10}(11x - 2\sqrt{x} + 11)}{x^2(x+1)},$$

$$f_{U_{13}}(x) = \frac{(\sqrt{x} - 1)^{10}}{x^2},$$

$$f_{U_{14}}(x) = \frac{(\sqrt{x} - 1)^{10}(5x + 22\sqrt{x} + 5)}{x^2(x + 1)}$$

and

$$f_{U_{15}}(x) = \frac{(\sqrt{x} - 1)^{10}(x + 170\sqrt{x} + 1)}{x^2(x + 1)}.$$

Theorem 8. *The following inequalities hold:*

$$U_{10} \leq \frac{1}{172}U_{15} \leq \frac{1}{32}U_{14} \leq \frac{1}{4}U_{11} \leq \frac{1}{2}U_{13} \leq \frac{1}{20}U_{12}. \tag{64}$$

The differences arising due (64) lead us to a single measure given by

$$U_{16}(a, b) = \frac{(\sqrt{a} - \sqrt{b})^{12}}{172(ab)^2(a + b)}, \quad a, b > 0, \ a \neq b. \tag{65}$$

Remarks 7. (i) In Stage 1 we have considered all possible differences. From Stage 2 onward we considered only the simple differences, just with previous measure in each case. This process continued up to Stage 4 until we reached to a single measure given in (65).

(ii) Details on the proofs of Theorems 5–8 are in author's work (Taneja, 2013c).

4.5 Equivalent Relations

As a consequence of above Theorems 6, 7 and 8, we have following equivalent relations:

- $V_1 = 14(K + 26\Delta - 48D_{CN}) = \frac{35}{4}(K + 30D_{CN} - 26D_{CG})$
  $= \frac{21}{4}(K + 14D_{CG} - 30D_{RG})$
  $= \frac{5}{2}(K + 12D_{RG} - 28hD_{RG}).$
- $V_2 = \Psi + 64h - 4\Delta - 8K.$
- $V_3 = 54(\Psi + 108\Delta - 192D_{CN}) = \frac{77}{2}(\Psi + 132D_{CN} - 108D_{CG})$
  $= \frac{51}{2}(\Psi + 68D_{CG} - 132D_{RG})$
  $= 15(\Psi + 72D_{RG} - 136h).$
- $V_4 = \Psi + 32h - 6K.$
- $V_5 = 2(F + 6K - 4\Delta - 3\Psi).$

- $V_6 = 82(F + 164\Delta - 288D_{CN}) = \frac{119}{2}(F + 204D_{CN} - 164D_{CG})$
  $= \frac{81}{2}(F + 108D_{CG} - 204D_{RG})$
  $= 25(F + 120D_{RG} - 216h)$.

- $V_7 = 2(F - 10K + 64h)$.

- $V_8 = 2(F + 2K - 2\Psi)$.

- $V_9 = L + 16K - 16\Delta - 8F$.

- $V_{10} = 440(L + 880\Delta - 1536D_{CN}) = 322(L + 1104D_{CN} - 880D_{CG})$
  $= 222(L + 592D_{CG} - 1104D_{RG})$
  $= 322(L + 6724D_{RG} - 1184h)$.

- $V_{11} = L + 384h - 56K$.

- $V_{12} = L + 12\Psi - 8K - 12F$.

- $V_{13} = L + 16K - 12\Psi$.

- $V_{14} = L + 4\Psi - 8F$.

- $U_1 = \Psi + 192D_{CN} - 100\Delta - 8K$.

- $U_2 = 2(F + 14K - 64h - 4\Psi)$.

- $U_3 = 4F + 576D_{CN} - 316\Delta - 9\Psi$.

- $U_4 = 9L + 4608D_{CN} - 2576\Delta - 64F$.

- $U_5 = \frac{1}{7}(7L + 6144h + 13824D_{CN} - 896K - 7920\Delta)$.

- $U_6 = \frac{2}{5}(5F + 576h + 1152D_{CN} - 90K - 656\Delta)$.

- $U_7 = L + 384h + 36\Psi - 92K - 18F$.

- $U_8 = L + 160K - 36\Psi - 768h$.

- $U_9 = L + 12\Psi - 8K - 12F$.

- $U_{10} = 2(F + 22K + 4\Delta - 5\Psi - 128h)$.

- $U_{11} = L + 16\Delta + 24\Psi - 16F - 32K$.

- $U_{12} = 392F + 8624K + 17696\Delta - 1960\Psi - 39424h - 32256D_{CN}$.

- $U_{13} = \frac{1}{196}(385L + 337920h + 183760\Delta + 17640\Psi - 49280K - 368640D_{CN} - 7840F)$.

- $U_{14} = \frac{1}{1449}(1400L + 279680\Delta + 103040K - 460800D_{CN} - 15680F - 706560h)$.

- $U_{15} = \frac{1}{9}(L + 12880\Delta + 320F + 7360K - 1656\Psi - 35328h - 23040D_{CN})$.

- $U_{16} := \frac{1}{172}(L + 64\Psi + 8704h + 11504\Delta - 208K - 23040D_{CN} - 24F)$.

5. PARAMETRIC GENERALIZED MEASURES

This section deals with generalizations of measures Δ, h, Ψ, K, and F in different situations. These measures are given in (6)–(8), (24) and (25). From now onward, instead of considering probability distributions P and Q, we shall consider real numbers a and b, $a, b > 0$.

5.1 Generalizations of Triangular Discrimination

In this subsection, we shall give two different generations of the measure given in (6), known by triangular discrimination. These generalizations are different from the one given in (13) and (23). Moreover, exponential representations are also presented.

5.1.1 First Generalization

In (13) and (23), there are two different generalizations of *triangular discrimination* along with other generalized measures. Here we shall give an alternative way to generalize *triangular discriminations*. For all $(a, b) \in \mathbb{R}_+^2$, let us consider

$$\Delta_t^1(a, b) = \frac{(a - b)^2 (\sqrt{a} - \sqrt{b})^{2t}}{(a + b)(\sqrt{ab})^t}, \qquad t = 0, 1, 2, 3, \ldots \tag{66}$$

In particular,

$$\Delta_0^1 = \Delta = \frac{(a - b)^2}{(a + b)},$$

$$\Delta_1^1 = D_{W_6 W_1}^{15} = K - 2\Delta = \frac{(a - b)^2 (\sqrt{a} - \sqrt{b})^2}{(a + b)\sqrt{ab}},$$

$$\Delta_2^1 = \Psi - 4K + 4\Delta = \frac{(a - b)^2 (\sqrt{a} - \sqrt{b})^4}{ab(a + b)}$$

and

$$\Delta_3^1 = V_5 = 2(F + 6K - 4\Delta - 3\Psi) = \frac{(a - b)^2 (\sqrt{a} - \sqrt{b})^6}{(ab)^{3/2}(a + b)}.$$

Consequently, we have

1. $\Delta \leq \frac{1}{2}K$.
2. $K \leq \frac{\Psi + 4\Delta}{4}$.
3. $4\Delta + 3\Psi \leq F + 6K$.

The expression (66) gives us first generalizations of the measure $\Delta(a, b)$. Let us write

$$\Delta_t^1(a, b) = b f_{\Delta_t^1}(a/b), \quad t \in \mathbb{N}, \tag{67}$$

where

$$f_{\Delta_t^1}(x) = \frac{(x-1)^2(\sqrt{x}-1)^{2t}}{(x+1)(\sqrt{x})^t}.$$

We can easily check that the measure $\Delta_t^1(a, b)$ given in (67) is convex for all $(a, b) \in \mathbb{R}_+^2$, $t \in \mathbb{N}$.

Now, we shall present *exponential representation* of the measure (66) using the function given in (67). Let us consider a linear combination of convex functions,

$$f_{\Delta^1}(x) = a_0 f_{\Delta_0^1}(x) + a_1 f_{\Delta_1^1}(x) + a_2 f_{\Delta_2^1}(x) + a_3 f_{\Delta_3^1}(x) + \dots$$

i.e.,

$$f_{\Delta^1}(x) = a_0 \frac{(x-1)^2}{x+1} + a_1 \frac{(x-1)^2(\sqrt{x}-1)^2}{(x+1)\sqrt{x}} + a_2 \frac{(x-1)^2(\sqrt{x}-1)^4}{x(x+1)} +$$
$$+ a_3 \frac{(x-1)^2(\sqrt{x}-1)^6}{(x)^{3/2}(x+1)} + \dots,$$

where $a_0, a_1, a_2, a_3, \dots$ are constants. For simplicity, let us choose,

$$a_0 = \frac{1}{0!}, \quad a_1 = \frac{1}{1!}, \quad a_2 = \frac{1}{2!}, \quad a_3 = \frac{1}{3!}, \quad \dots$$

Thus, we have

$$f_{\Delta^1}(x) = \frac{1}{0!}\frac{(x-1)^2}{x+1} + \frac{1}{1!}\frac{(x-1)^2(\sqrt{x}-1)^2}{(x+1)\sqrt{x}} + \frac{1}{2!}\frac{(x-1)^2(\sqrt{x}-1)^4}{x(x+1)} +$$
$$+ \frac{1}{3!}\frac{(x-1)^2(\sqrt{x}-1)^6}{(x)^{3/2}(x+1)} + \dots$$
$$= \frac{(x-1)^2}{x+1}\left[\frac{1}{0!} + \frac{1}{1!}\left(\frac{(\sqrt{x}-1)^2}{\sqrt{x}}\right)^1 + \frac{1}{2!}\left(\frac{(\sqrt{x}-1)^2}{\sqrt{x}}\right)^2 + \right.$$
$$\left. + \frac{1}{3!}\left(\frac{(\sqrt{x}-1)^2}{\sqrt{x}}\right)^3 + \dots\right].$$

Which will give us

$$f_{\Delta^1}(x) = \frac{(x-1)^2}{x+1}\exp\left(\frac{(x-1)^2}{\sqrt{x}}\right). \tag{68}$$

Because of (68), we have the following *exponential triangular discrimination*:

$$E_{\Delta^1}(a, b) = bf_{\Delta^1}(a/b) = \frac{(a-b)^2}{a+b} \exp\left(\frac{(a-b)^2}{\sqrt{ab}}\right), \quad (a, b) \in \mathbb{R}_+^2. \tag{69}$$

5.1.2 Second Generalization

For all $(a, b) \in \mathbb{R}_+^2$, let us consider the following measures

$$\Delta_t^2(a, b) = \frac{(a-b)^{2(t+1)}}{(a+b)(ab)^t}, \quad t = 0, 1, 2, 3, \dots \tag{70}$$

In particular,

$$\Delta_0^2 = \Delta = \frac{(a-b)^2}{a+b}$$

and

$$\Delta_1^2 = 2D_{W_7 W_1}^{21} = \Psi - 4\Delta = \frac{(a-b)^4}{ab(a+b)}.$$

The expression (70) gives us the second generalization of the measure $\Delta(a, b)$. Again the measure $\Delta_t^2(a, b)$ given in (70) is convex for all $(a, b) \in \mathbb{R}_+^2$, $t \in \mathbb{N}$. Following the similar lines of (68) and (69), the exponential representation of the measure $\Delta_t^2(a, b)$ is given by

$$E_{\Delta^2}(a, b) = \frac{(a-b)^2}{a+b} \exp\left(\frac{(a-b)^2}{ab}\right), \quad (a, b) \in \mathbb{R}_+^2. \tag{71}$$

The *generalized triangular discrimination measures* (13), (23), (66) and (70) considered here are different from Topsoe (2000):

$$\Delta_t(a, b) = \frac{(a-b)^{2t}}{(a+b)^{2t-1}}, \quad t \in \mathbb{N}_+.$$

5.2 Generalizations of Measure K

This subsection deals with different generalizations of the measure K given in (24). Exponential generalizations are also given.

5.2.1 First Generalization

For all $(a, b) \in \mathbb{R}_+^2$, let us consider the following measures

$$K_t^1(a, b) = \frac{(a-b)^2(\sqrt{a}-\sqrt{b})^{2t}}{(\sqrt{ab})^{t+1}}, \quad t = 0, 1, 2, 3, \dots \tag{72}$$

In particular,

$$K_0^1 = K = \frac{(a-b)^2}{\sqrt{ab}},$$

$$K_1^1 = 2D_{W_7 W_6}^{16} = \Psi - 2K = \frac{(a-b)^2(\sqrt{a}-\sqrt{b})^2}{ab},$$

$$K_2^1 = V_8 = 2(F + 2K - 2\Psi) = \frac{(a-b)^2(\sqrt{a}-\sqrt{b})^4}{(ab)^{3/2}}$$

and

$$K_3^1 = V_{12} = L + 12\Psi - 8K - 12F = \frac{(a-b)^2(\sqrt{a}-\sqrt{b})^6}{(ab)^2}.$$

Consequently, we have

1. $\Delta \leq \frac{1}{2}\Psi$.
2. $\Psi \leq \frac{F+2K}{2}$.
3. $2K + 3F \leq \frac{L+12\Psi}{4}$.

The expression (72) gives the first parametric generalization of the measure $K(a, b)$ given by (24). Again the measure $K_t^1(a, b)$ given in (72) is convex for all $(a, b) \in \mathbb{R}_+^2$, $t \in \mathbb{N}$. Following the similar lines of (68) and (69), the exponential representation of the measure $K_t^1(a, b)$ is given by

$$E_K^1(a, b) = \frac{(\sqrt{a}-\sqrt{b})^2}{\sqrt{ab}} \exp\left(\frac{(a-b)^2}{\sqrt{ab}}\right), \quad (a, b) \in \mathbb{R}_+^2. \tag{73}$$

5.2.2 Second Generalization

For all $(a, b) \in \mathbb{R}_+^2$, let us consider the following measures

$$K_t^2(a, b) = \frac{(a-b)^{2(t+1)}}{(\sqrt{ab})^{2t+1}}, \quad t = 0, 1, 2, 3, \dots \tag{74}$$

In particular,

$$K_0^2 = K = \frac{(a-b)^2}{\sqrt{ab}}$$

and

$$K_1^2 = F - 2K = \frac{(a-b)^4}{(ab)^{3/2}}.$$

The expression (74) gives the second generalization of the measure $K(a, b)$ given by (24). The measure $K_t^2(a, b)$ given by (74) is convex for all $(a, b) \in$

$\mathbb{R}^2_+$, $t \in \mathbb{N}$. Following the similar lines of (68) and (69), the exponential representation of the measure $K_t^2(a, b)$ is given by

$$E_{K^2}(a, b) = \frac{(a-b)^2}{\sqrt{ab}} \exp\left(\frac{(a-b)^2}{ab}\right), \quad (a, b) \in \mathbb{R}^2_+. \tag{75}$$

5.3 Generalizations of Hellinger's Discrimination

This subsection deals with generalizations of Hellinger's discrimination given by (7). For all $(a, b) \in \mathbb{R}^2_+$, let us consider the following measures

$$h_t(a, b) = \frac{(\sqrt{a} - \sqrt{b})^{2(t+1)}}{(\sqrt{ab})^t}, \quad t \in \mathbb{N} \tag{76}$$

In particular, we have

$$h_0 = 2h = (\sqrt{a} - \sqrt{b})^2,$$

$$h_1 = D_{W_6 W_5}^{11} = K - 8h = \frac{(\sqrt{a} - \sqrt{b})^4}{\sqrt{ab}},$$

$$h_2 = V_4 = \Psi + 32h - 6K = \frac{(\sqrt{a} - \sqrt{b})^6}{ab},$$

$$h_3 = U_2 = 2(F + 14K - 4\Psi - 64h) = \frac{(\sqrt{a} - \sqrt{b})^8}{(ab)^{3/2}}$$

and

$$h_4 = U_{13} = L + 44\Psi - 120K + 512h - 20F = \frac{(\sqrt{a} - \sqrt{b})^{10}}{(ab)^2}.$$

Consequently, we have
1. $h \le \frac{1}{8}K$.
2. $K \le \frac{\Psi + 32h}{6}$.
3. $4\Psi + 64h \le F + 14K$.
4. $F + 6K \le \frac{L + 44\Psi + 512h}{20}$.

The measure (76) gives generalized Hellinger's discrimination. The measure $h_t(a, b)$ given in (76) is convex for all $(a, b) \in \mathbb{R}^2_+$, $t \in \mathbb{N}$. Following the similar lines of (68) and (69), the exponential representation of the measure $h_t(a, b)$ is given by

$$E_h(a, b) = (\sqrt{a} - \sqrt{b})^2 \exp\left(\frac{(\sqrt{a} - \sqrt{b})^2}{\sqrt{ab}}\right), \quad (a, b) \in \mathbb{R}_+^2. \qquad (77)$$

5.4 New Parametric Measure

For all $(a, b) \in \mathbb{R}_+^2$, let us consider the following parametric measure

$$Z_t(a, b) = \frac{(\sqrt{a} - \sqrt{b})^{2(t+2)}}{(a + b)(\sqrt{ab})^t}, \quad t = 0, 1, 2, 3, \dots \qquad (78)$$

In particular, we will have

$$Z_0 = \frac{7}{2}D^1_{W_2 W_1} = 12D_{CN} - 7\Delta = \frac{(\sqrt{a} - \sqrt{b})^4}{(a + b)},$$

$$Z_1 = V_1 = K + 26\Delta - 48D_{CN} = \frac{(\sqrt{a} - \sqrt{b})^6}{(a + b)\sqrt{ab}},$$

$$Z_2 = U_1 = \Psi + 192D_{CN} - 100\Delta - 8K = \frac{(\sqrt{a} - \sqrt{b})^8}{ab(a + b)},$$

$$Z_3 = U_{10} = 2(F + 22K + 4\Delta - 5\Psi - 128h) = \frac{(\sqrt{a} - \sqrt{b})^{10}}{(a + b)(ab)^{3/2}}$$

and

$$Z_4 = 172U_{16} = \frac{1}{7}\left(\begin{array}{l} 7L + 448\Delta - 1456K + 3728\Delta + \\ +9728h - 7680D_{CN} - 168F \end{array}\right) = \frac{(\sqrt{a} - \sqrt{b})^{12}}{(a + b)(ab)^2}.$$

Consequently, we have
1. $\Delta \leq \frac{12}{7}(C - N)$.
2. $C \leq \frac{48N + K + 26\Delta}{48}$.
3. $192N + 100\Delta + 8K \leq \Psi + 192C$.
4. $5\Psi + 128h \leq F + 22K + 4\Delta$.
5. $7680C + 168F + 1456K \leq 7680N + 7L + 448\Delta + 3728\Delta + 9728h$.

The measure $Z_t(a, b)$ given in (78) is convex for all $(a, b) \in \mathbb{R}_+^2$, $t \in \mathbb{N}$. Following the similar lines of (68) and (69), the exponential representation of the measure $Z_t(a, b)$ is given by

$$E_Z(a, b) = \frac{(a - b)^4}{a + b} \exp\left(\frac{(\sqrt{a} - \sqrt{b})^2}{\sqrt{ab}}\right), \quad (a, b) \in \mathbb{R}_+^2. \tag{79}$$

Remarks 8. (i) Following the similar lines of (68) and (69), the exponential representation of the principal measure $L_t(a, b)$ appearing in (26) is given by

$$E_\Delta(a, b) = \frac{2(a - b)^2}{a + b} \exp\left(\frac{a + b}{2\sqrt{ab}}\right), \quad (a, b) \in \mathbb{R}_+^2. \tag{80}$$

(ii) The exponential expressions (71), (73), (75), (77), (79) and (80) promise new studies in future.

6. COMBINED INEQUALITIES

In Sections 1 and 2, we studied divergence measures, and then in Theorem 1, these measures are connected through inequalities (34), (35) and (36). In Section 3, we studied *seven means* given in (40). In Theorem 3, these are connected through inequalities by considering *difference of means*. Inequalities (49) connect *difference of mean* with some *divergence measures* given in (17). Still in the literature, there exists another famous measure, known by *generalized Gini-mean* based on two parameters. This mean in particular contains well known power mean (Czinder & Pales, 2005). Aim of this section is to study extensively generalized Gini-mean with parameters.

6.1 Gini Mean of Order r and s

The *Gini mean* (Gini, 1938) *of order r and s* is given by

$$E_{r,s}(a, b) = \begin{cases} \left(\dfrac{a^r + b^r}{a^s + b^s}\right)^{\frac{1}{r-s}}, & r \neq s \\[2ex] \exp\left(\dfrac{a^r \ln a + b^r \ln b}{a^r + b^r}\right), & r = s \neq 0 \\[2ex] \sqrt{ab}, & r = s = 0 \end{cases} \tag{81}$$

In particular, when $s = 0$ (Czinder & Pales, 2005) in (81), we have

$$E_{r,0}(a, b) := B_r(a, b) = \begin{cases} \left(\dfrac{a^r + b^r}{2} \right)^{\frac{1}{r}}, & r \neq 0 \\ \sqrt{ab}, & r = 0 \end{cases} \qquad (82)$$

Again, when $s = r - 1$ in (81), we have

$$E_{r,r-1}(a, b) := L_r(a, b) = \frac{a^r + b^r}{a^{r-1} + b^{r-1}}, \qquad r \in \mathbb{R} \qquad (83)$$

The expression (82) is famous as *mean of order r* or *power-mean*. The expression (83) is known as Lehmer mean (Lehmer, 1971). Both the means given in (82) and (83) are monotonically increasing in r (Chen, 2008). Moreover, these two have the following inequality (Chen, 2008) among each other:

$$B_r(a, b) \begin{cases} < L_r(a, b), & r > 1 \\ > L_r(a, b), & r < 1 \end{cases} \qquad (84)$$

Since $E_{r,s} = E_{s,r}$, the Gini-mean $E_{r,s}(a, b)$ given by (6.1) is an increasing function in r or s. Using the monotonic property (Sánder, 2004; Simic, 2009a) we have the following inequalities among particular cases of (81):

$$E_{-3,-2} \leq E_{-2,-1} \leq E_{-3/2,-1/2} \leq E_{-1,0} \leq E_{-1/2,0} \leq$$
$$\leq E_{-1/2,1/2} \leq E_{0,1/2} \leq E_{0,1} \leq \{E_{0,2} \text{ or } E_{1/2,1}\} \leq E_{1,2}. \qquad (85)$$

Let us re-write the expression (85) as

$$P_1 \leq P_2 \leq P_3 \leq H \leq P_4 \leq G \leq N_1 \leq A \leq (P_5 \text{ or } S) \leq P_6, \qquad (86)$$

where $P_1 = E_{-3,-2} = L_{-2}$, $P_2 = E_{-2,-1} = L_{-1}$, $P_3 = E_{-3/2,-1/2} = L_{-1/2}$, $H = E_{-1,0} = L_0 = B_{-1}$, $P_4 = E_{-1/2,0} = B_{-1/2}$, $G = E_{-1/2,1/2} = L_{1/2} = B_0$, $N_1 = E_{0,1/2} = B_{1/2}$, $A = E_{0,1} = L_1 = B_1$, $P_5 = E_{1/2,1}$, $B_2 = E_{0,2} = S$ and $P_6 = E_{1,2} = L_2$.

The means H, G, A and S are well known as, *harmonic, geometric, arithmetic* and the *square-root means* respectively.

6.2 Combined Inequalities

Taneja (2005a) studied the following inequalities:

$$H \leq G \leq N_1 \leq N \leq N_2 \leq A \leq S, \qquad (87)$$

where

$$N_2(a, b) = \left(\frac{\sqrt{a} + \sqrt{b}}{2} \right) \left(\sqrt{\frac{a+b}{2}} \right)$$

and

$$N(a, b) = \frac{a + \sqrt{ab} + b}{3}.$$

$N(a, b)$ is famous as *Heron's mean* studied in Section 3. Some applications of the inequalities (87) in terms of probability distributions are done in (Shi, Zhang, & Li, 2010; Simic, 2009b). Combining (86) and (87), we have the following sequence of inequalities:

$$P_1 \leq P_2 \leq P_3 \leq H \leq P_4 \leq G \leq N_1 \leq N_3 \leq N_2 \leq A \leq \{P_5 \text{ or } S\} \leq P_6. \quad (88)$$

Comparing (88) with inequalities (40) for *seven means*, we observe that except R all other means are already in expression (88), where $P_6 = C$. After verifications, we can extend the inequalities (88) with *seven means* and rewrite them as:

$$P_1 \leq P_2 \leq P_3 \leq H \leq P_4 \leq G \leq N_1 \leq N \leq N_2 \leq A \leq \begin{Bmatrix} P_5 \\ R \leq S \end{Bmatrix} \leq C \quad (89)$$

Expression (89) combines three expressions given by (40), (87) and (88). The inequalities appearing in (89) admit many non-negative differences. Let us write them as:

$$D_{tp}(a, b) = b f_{tp}\left(\frac{a}{b} \right) = b \left[f_t\left(\frac{a}{b} \right) - f_p\left(\frac{a}{b} \right) \right], \quad (90)$$

where

$$f_{tp}(x) = f_t(x) - f_p(x), \quad f_t(x) \geq f_p(x), \quad \forall x > 0.$$

More precisely, the function $f : (0, \infty) \rightarrow \mathbb{R}$ appearing in (90) leads us to the following inequalities:

$$f_{P_1}(x) \leq f_{P_2}(x) \leq f_{P_3}(x) \leq f_H(x) \leq f_{P_4}(x) \leq f_G(x) \leq f_{N_1}(x) \leq$$

$$\leq f_N(x) \leq f_{N_2}(x) \leq f_A(x) \leq \begin{Bmatrix} f_{P_5}(x) \\ f_R(x) \leq f_S(x) \end{Bmatrix} \leq f_C(x). \quad (91)$$

Equivalently,

$$\frac{x(x^2+1)}{x^3+1} \le \frac{x(x+1)}{x^2+1} \le \frac{x(\sqrt{x}+1)}{x^{3/2}+1} \le \frac{2x}{1+x} \le \frac{4x}{(\sqrt{x}+1)^2} \le$$

$$\le \sqrt{x} \le \left(\frac{\sqrt{x}+1}{2}\right)^2 \le$$

$$\le \frac{x+\sqrt{x}+1}{3} \le \left(\frac{\sqrt{x}+1}{2}\right)\left(\sqrt{\frac{x+1}{2}}\right) \le \frac{x+1}{2} \le$$

$$\le \left\{ \begin{array}{c} (\frac{x+1}{\sqrt{x}+1})^2 \\ \frac{2(x^2+x+1)}{3(x+1)} \end{array} \le \sqrt{\frac{x^2+1}{2}} \right\} \le \frac{x^2+1}{x+1}.$$

Below are the *first* and *second order derivatives* of the functions given in (91). These shall be used later to prove important results.

1. $f'_{P_1}(x) = \frac{2x^3-3x^2-1}{(x^3+1)^2}$ and $f''_{P_1}(x) = \frac{6x(x^4-2x^3-2x+1)}{(x^3+1)^3}$.

2. $f'_{P_2}(x) = -\frac{x^2-2x-1}{(x^2+1)^2}$ and $f''_{P_2}(x) = \frac{2(x+1)(x^2-4x+1)}{(x^2+1)^3}$.

3. $f'_{P_3}(x) = -\frac{\sqrt{x}-2}{(x-\sqrt{x}+1)^2}$ and $f''_{P_3}(x) = \frac{3(x^2-x^{3/2}-4x-\sqrt{x}+1)}{4\sqrt{x}(\sqrt{x}+1)^2(x-\sqrt{x}+1)^3}$.

4. $f'_H(x) = \frac{2}{(x+1)^2}$ and $f''_H(x) = -\frac{4}{(x+1)^3}$.

5. $f'_{P_4}(x) = \frac{4}{(\sqrt{x}+1)^3}$ and $f''_{P_4}(x) = -\frac{6}{\sqrt{x}(\sqrt{x}+1)^6}$.

6. $f'_G(x) = \frac{1}{2\sqrt{x}}$ and $f''_G(x) = -\frac{1}{4x^{3/2}}$.

7. $f'_{N_1}(x) = \frac{\sqrt{x}+1}{4\sqrt{x}}$ and $f''_{N_1}(x) = -\frac{1}{8x^{3/2}}$.

8. $f'_N(x) = \frac{2\sqrt{x}+1}{6\sqrt{x}}$ and $f''_N(x) = -\frac{1}{12x^{3/2}}$.

9. $f'_{N_2}(x) = \frac{2x+\sqrt{x}+1}{4\sqrt{x(2x+2)}}$ and $f''_{N_2}(x) = -\frac{x^{3/2}+1}{4x^{3/2}(2x+2)^{3/2}}$.

10. $f'_A(x) = \frac{1}{2}$ and $f''_A(x) = 0$.

11. $f'_{P_5}(x) = \frac{(x+1)(x+2\sqrt{x}-1)}{\sqrt{x}(\sqrt{x}+1)^3}$ and $f''_{P_5}(x) = \frac{(\sqrt{x}+1)^4-12x}{2x^{3/2}(\sqrt{x}+1)^4}$.

12. $f'_R(x) = \frac{2}{3}\frac{x(x+2)}{(x+1)^2}$ and $f''_R(x) = \frac{4}{3(x+1)^3}$.

13. $f'_S(x) = \frac{x}{\sqrt{2x^2+2}}$ and $f''_S(x) = \frac{4}{(2x^2+2)^{3/2}}$.

14. $f'_C(x) = \frac{x^2+2x-1}{(x+1)^2}$ and $f''_C(x) = \frac{4}{(x+1)^3}$.

From the values of second derivatives we observe that not all the measures are convex. Values of second order derivatives at $x = 1$ are given by

1. $f''_{P_1}(1) = -\frac{3}{2}.$ **6.** $f''_G(1) = -\frac{1}{4}.$ **11.** $f''_{P_5}(1) = \frac{1}{8}.$

2. $f''_{P_2}(1) = -1.$ **7.** $f''_{N_1}(1) = -\frac{1}{8}.$ **12.** $f''_R(1) = \frac{1}{6}.$

3. $f''_{P_3}(1) = -\frac{3}{4}.$ **8.** $f''_N(x) = -\frac{1}{12}.$ **13.** $f''_S(1) = \frac{1}{4}.$

4. $f''_H(1) = -\frac{1}{2}.$ **9.** $f''_{N_2}(1) = -\frac{1}{16}.$ **14.** $f''_C(1) = \frac{1}{2}.$

5. $f''_{P_4}(x) = -\frac{3}{8}.$ **10.** $f''_A(x) = 0.$

$$\dots \quad (92)$$

6.3 Main Results

Inequalities (89) admit many nonnegative measures. These measures satisfy some obvious inequalities given below.

- $D^1_{P_2 P_1}.$

- $D^2_{P_3 P_2} \le D^3_{P_3 P_1}.$

- $D^4_{HP_3} \le D^5_{HP_2} \le D^6_{HP_1}.$

- $D^7_{P_4 H} \le D^8_{P_4 P_3} \le D^9_{P_4 P_2} \le D^{10}_{P_4 P_1}.$

- $D^{11}_{GP_4} \le D^{12}_{GH} \le D^{13}_{GP_3} \le D^{14}_{GP_2} \le D^{15}_{GP_1}.$

- $D^{16}_{N_1 G} \le D^{17}_{N_1 P_4} \le D^{18}_{N_1 H} \le D^{19}_{N_1 P_3} \le D^{20}_{N_1 P_2} \le D^{21}_{N_1 P_1}.$

- $D^{22}_{NN_1} \le D^{23}_{NG} \le D^{24}_{NP_4} \le D^{25}_{NH} \le D^{26}_{NP_3} \le D^{27}_{NP_2} \le D^{28}_{NP_1}.$

- $D^{29}_{N_2 N} \le D^{30}_{N_2 N_1} \le D^{31}_{N_2 G} \le D^{32}_{N_2 P_4} \le D^{33}_{N_2 H} \le D^{34}_{N_2 P_3} \le D^{35}_{N_2 P_2} \le D^{36}_{N_2 P_1}.$

- $D^{37}_{AN_2} \le D^{38}_{AN} \le D^{39}_{AN_1} \le D^{40}_{AG} \le D^{41}_{AP_4} \le D^{42}_{AH} \le D^{43}_{AP_3} \le D^{44}_{AP_2} \le D^{45}_{AP_1}.$

- $D^{46}_{P_5 A} \le D^{47}_{P_5 N_2} \le D^{48}_{P_5 N} \le D^{49}_{P_5 N_1} \le D^{50}_{P_5 G} \le D^{51}_{P_5 P_4} \le D^{52}_{P_5 H} \le D^{53}_{P_5 P_3} \le$
  $\le D^{54}_{P_5 P_2} \le D^{55}_{P_5 P_1}.$

- $D^{56}_{RA} \le D^{57}_{RN_2} \le D^{58}_{RN} \le D^{59}_{RN_1} \le D^{60}_{RG} \le D^{61}_{RP_4} \le D^{62}_{RH} \le D^{63}_{RP_3} \le$
  $\le D^{64}_{RP_2} \le D^{65}_{RP_1}.$

- $D^{66}_{SR} \le D^{67}_{SA} \le D^{68}_{SN_2} \le D^{69}_{SN} \le D^{70}_{SN_1} \le D^{71}_{SG} \le D^{72}_{SP_4} \le D^{73}_{SH} \le D^{74}_{SP_3} \le$
  $\le D^{75}_{SP_2} \le D^{76}_{SP_1}.$

- $\left.\begin{matrix} D^{79}_{CP_5} \\ D^{77}_{CS} \le D^{78}_{CR} \end{matrix}\right\} \le D^{80}_{CA} \le D^{81}_{CN_2} \le D^{82}_{CN} \le D^{83}_{CN_1} \le D^{84}_{CG} \le$
  $\le D^{85}_{CP_4} \le D^{86}_{CH} \le D^{87}_{CP_3} \le D^{88}_{CP_2} \le D^{89}_{CP_1}.$

$$\dots \quad (93)$$

The difference measures appearing above are understood as, $D^1_{P_2 P_1} = P_2 - P_1$, $D^{26}_{NP_3} = N - P_3$, $D^{78}_{CR} = C - R$, etc. After simplifications, we can easily check the following equality relations.

1. $2D^{42}_{AH} = 6D^{56}_{RA} = \frac{3}{2}D^{62}_{RH} = 3D^{78}_{CR} = 2D^{80}_{CA} = D^{86}_{CH}$.
2. $2D^{16}_{N_1 G} = 6D^{22}_{NN_1} = \frac{3}{2}D^{23}_{NG} = \frac{1}{2}D^{51}_{P_5 P_4} = 3D^{38}_{AN} = 2D^{39}_{AN_1} = D^{40}_{AG}$.
3. $3D^{58}_{RN} = D^{84}_{CG}$.

Because of above equalities, out of 89 nonnegative differences given in (93), we are left only with 77. These are connected with each other in three different inequalities given in theorem below.

Theorem 9. *The nonnegative difference arising due to (89), given in (93) satisfies the following inequalities:*

$$\frac{1}{2}D^5_{HP_2} \leq \begin{cases} \left.\begin{matrix} \frac{1}{3}D^{14}_{GP_2} \leq \frac{2}{7}D^{20}_{N_1 P_2} \leq \frac{3}{11}D^{27}_{NP_2} \\ \frac{1}{8}D^{89}_{CP_1} \end{matrix}\right\} \leq \begin{cases} \frac{1}{6}D^{88}_{CP_2} \\ \frac{2}{9}D^{54}_{P_5 P_2} \end{cases} \\[4em] \left.\begin{matrix} D^4_{HP_3} \\ \frac{2}{5}D^9_{P_4 P_2} \end{matrix}\right\} \leq \frac{2}{3}D^8_{P_4 P_3} \leq \begin{cases} \frac{2}{9}D^{54}_{P_5 P_2} \\ 2D^7_{P_4 H} \leq D^{12}_{GH} \leq \begin{cases} \frac{2}{3}D^{79}_{CP_5} \\ \frac{1}{5}D^{87}_{CP_3} \leq f^{67}_{SA} \end{cases} \\ \frac{1}{2}D^{13}_{GP_3} \leq 3D^{66}_{SR} \leq \frac{1}{6}D^{88}_{CP_2} \leq \\ \qquad \leq \frac{1}{5}D^{87}_{CP_3} \leq f^{67}_{SA} \end{cases} \end{cases} \leq$$

$$\leq \begin{cases} \left.\begin{matrix} \frac{1}{3}D^{73}_{SH} \leq \frac{1}{2}D^{42}_{AH} \\ \frac{4}{5}D^{68}_{SN_2} \\ \frac{3}{4}D^{69}_{SN} \end{matrix}\right\} \leq \begin{cases} \frac{2}{5}D^{72}_{SP_4} \\ \frac{3}{7}D^{82}_{CN} \end{cases} \leq \begin{cases} \frac{2}{5}D^{83}_{CN_1} \\ \frac{2}{7}D^{85}_{CP_4} \end{cases} \leq \\[2em] \qquad \leq \frac{1}{3}D^{84}_{CG} \leq \frac{12}{11}D^{57}_{RN_2} \\[1em] \left.\begin{matrix} \frac{4}{5}D^{68}_{SN_2} \\ \frac{3}{4}D^{69}_{SN} \end{matrix}\right\} \leq \frac{2}{3}D^{70}_{SN_1} \leq \begin{cases} \frac{1}{3}D^{84}_{CG} \leq \frac{12}{11}D^{57}_{RN_2} \\ \frac{1}{2}D^{71}_{SG} \end{cases} \end{cases} \leq$$

$$\leq \begin{cases} \frac{2}{5}D^{52}_{P_5 H} \leq \frac{3}{5}D^{60}_{RG} \\ \frac{6}{7}D^{59}_{RN_1} \end{cases} \leq 4D^{30}_{N_2 N_1} \leq \frac{4}{3}D^{31}_{N_2 G} \leq D^{40}_{AG} \leq 4D^{37}_{AN_2} \leq$$

$$\leq \frac{2}{3}D^{50}_{P_5 G} \leq D^{49}_{P_5 N_1} \leq \frac{6}{5}D^{48}_{P_5 N_3} \leq \frac{4}{3}D^{47}_{P_5 N_2} \leq 2D^{46}_{P_5 A}.$$

$$(94)$$

$$\frac{1}{2}D^1_{P_2P_1} \le \frac{1}{3}D^3_{P_3P_1} \le$$

$$\le \left\{ \begin{array}{l} \frac{1}{4}D^6_{HP_1} \le \frac{2}{9}D^{10}_{P_4P_1} \le \frac{1}{5}D^{15}_{GP_1} \le \\[2mm] \quad \le \left\{ \begin{array}{l} \frac{2}{11}D^{21}_{N_1P_1} \le \frac{3}{17}D^{28}_{N_3P_1} \le \frac{4}{23}D^{36}_{N_2P_1} \le \left\{ \begin{array}{l} \frac{1}{6}D^{45}_{AP_1} \\[2mm] \frac{2}{7}D^{20}_{N_1P_2} \end{array} \right. \\[6mm] \frac{1}{3}D^{14}_{GP_2} \le \frac{2}{7}D^{20}_{N_1P_2} \end{array} \right. \\[8mm] \left. \begin{array}{l} \frac{1}{4}D^6_{HP_1} \\[2mm] D^2_{P_3P_2} \end{array} \right\} \le \frac{1}{2}D^5_{HP_2} \end{array} \right\} \le$$

$$\le \frac{3}{11}D^{27}_{NP_2} \le \frac{1}{15}D^{35}_{N_2P_2} \le \left\{ \begin{array}{l} \left. \begin{array}{l} 3D^{66}_{SR} \\[2mm] \frac{2}{5}D^{19}_{N_1P_3} \end{array} \right\} \le \frac{3}{8}D^{26}_{NP_3} \le \frac{4}{11}D^{34}_{N_2P_3} \\[6mm] \frac{1}{4}D^{44}_{AP_2} \le \left\{ \begin{array}{l} \frac{1}{5}D^{75}_{SP_2} \\[2mm] \frac{3}{14}D^{64}_{RP_2} \end{array} \right. \end{array} \right\} \le$$

$$\le \frac{1}{3}D^{43}_{AP_3} \le \left\{ \begin{array}{l} \frac{2}{3}D^{18}_{N_1H} \\[2mm] \frac{1}{4}D^{74}_{SP_3} \le \frac{3}{11}D^{63}_{RP_3} \end{array} \right\} \le \frac{3}{5}D^{25}_{NH} \le \frac{4}{7}D^{33}_{N_2H} \le$$

$$\le \left\{ \begin{array}{l} \left. \begin{array}{l} 12D^{29}_{N_2N} \\[2mm] \frac{1}{2}D^{42}_{AH} \end{array} \right\} \le \left\{ \begin{array}{l} D^{17}_{N_1P_4} \le \frac{4}{5}D^{32}_{N_2P_4} \\[2mm] \frac{2}{7}D^{85}_{CP_4} \end{array} \right\} \le \frac{6}{7}D^{24}_{NP_4} \\[8mm] \frac{1}{2}D^{42}_{AH} \le \frac{4}{9}D^{81}_{CN_2} \le \left\{ \begin{array}{l} \frac{1}{3}D^{84}_{CG} \\[2mm] \frac{6}{13}D^{61}_{RP_4} \end{array} \right. \end{array} \right\} \le \frac{2}{3}D^{41}_{AP_4} \le D^{40}_{AG}$$

$$\tag{95}$$

and

$$\left\{ \begin{array}{l} \frac{1}{6}D^{45}_{AP_1} \le \left\{ \begin{array}{l} \frac{3}{20}D^{65}_{RP_1} \le \frac{1}{7}D^{76}_{SP_1} \le \frac{1}{4}D^{44}_{AP_2} \\[2mm] \frac{2}{13}D^{55}_{P_5P_1} \end{array} \right\} \le \frac{1}{3}D^{43}_{AP_3} \le \frac{3}{11}D^{63}_{RP_3} \le \frac{2}{7}D^{53}_{P_5P_3} \\[6mm] \frac{1}{3}D^{14}_{GP_2} \le \frac{2}{3}D^{79}_{CP_5} \le 2D^{11}_{GP_4} \le \frac{1}{2}D^{42}_{AH} \end{array} \right\} \le$$

$$\le \left\{ \begin{array}{l} \frac{3}{7}D^{82}_{CN} \le \frac{2}{7}D^{85}_{CP_4} \\[2mm] \frac{4}{9}D^{81}_{CN_2} \end{array} \right\} \le D^{77}_{CS} \le D^{40}_{AG}.$$

$$\tag{96}$$

Proof. It is based on the following results. Since Lemma 4, Note 1 and Note 2 are frequently used, their citations are excluded. The same is with the derivatives given in (92).

1. For $D^5_{HP_2} \leq \frac{1}{6}D^{14}_{GP_2}$: We have, $\beta_1 = \frac{f''_{HP_2}(1)}{f''_{GP_2}(1)} = \frac{f''_H(1)-f''_{P_2}(1)}{f''_G(1)-f''_{P_2}(1)} = \frac{1}{6}$. In order to show $\frac{1}{6}D^{14}_{GP_2} - D^5_{HP_2} \geq 0$, let us consider, $\frac{1}{6}D^{14}_{GP_2} - D^5_{HP_2} = bg_1(a/b)$, where

$$g_1(x) = \frac{1}{6}f^{14}_{GP_2}(x) - f^5_{HP_2}(x)$$
$$= \frac{\sqrt{x}(2x^3 - 5x^{5/2} + 2x^2 + 2x^{3/2} + 2x - 5\sqrt{x} + 2)}{3(x+1)(x^2+1)}$$
$$= \frac{\sqrt{x}(\sqrt{x}-1)^4(2x+3\sqrt{x}+2)}{3(x+1)(x^2+1)}.$$

Since $g_1(x) \geq 0$, $\forall x > 0$, hence proving the required result.

2. For $D^5_{HP_2} \leq \frac{1}{4}D^{89}_{CP_1}$: We have, $\beta_2 = \frac{f''_{HP_2}(1)}{f''_{CP_1}(1)} = \frac{f''_H(1)-f''_{P_2}(1)}{f''_C(1)-f''_{P_1}(1)} = \frac{1}{4}$. In order to show $\frac{1}{4}D^{89}_{CP_1} - D^5_{HP_2} \geq 0$, let us consider, $\frac{1}{4}D^{89}_{CP_1} - D^5_{HP_2} = bg_2(a/b)$, where

$$g_2(x) = \frac{1}{4}f^{89}_{CP_1}(x) - f^5_{HP_2}(x) = \frac{(x-1)^6}{4(x^2+1)(x^3+1)} \geq 0, \quad \forall x > 0.$$

3. For $D^5_{HP_2} \leq 2D^4_{HP_3}$: We have, $\beta_3 = \frac{f''_{HP_2}(1)}{f''_{HP_3}(1)} = \frac{f''_H(1)-f''_{P_2}(1)}{f''_H(1)-f''_{P_3}(1)} = 2$. In order to show $2f^4_{HP_3} - D^5_{HP_2} \geq 0$, let us consider, $2D^4_{HP_3} - D^5_{HP_2} = bg_3(a/b)$, where

$$g_3(x) = 2f^4_{HP_3}(x) - f^5_{HP_2}(x)$$
$$= \frac{x(x^3 - 3x^{5/2} + 3x^2 - 2x^{3/2} + 3x - 3\sqrt{x} + 1)}{(x+1)(x^2+1)(x-\sqrt{x}+1)}$$
$$= \frac{x(\sqrt{x}-1)^4(x+\sqrt{x}+1)}{(x+1)(x^2+1)(x-\sqrt{x}+1)} \geq 0, \quad \forall x > 0.$$

4. For $D^5_{HP_2} \leq \frac{1}{5}D^9_{P_4P_2}$: We have, $\beta_4 = \frac{f''_{HP_2}(1)}{f''_{P_4P_2}(1)} = \frac{f''_H(1)-f''_{P_2}(1)}{f''_{P_4}(1)-f''_{P_2}(1)} = \frac{1}{5}$. In order to show $\frac{1}{5}D^9_{P_4P_2} - D^5_{HP_2} \geq 0$, let us consider, $\frac{1}{5}D^9_{P_4P_2} - D^5_{HP_2} = bg_4(a/b)$, where

$$g_4(x) = \frac{1}{5}f^9_{P_4P_2}(x) - f^5_{HP_2}(x)$$
$$= \frac{x(7x^3 + 9x^2 - 18x^{5/2} + 9x + 4x^{3/2} - 18\sqrt{x} + 7)}{5(x+1)(\sqrt{x}+1)^2(x^2+1)}$$
$$= \frac{x(7x + 10\sqrt{x} + 7)(\sqrt{x}-1)^4}{5(x+1)(\sqrt{x}+1)^2(x^2+1)} \geq 0, \quad \forall x > 0.$$

5. For $D_{GP_2}^{14} \leq \frac{6}{7}D_{N_1P_2}^{20}$: We have, $\beta_5 = \frac{f_{GP_2}''(1)}{f_{N_1P_2}''(1)} = \frac{f_G''(1)-f_{P_2}''(1)}{f_{N_1}''(1)-f_{P_2}''(1)} = \frac{6}{7}$. In order to show $\frac{6}{7}D_{N_1P_2}^{20} - D_{GP_2}^{14} \geq 0$, let us consider, $\frac{6}{7}D_{N_1P_2}^{20} - D_{GP_2}^{14} = bg_5(a/b)$, where

$$g_5(x) = \frac{6}{7}f_{N_1P_2}^{20}(x) - f_{GP_2}^{14}(x)$$

$$= \frac{(3x^3 + 5x - 8x^{5/2} - 8\sqrt{x} + 5x^2 + 3)}{14(x^2 + 1)}$$

$$= \frac{(3x + 4\sqrt{x} + 3)(\sqrt{x} - 1)^4}{14(x^2 + 1)} \geq 0, \quad \forall x > 0.$$

6. For $D_{N_1P_2}^{20} \leq \frac{21}{22}D_{NP_2}^{27}$: We have, $\beta_6 = \frac{f_{N_1P_2}''(1)}{f_{NP_2}''(1)} = \frac{f_{N_1}''(1)-f_{P_2}''(1)}{f_N''(1)-f_{P_2}''(1)} = \frac{21}{22}$. In order to show $\frac{21}{22}D_{NP_2}^{27} - D_{N_1P_2}^{20} \geq 0$, let us consider, $\frac{21}{22}D_{NP_2}^{27} - D_{N_1P_2}^{20} = bg_6(a/b)$, where

$$g_6(x) = \frac{21}{22}f_{NP_2}^{27}(x) - f_{N_1P_2}^{20}(x)$$

$$= \frac{(3x^3 + 5x - 8x^{5/2} - 8\sqrt{x} + 5x^2 + 3)}{44(x^2 + 1)}$$

$$= \frac{(3x + 4\sqrt{x} + 3)(\sqrt{x} - 1)^4}{44(x^2 + 1)} \geq 0, \quad \forall x > 0.$$

7. For $D_{NP_2}^{27} \leq \frac{11}{12}D_{CP_2}^{88}$: We have, $\beta_7 = \frac{f_{NP_2}''(1)}{f_{CP_2}''(1)} = \frac{f_N''(1)-f_{P_2}''(1)}{f_C''(1)-f_{P_2}''(1)} = \frac{11}{12}$. In order to show $\frac{11}{12}D_{CP_2}^{88} - D_{NP_2}^{27} \geq 0$, let us consider, $\frac{11}{12}D_{CP_2}^{88} - D_{NP_2}^{27} = bg_7(a/b)$, where

$$g_7(x) = \frac{11}{12}f_{CP_2}^{88}(x) - f_{NP_2}^{27}(x)$$

$$= \frac{(5x^4 - 6x^{7/2} - 5x^3 - 6x^{5/2} + 24x^2 - 6x^{3/2} - 5x - 6\sqrt{x} + 5)}{18(x + 1)(x^2 + 1)}$$

$$= \frac{(5x^2 + 14x^{3/2} + 21x + 14\sqrt{x} + 5)(\sqrt{x} - 1)^4}{18(x + 1)(x^2 + 1)} \geq 0, \quad \forall x > 0.$$

8. For $D_{NP_2}^{27} \leq \frac{22}{27}D_{P_5P_2}^{54}$: We have, $\beta_8 = \frac{f_{NP_2}''(1)}{f_{P_5P_2}''(1)} = \frac{f_N''(1)-f_{P_2}''(1)}{f_{P_5}''(1)-f_{P_2}''(1)} = \frac{22}{27}$. In order to show $\frac{22}{27}D_{P_5P_2}^{54} - D_{NP_2}^{27} \geq 0$, let us consider, $\frac{22}{27}D_{P_5P_2}^{54} - D_{NP_2}^{27} = bg_8(a/b)$, where

$$g_8(x) = \frac{22}{27}f_{P_5P_2}^{54}(x) - f_{NP_2}^{27}(x) = \frac{u_8(x)}{27(\sqrt{x} + 1)^2(x^2 + 1)}$$

with

$$u_8(x) = 13x^4 - 27x^{7/2} + 13x^3 - 17x^{5/2} +$$
$$+ 36x^2 - 17x^{3/2} + 13x - 27\sqrt{x} + 13$$
$$= (\sqrt{x} - 1)^4(13x^2 + 25x^{3/2} + 35x + 25\sqrt{x} + 13) \geq 0, \quad \forall x > 0.$$

9. For $D_{CP_1}^{89} \leq \frac{4}{3}D_{CP_2}^{88}$: We have, $\beta_9 = \frac{f_{CP_1}''(1)}{f_{CP_2}''(1)} = \frac{f_C''(1) - f_{P_1}''(1)}{f_C''(1) - f_{P_2}''(1)} = \frac{4}{3}$. In order to show $\frac{4}{3}D_{CP_2}^{88} - D_{CP_1}^{89} \geq 0$, let us consider, $\frac{4}{3}D_{CP_2}^{88} - D_{CP_1}^{89} = bg_9(a/b)$, where

$$g_9(x) = \frac{4}{3}f_{CP_2}^{88}(x) - f_{CP_1}^{89}(x) = \frac{(x+1)^2(x-1)^4}{3(x^3+1)(x^2+1)} \geq 0, \quad \forall x > 0.$$

10. For $D_{CP_1}^{89} \leq \frac{16}{9}D_{P_5P_2}^{54}$: We have, $\beta_{10} = \frac{f_{CP_1}''(1)}{f_{P_5P_2}''(1)} = \frac{f_C''(1) - f_{P_1}''(1)}{f_{P_5}''(1) - f_{P_2}''(1)} = \frac{16}{9}$. In order to show $\frac{16}{9}D_{P_5P_2}^{54} - D_{CP_1}^{89} \geq 0$, let us consider, $\frac{16}{9}D_{P_5P_2}^{54} - D_{CP_1}^{89} = bg_{10}(a/b)$, where

$$g_{10}(x) = \frac{16}{9}f_{P_5P_2}^{54}(x) - f_{CP_1}^{89}(x) = \frac{u_{10}(x)}{9(\sqrt{x}+1)^2(x^3+1)(x^2+1)},$$

with

$$u_{10}(x) = 7x^7 - 18x^{13/2} + 25x^6 + 4x^{11/2} - 9x^5 - 86x^{9/2} + 41x^4 +$$
$$+ 72x^{7/2} + 41x^3 - 86x^{5/2} - 9x^2 + 4x^{3/2} + 25x - 18\sqrt{x} + 7$$
$$= (\sqrt{x} - 1)^4(7x^5 + 10x^{9/2} + 23x^4 + 64x^{7/2} + 142x^3 +$$
$$+ 180x^{5/2} + 142x^2 + 64x^{3/2} + 23x + 10\sqrt{x} + 7) \geq 0, \quad \forall x > 0.$$

11. For $D_{HP_3}^4 \leq \frac{2}{3}D_{P_4P_3}^8$: We have, $\beta_{11} = \frac{f_{HP_3}''(1)}{f_{P_4P_3}''(1)} = \frac{f_H''(1) - f_{P_3}''(1)}{f_{P_4}''(1) - f_{P_3}''(1)} = \frac{2}{3}$. In order to show $\frac{2}{3}D_{P_4P_3}^8 - D_{HP_3}^4 \geq 0$, let us consider, $\frac{2}{3}D_{P_4P_3}^8 - D_{HP_3}^4 = bg_{11}(a/b)$, where

$$g_{11}(x) = \frac{2}{3}f_{P_4P_3}^8(x) - f_{HP_3}^4(x) = \frac{x^2 - 4x^{3/2} + 6x - 4\sqrt{x} + 1}{(x+1)(\sqrt{x}+1)^2(x - \sqrt{x}+1)}$$
$$= \frac{(\sqrt{x}-1)^4}{(x+1)(\sqrt{x}+1)(x^{3/2}+1)} \geq 0, \quad \forall x > 0.$$

12. For $D_{P_4P_2}^9 \leq \frac{5}{3}D_{P_4P_3}^8$: We have, $\beta_{12} = \frac{f_{P_4P_2}''(1)}{f_{P_4P_3}''(1)} = \frac{f_{P_4}''(1) - f_{P_2}''(1)}{f_{P_4}''(1) - f_{P_3}''(1)} = \frac{5}{3}$. In order to show $\frac{5}{3}D_{P_4P_3}^8 - D_{P_4P_2}^9 \geq 0$, let us consider, $\frac{5}{3}D_{P_4P_3}^8 - D_{P_4P_2}^9 =$

$bg_{12}(a/b)$, where

$$g_{12}(x) = \frac{5}{3} f^8_{P_4 P_3}(x) - f^9_{P_4 P_2}(x)$$

$$= \frac{x(2x^3 + 2x^2 - 5x^{5/2} + 2x^{3/2} + 2x - 5\sqrt{x} + 2)}{(x^2 + 1)(\sqrt{x} + 1)^2 (x - \sqrt{x} + 1)}$$

$$= \frac{(2x + 3\sqrt{x} + 2)(\sqrt{x} - 1)^4}{(x^2 + 1)(\sqrt{x} + 1)(x^{3/2} + 1)} \geq 0, \quad \forall x > 0.$$

13. For $D^8_{P_4 P_3} \leq \frac{1}{3} D^{54}_{P_5 P_2}$: We have, $\beta_{13} = \frac{f''_{P_4 P_3}(1)}{f''_{P_5 P_2}(1)} = \frac{f''_{P_4}(1) - f''_{P_3}(1)}{f''_{P_5}(1) - f''_{P_2}(1)} = \frac{1}{3}$.
In order to show $\frac{1}{3} D^{54}_{P_5 P_2} - D^8_{P_4 P_3} \geq 0$, let us consider, $\frac{1}{3} D^{54}_{P_5 P_2} - D^8_{P_4 P_3} = bg_{13}(a/b)$, where

$$g_{13}(x) = \frac{1}{3} f^{54}_{P_5 P_2}(x) - f^8_{P_4 P_3}(x) = \frac{u_{13}(x)}{3(x^2 + 1)(\sqrt{x} + 1)(x^{3/2} + 1)}$$

with

$$u_{13}(x) = x^5 - x^{9/2} - 7x^4 + 15x^{7/2} - 6x^3 - 4x^{5/2} - 6x^2 + 15x^{3/2} - 7x -$$
$$- \sqrt{x} + 1$$
$$= (\sqrt{x} - 1)^4 \big[(x + \sqrt{x} + 1)(x - 1)^2 + 2\sqrt{x}(x^2 - x + 1) + x^{3/2} \big] \geq 0,$$
$$\forall x > 0.$$

14. For $D^8_{P_4 P_3} \leq 3 D^7_{P_4 H}$: We have, $\beta_{14} = \frac{f''_{P_4 P_3}(1)}{f''_{P_4 H}(1)} = \frac{f''_{P_4}(1) - f''_{P_3}(1)}{f''_{P_4}(1) - f''_{H}(1)} = \frac{1}{3}$.
In order to show $3 D^7_{P_4 H} - D^8_{P_4 P_3} \geq 0$, let us consider, $3 D^7_{P_4 H} - D^8_{P_4 P_3} = bg_{14}(a/b)$, where

$$g_{14}(x) = 3 f^7_{P_4 H}(x) - f^8_{P_4 P_3}(x)$$

$$= \frac{3x(x^2 - 4x^{3/2} + 6x - 4\sqrt{x} + 1)}{(\sqrt{x} + 1)(x^{3/2} + 1)(x + 1)}$$

$$= \frac{3x(\sqrt{x} - 1)^4}{(\sqrt{x} + 1)(x^{3/2} + 1)(x + 1)} \geq 0, \quad \forall x > 0.$$

15. For $D^8_{P_4 P_3} \leq \frac{3}{4} D^{13}_{G P_3}$: We have, $\beta_{15} = \frac{f''_{P_4 P_3}(1)}{f''_{G P_3}(1)} = \frac{f''_{P_4}(1) - f''_{P_3}(1)}{f''_{G}(1) - f''_{P_3}(1)} = \frac{3}{4}$.
In order to show $\frac{3}{4} D^{13}_{G P_3} - D^8_{P_4 P_3} \geq 0$, let us consider, $\frac{3}{4} D^{13}_{G P_3} - D^8_{P_4 P_3} =$

$bg_{15}(a/b)$, where

$$g_{15}(x) = \frac{3}{4}f^{13}_{GP_3}(x) - f^8_{P_4P_3}(x)$$

$$= \frac{3\sqrt{x}(x^2 - 4x^{3/2} + 6x - 4\sqrt{x} + 1)}{4(\sqrt{x} + 1)(x^{3/2} + 1)}$$

$$= \frac{3\sqrt{x}(\sqrt{x} - 1)^4}{4(\sqrt{x} + 1)(x^{3/2} + 1)} \geq 0, \quad \forall x > 0.$$

16. For $D^{13}_{GP_3} \leq 6D^{66}_{SR}$: We have, $\beta_{16} = \frac{f''_{GP_3}(1)}{f''_{SR}(1)} = \frac{f''_G(1) - f''_{P_3}(1)}{f''_S(1) - f''_R(1)} = 6$. In order to show $6D^{66}_{SR} - D^{13}_{GP_3} \geq 0$, let us consider, $6D^{66}_{SR} - D^{13}_{GP_3} = bg_{16}(a/b)$, where

$$g_{16}(x) = 6f^{66}_{SR}(x) - f^{13}_{GP_3}(x) = \frac{u_{16}(x)}{(x - \sqrt{x} + 1)(x + 1)}$$

with

$$u_{16}(x) = 3\sqrt{2x^2 + 2}(x - \sqrt{x} + 1)(x + 1) -$$
$$- \left(4x^3 - 3x^{5/2} + 6x^2 - 2x^{3/2} + 6x - 3\sqrt{x} + 4\right)$$
$$= 3\sqrt{2x^2 + 2}(x - \sqrt{x} + 1)(x + 1) - \left[x^2(2\sqrt{x} - 1)^2 +\right.$$
$$\left. + x^{5/2} + 4x^2 + x(\sqrt{x} - 1)^2 + 4x + (\sqrt{x} - 2)^2 + \sqrt{x}\right].$$

Now we shall show that $u_{16}(x) \geq 0$, $\forall x > 0$. Let us consider

$$u_{16}(x) = \left[3\sqrt{2x^2 + 2}(x - \sqrt{x} + 1)(x + 1)\right]^2 - \left[x^2(2\sqrt{x} - 1)^2 +\right.$$
$$\left. + x^{5/2} + 4x^2 + x(\sqrt{x} - 1)^2 + 4x + (\sqrt{x} - 2)^2 + \sqrt{x}\right]^2$$
$$= (\sqrt{x} - 1)^6 \left(2x^3 + 3x^2 + 2x^{3/2} + 3x + 2\right) \geq 0, \quad \forall x > 0.$$

17. For $D^{66}_{SR} \leq \frac{1}{18}D^{88}_{CP_2}$: We have, $\beta_{17} = \frac{f''_{SR}(1)}{f''_{CP_2}(1)} = \frac{f''_S(1) - f''_R(1)}{f''_G(1) - f''_{P_2}(1)} = \frac{1}{18}$. In order to show $\frac{1}{18}D^{88}_{CP_2} - D^{66}_{SR} \geq 0$, let us consider, $\frac{1}{18}D^{88}_{CP_2} - D^{66}_{SR} = bg_{17}(a/b)$, where

$$g_{17}(x) = \frac{1}{18}f^{88}_{CP_2}(x) - f^{66}_{SR}(x) = \frac{u_{17}(x)}{18(x^2 + 1)(x + 1)}$$

with

$$u_{17}(x) = \left(x^2 + x + 1\right)\left(12(x^2 + 1) + (x - 1)^2\right) - 9\sqrt{2x^2 + 2}(x^2 + 1)(x + 1).$$

Now we shall show that $u_{17}(x) \geq 0$, $\forall x > 0$. Let us consider

$$
\begin{aligned}
v_{17}(x) &= \left[(x^2 + x + 1)(12(x^2 + 1) + (x - 1)^2) \right]^2 \\
&\quad - \left[9\sqrt{2x^2 + 2}(x^2 + 1)(x + 1) \right]^2 \\
&= (x - 1)^4 (7x^4 - 10x^3 + 15x^2 - 10x + 7) \geq 0, \quad \forall x > 0.
\end{aligned}
$$

18. For $D_{CP_2}^{88} \leq \frac{6}{5} D_{CP_3}^{87}$: We have, $\beta_{17} = \frac{g_{CP_2}''(1)}{g_{CP_3}''(1)} = \frac{f_C''(1) - f_{P_2}''(1)}{f_C''(1) - f_{P_3}''(1)} = \frac{6}{5}$. In order to show $\frac{6}{5} D_{CP_3}^{87} - D_{CP_2}^{88} \geq 0$, let us consider, $\frac{6}{5} D_{CP_3}^{87} - D_{CP_2}^{88} = b\eta_{17}(a/b)$, where

$$
\eta_{17}(x) = \frac{6}{5} g_{CP_3}^{87}(x) - g_{CP_2}^{88}(x) = \frac{u_{17}(x)}{5(x - \sqrt{x} + 1)(x^2 + 1)(x + 1)}
$$

with

$$
\begin{aligned}
u_{17}(x) &= x^5 - x^{9/2} - 5x^{7/2} + 11x^3 - 12x^{5/2} + 11x^2 - 5x^{3/2} - \sqrt{x} + 1 \\
&= (\sqrt{x} - 1)^4 (x^3 + 3x^{5/2} + 6x^2 + 5x^{3/2} + 6x + 3\sqrt{x} + 1) \geq 0, \\
&\quad \forall x > 0.
\end{aligned}
$$

19. For $D_{CP_3}^{87} \leq 5D_{SA}^{67}$: We have, $\beta_{19} = \frac{f_{CP_3}''(1)}{f_{SA}''(1)} = \frac{f_C''(1) - f_{P_3}''(1)}{f_C''(1) - f_A''(1)} = 5$. In order to show $5D_{SA}^{67} - D_{CP_3}^{87} \geq 0$, let us consider, $5D_{SA}^{67} - D_{CP_3}^{87} = bg_{19}(a/b)$, where

$$
g_{19}(x) = 5f_{SA}^{67}(x) - f_{CP_3}^{87}(x) = \frac{u_{19}(x)}{2(x - \sqrt{x} + 1)(x + 1)}
$$

with

$$
\begin{aligned}
u_{19}(x) &= 5\sqrt{2x^2 + 2}(x - \sqrt{x} + 1)(x + 1) - \\
&\quad - \left[(7x^2 + 7)(x - \sqrt{x} + 1) + 8x(\sqrt{x} - 1)^2 + 6x^{3/2} \right].
\end{aligned}
$$

Now we shall show that $u_{19}(x) \geq 0$, $\forall x > 0$. Let us consider

$$
\begin{aligned}
v_{19}(x) &= \left[5\sqrt{2x^2 + 2}(x - \sqrt{x} + 1)(x + 1) \right]^2 - \\
&\quad - \left[(7x^2 + 7)(x - \sqrt{x} + 1) + 8x(\sqrt{x} - 1)^2 + 6x^{3/2} \right]^2 \\
&= (\sqrt{x} - 1)^6 (x^3 + 4x^{5/2} + 10x^{3/2} + 4\sqrt{x} + 1) \geq 0, \quad \forall x > 0.
\end{aligned}
$$

20. For $D_{P_4 H}^{7} \leq \frac{1}{2} D_{GH}^{12}$: We have, $\beta_{20} = \frac{f_{P_4 H}''(1)}{f_{GH}''(1)} = \frac{f_{P_4}''(1) - f_H''(1)}{f_G''(1) - f_H''(1)} = \frac{1}{2}$. In order to show $\frac{1}{2} D_{GH}^{12} - D_{P_4 H}^{7} \geq 0$, let us consider, $\frac{1}{2} D_{GH}^{12} - D_{P_4 H}^{7} = bg_{20}(a/b)$,

where

$$g_{20}(x) = \frac{1}{2}f_{GH}^{12}(x) - f_{P_4H}^7(x)$$
$$= \frac{\sqrt{x}(x^2 - 4x^{3/2} + 6x - 4\sqrt{x} + 1)}{2(\sqrt{x} + 1)^2(x + 1)}$$
$$= \frac{\sqrt{x}(\sqrt{x} - 1)^2}{2(\sqrt{x} + 1)^2(x + 1)} \geq 0, \quad \forall x > 0.$$

21. For $D_{GH}^{12} \leq \frac{2}{3}D_{CP_5}^{79}$: We have, $\beta_{21} = \frac{f_{GH}''(1)}{f_{CP_5}''(1)} = \frac{f_G''(1) - f_H''(1)}{f_C''(1) - f_{P_5}''(1)} = \frac{2}{3}$. In order to show $\frac{2}{3}D_{CP_5}^{79} - D_{GH}^{12} \geq 0$, let us consider, $\frac{2}{3}D_{CP_5}^{79} - D_{GH}^{12} = bg_{21}(a/b)$, where

$$g_{21}(x) = \frac{2}{3}f_{CP_5}^{79}(x) - f_{GH}^{12}(x)$$
$$= \frac{\sqrt{x}(x^2 - 4x^{3/2} + 6x - 4\sqrt{x} + 1)}{3(\sqrt{x} + 1)^2(x + 1)}$$
$$= \frac{\sqrt{x}(\sqrt{x} - 1)^2}{3(\sqrt{x} + 1)^2(x + 1)} \geq 0, \quad \forall x > 0.$$

22. For $D_{GH}^{12} \leq \frac{1}{5}D_{CP_3}^{87}$: We have, $\beta_{22} = \frac{f_{GH}''(1)}{f_{CP_3}''(1)} = \frac{f_G''(1) - f_H''(1)}{f_C''(1) - f_{P_3}''(1)} = \frac{1}{5}$. In order to show $\frac{1}{5}D_{CP_3}^{87} - D_{GH}^{12} \geq 0$, let us consider, $\frac{1}{5}D_{CP_3}^{87} - D_{GH}^{12} = bg_{20}(a/b)$, where

$$g_{20}(x) = \frac{1}{5}f_{CP_3}^{87}(x) - f_{GH}^{12}(x)$$
$$= \frac{x^3 - 6x^{5/2} + 15x^2 - 20x^{3/2} + 15x - 6\sqrt{x} + 1}{5(x - \sqrt{x} + 1)(x + 1)}$$
$$= \frac{(\sqrt{x} - 1)^6}{5[(\sqrt{x} - 1)^2 + \sqrt{x}](x + 1)} \geq 0, \quad \forall x > 0.$$

23. For $D_{P_5P_2}^{54} \leq \frac{3}{2}D_{SH}^{73}$: We have, $\beta_{23} = f_{P_5P_2}''(1)/f_{SH}''(1) = \frac{3}{2}$. In order to show $\frac{3}{2}D_{SH}^{73} - D_{P_5P_2}^{54} \geq 0$, let us consider, $\frac{3}{2}D_{SH}^{73} - D_{P_5P_2}^{54} = bg_{23}(a/b)$, where

$$g_{23}(x) = \frac{3}{2}f_{SH}^{73}(x) - f_{P_5P_2}^{54}(x) = \frac{u_{23}(x)}{2(\sqrt{x} + 1)^2(x^2 + 1)(x + 1)}$$

with

$$u_{23}(x) = \left(3x^5 + 10x^4 + 2x^{7/2} + x^3 + 6x^2(\sqrt{x} - 1)^2 + x^2 + 2x^{3/2} + \right.$$
$$\left. + 10x + 3\right) - \sqrt{2x^2 + 2}(\sqrt{x} + 1)^2(x + 1)(x^2 + 1).$$

Now we shall show that $u_{23}(x) \geq 0$, $\forall x > 0$. Let us consider

$$v_{23}(x) = \left(3x^5 + 10x^4 + 2x^{7/2} + x^3 + 6x^2(\sqrt{x} - 1)^2 + x^2 + 2x^{3/2} + \right.$$
$$\left. + 10x + 3\right)^2 - \left[\sqrt{2x^2 + 2}(\sqrt{x} + 1)^2(x + 1)(x^2 + 1)\right]^2.$$

After simplifications, we have

$$v_{23}(x) = (\sqrt{x} - 1)^2 \left(7x^9 + 6x^{17/2} + 49x^8 + 80x^{15/2} + 219x^7 + \right.$$
$$+ 278x^{13/2} + 459x^6 + 360x^{11/2} + 370x^5 +$$
$$+ 184x^{9/2} + 370x^4 + 360x^{7/2} + 459x^3 +$$
$$\left. + 278x^{5/2} + 219x^2 + 80x^{3/2} + 49x + 6\sqrt{x} + 7\right) \geq 0, \quad \forall x > 0.$$

24. For $D_{P_5 P_2}^{54} \leq \frac{18}{5} D_{SN_2}^{68}$: We have, $\beta_{24} = f''_{P_5 P_2}(1)/f''_{SN_2}(1) = \frac{18}{5}$. In order to show $\frac{18}{5} D_{SN_2}^{68} - D_{P_5 P_2}^{54} \geq 0$, let us consider, $\frac{18}{5} D_{SN_2}^{68} - D_{P_5 P_2}^{54} = b g_{24}(a/b)$, where

$$g_{24}(x) = \frac{18}{5} f_{SN_2}^{68}(x) - f_{P_5 P_2}^{54}(x) = \frac{u_{24}(x)}{10(\sqrt{x} + 1)^2(x^2 + 1)}$$

with

$$u_{24}(x) = 18\sqrt{2x^2 + 2}(\sqrt{x} + 1)^2(x^2 + 1) -$$
$$- \left[10(x^{3/2} - 1)^2(x + 1) + 9\sqrt{2x + 2}(\sqrt{x} + 1)^3(x^2 + 1)\right].$$

Now we shall show that $u_{24}(x) \geq 0$, $\forall x > 0$. Let us consider

$$v_{24}(x) = \left[18\sqrt{2x^2 + 2}(\sqrt{x} + 1)^2(x^2 + 1)\right]^2 -$$
$$- \left[10(x^{3/2} - 1)^2(x + 1) + 9\sqrt{2x + 2}(\sqrt{x} + 1)^3(x^2 + 1)\right]^2.$$

After simplifications, we have

$$v_{24}(x) = 2(\sqrt{x} - 1)^2 \left[193x^7 + 1196x^{13/2} + 2747x^6 + 3688x^{11/2} + \right.$$
$$+ 3283x^5 + 4088x^{9/2} + 6537x^4 + 8376x^{7/2} + 6537x^3 +$$
$$+ 4088x^{5/2} + 3283x^2 + 3688x^{3/2} + 2747x + 1196\sqrt{x} +$$
$$\left. + 193 - 90\sqrt{2x + 2}(x + \sqrt{x} + 1)^2(\sqrt{x} + 1)^3(x + 1)(x^2 + 1)\right].$$

Let us apply again the procedure (1), we get

$$p_{24}(x) = \left[193x^7 + 1196x^{13/2} + 2747x^6 + 3688x^{11/2} + 3283x^5 + \right.$$
$$+ 4088x^{9/2} + 6537x^4 + 8376x^{7/2} + 6537x^3 + 4088x^{5/2} +$$
$$+ 3283x^2 + 3688x^{3/2} + 2747x + 1196\sqrt{x} + 193\big]^2$$
$$\left. - \left[90\sqrt{2x+2}(x+\sqrt{x}+1)^2(\sqrt{x}+1)^3(x+1)(x^2+1)\right]^2.\right.$$

After simplifications, we get

$$p_{24}(x) = (\sqrt{x}-1)^2\big(21049 + 2310817x + 25551870x^2 + 341754\sqrt{x} +$$
$$+ 9261072x^{3/2} + 232560190x^{9/2} + 175611907x^9 + 290764347x^5 +$$
$$+ 88137742x^3 + 175611907x^4 + 128294160x^{7/2} + 52840644x^{5/2} +$$
$$+ 356742600x^{13/2} + 333277120x^{11/2} + 25551870x^{11} +$$
$$+ 52840644x^{21/2} + 88137742x^{10} + 128294160x^{19/2} + 21049x^{13} +$$
$$+ 341754x^{25/2} + 2310817x^{12} + 9261072x^{23/2} + 232560190x^{17/2} +$$
$$+ 333277120x^{15/2} + 290764347x^8 + 352628508x^6 + 352628508x^7\big).$$

Here we applied Note 2 twice to get the required result. Since $p_{24}(x) \geq 0$, giving $v_{24}(x) \geq 0$ and $u_{24}(x) \geq 0$, $\forall x > 0$, hence proving the required result.

25. For $D^{54}_{P_5 P_2} \leq \frac{27}{8}D^{69}_{SN_3}$: We have, $\beta_{25} = f''_{P_5 P_2}(1)/f''_{SN_3}(1) = \frac{27}{8}$. In order to show $\frac{27}{8}D^{69}_{SN_3} - D^{54}_{P_5 P_2} \geq 0$, let us consider, $\frac{27}{8}D^{69}_{SN_3} - D^{54}_{P_5 P_2} = bg_{25}(a/b)$, where

$$g_{25}(x) = \frac{27}{8}f^{69}_{SN_3}(x) - f^{54}_{P_5 P_2}(x) = \frac{u_{25}(x)}{16(\sqrt{x}+1)^2(x^2+1)}$$

with

$$u_{25}(x) = 27\sqrt{2x^2+2}(\sqrt{x}+1)^2(x^2+1) - 2(x+\sqrt{x}+1)\times$$
$$\times \left(17x^3 + 10x^{5/2} + 9x^2 + 8x(\sqrt{x}-1)^2 + 9x + 10\sqrt{x} + 17\right).$$

Now we shall show that $u_{25}(x) \geq 0$, $\forall x > 0$. Let us consider

$$v_{25}(x) = \left[27\sqrt{2x^2+2}(\sqrt{x}+1)^2(x^2+1)\right]^2 - \left[2(x+\sqrt{x}+1)\times\right.$$
$$\left.\times \left(17x^3 + 10x^{5/2} + 9x^2 + 8x(\sqrt{x}-1)^2 + 9x + 10\sqrt{x} + 17\right)\right]^2.$$

After simplifications, we have

$$v_{25}(x) = (\sqrt{x} - 1)^4 \big(302 + 3368\sqrt{x} + 11508x + 21864x^{3/2} + 24842x^2 + \\ + 19088x^{5/2} + 14312x^3 + 19088x^{7/2} + 24842x^4 + \\ + 21864x^{9/2} + 11508x^5 + 3368x^{11/2} + 302x^6\big) \geq 0, \quad \forall x > 0.$$

26. For $D_{CP_5}^{79} \leq \frac{1}{2}D_{SH}^{73}$: We have, $\beta_{26} = f_{CP_5}''(1)/f_{SH}''(1) = \frac{1}{2}$. In order to show $\frac{1}{2}D_{SH}^{73} - D_{CP_5}^{79} \geq 0$, let us consider, $\frac{1}{2}D_{SH}^{73} - D_{CP_5}^{79} = bg_{26}(a/b)$, where

$$g_{26}(x) = \frac{1}{2}f_{SH}^{73}(x) - f_{CP_5}^{79}(x) = \frac{u_{26}(x)}{4(\sqrt{x} + 1)^2(x + 1)}$$

with

$$u_{26}(x) = \sqrt{2x^2 + 2}(\sqrt{x} + 1)^2(x + 1) - 4\sqrt{x}\big(2x^2 - x^{3/2} + 2x - \sqrt{x} + 2\big)$$
$$= \sqrt{2x^2 + 2}(\sqrt{x} + 1)^2(x + 1) - 4\sqrt{x}\big[x^2 + (x + 1)(x - \sqrt{x} + 1) + 1\big].$$

Now we shall show that $u_{26}(x) \geq 0$, $\forall x > 0$. Let us consider

$$v_{26}(x) = \big[\sqrt{2x^2 + 2}(\sqrt{x} + 1)^2(x + 1)\big]^2 - 16x\big[x^2 + (x + 1)(x - \sqrt{x} + 1) + 1\big]^2.$$

After simplifications, we have

$$v_{26}(x) = (\sqrt{x} - 1)^4\big[2x^4 + 16x^{7/2} + 4x^3 + 10x^{5/2} + \\ + 6x^{3/2}(x - \sqrt{x} + 1) + 10x^{3/2} + 4x + 16\sqrt{x} + 2\big] \geq 0, \quad \forall x > 0.$$

27. For $D_{SN_2}^{68} \leq \frac{6}{5}D_{CP_5}^{79}$: We have, $\beta_{27} = f_{SN_2}''(1)/f_{CP_5}''(1) = \frac{6}{5}$. In order to show $\frac{6}{5}D_{CP_5}^{79} - D_{SN_2}^{68} \geq 0$, let us consider, $\frac{6}{5}D_{CP_5}^{79} - D_{SN_2}^{68} = bg_{27}(a/b)$, where

$$g_{27}(x) = \frac{6}{5}f_{CP_5}^{79}(x) - f_{SN_2}^{68}(x) = \frac{u_{27}(x)}{10(\sqrt{x} + 1)^2(x + 1)}$$

with

$$u_{27}(x) = 6\sqrt{2x^2 + 2}(\sqrt{x} + 1)^2(x + 1) - \\ - \big[3\sqrt{2x + 2}(\sqrt{x} + 1)^3(x + 1) + 20\sqrt{x}(x + \sqrt{x} + 1)(\sqrt{x} - 1)^2\big].$$

Now we shall show that $u_{26}(x) \geq 0$, $\forall x > 0$. Let us consider

$$v_{27}(x) = \big[6\sqrt{2x^2 + 2}(\sqrt{x} + 1)^2(x + 1)\big]^2 - \\ - \big[3\sqrt{2x + 2}(\sqrt{x} + 1)^3(x + 1) + 20\sqrt{x}(x + \sqrt{x} + 1)(\sqrt{x} - 1)^2\big]^2.$$

After simplifications, we have

$$v_{27}(x) = 2(\sqrt{x} - 1)^2 \big[27x^5 + 144x^{9/2} + 187x^4 + 720x^{7/2} + 1026x^3 +$$
$$+ 1552x^{5/2} + 1026x^2 + 720x^{3/2} + 187x + 144\sqrt{x} + 27 -$$
$$- 60\sqrt{x}\sqrt{2x + 2}(x + \sqrt{x} + 1)(\sqrt{x} + 1)^3(x + 1)\big]$$

Let us apply again the procedure (1), we get

$$p_{27}(x) = \big(27x^5 + 144x^{9/2} + 187x^4 + 720x^{7/2} + 1026x^3 +$$
$$+ 1552x^{5/2} + 1026x^2 + 720x^{3/2} + 187x + 144\sqrt{x} + 27\big)^2 -$$
$$- \big[60\sqrt{x}\sqrt{2x + 2}(x + \sqrt{x} + 1)(\sqrt{x} + 1)^3(x + 1)\big]^2$$

After simplifications, we get

$$p_{27}(x) = (\sqrt{x} - 1)^2 \big[729 + 9234\sqrt{x} + 41373x + 108648x^{3/2} +$$
$$+ 236056x^2 + 335240x^{5/2} + 355728x^3 + 133272x^{7/2} -$$
$$- 110046x^4 - 285108x^{9/2} - 110046x^5 + 133272x^{11/2} +$$
$$+ 355728x^6 + 335240x^{13/2} + 236056x^7 +$$
$$+ 108648x^{15/2} + 41373x^8 + 9234x^{17/2} + 729x^9\big] \geq 0, \quad \forall x > 0.$$

Here also we applied Note 2 twice to get the required result.

28. For $D_{CP_5}^{79} \leq \frac{9}{8}D_{SN_3}^{69}$: We have, $\beta_{28} = f_{CP_5}''(1)/f_{SN_3}''(1) = \frac{9}{8}$. In order to show $\frac{1}{2}D_{SH}^{73} - D_{CP_5}^{79} \geq 0$, let us consider, $\frac{9}{8}D_{SN_3}^{69} - D_{CP_5}^{79} = bg_{28}(a/b)$, where

$$g_{28}(x) = \frac{9}{8}D_{SN_3}^{69} - D_{CP_5}^{79} = \frac{u_{28}(x)}{16(\sqrt{x} + 1)^2(x + 1)},$$

with

$$u_{28}(x) = 9\sqrt{2x^2 + 2}(\sqrt{x} + 1)^2(x + 1) - 2(x + \sqrt{x} + 1) \times$$
$$\times \big(3x^2 + 9x^{3/2} + 13\sqrt{x}(\sqrt{x} - 1)^2 + 9\sqrt{x} + 3\big).$$

Now we shall show that $u_{28}(x) \geq 0$, $\forall x > 0$. Let us consider

$$v_{27}(x) = 9\sqrt{2x^2 + 2}(\sqrt{x} + 1)^2(x + 1) - \big[2(x + \sqrt{x} + 1) \times$$
$$\times \big(3x^2 + 9x^{3/2} + 13\sqrt{x}(\sqrt{x} - 1)^2 + 9\sqrt{x} + 3\big)\big]^2.$$

After simplifications, we have

$$v_{28}(x) = 2(\sqrt{x} - 1)^4\big(63x^4 + 276x^{7/2} + 136x^3 -$$
$$- 4x^{5/2} - 366x^2 - 4x^{3/2} + 136x + 276\sqrt{x} + 63\big).$$

Since there are negative terms in the above expression, reorganizing again, we get

$$v_{28}(x) = 2(\sqrt{x}-1)^4\big[(2x+183\sqrt{x}+2)(\sqrt{x}-1)^2(x+\sqrt{x}+1)^2 + \\ + 61(x^4+1) + 93\sqrt{x}(x^3+1) + 134x(x^2+1)\big] \geq 0, \quad \forall x > 0.$$

29. For $D_{SA}^{67} \leq \frac{1}{3}D_{SH}^{73}$: We have, $\beta_{29} = f_{SA}''(1)/f_{SH}''(1) = \frac{1}{3}$. In order to show $\frac{1}{3}D_{SH}^{73} - D_{SA}^{67} \geq 0$, let us consider, $\frac{1}{3}D_{SH}^{73} - D_{SA}^{67} = bg_{29}(a/b)$, where

$$g_{29}(x) = \frac{1}{3}f_{SH}^{73}(x) - f_{SA}^{67}(x) = \frac{3x^2 + 2x + 3 - 2(x+1)\sqrt{2x^2+2}}{6(x+1)}.$$

In order to prove $g_{29}(x) \geq 0$, $\forall x > 0$, let us consider

$$v_{29}(x) = \big[3x^2+2x+3\big]^2 - \big[2(x+1)\sqrt{2x^2+2}\big]^2 = (x-1)^4 \geq 0, \quad \forall x > 0.$$

30. For $D_{SA}^{67} \leq \frac{4}{5}D_{SN_2}^{68}$: We have, $\beta_{30} = f_{SA}''(1)/f_{SN_2}''(1) = \frac{4}{5}$. In order to show $\frac{4}{5}D_{SN_2}^{68} - D_{SA}^{67} \geq 0$, let us consider, $\frac{4}{5}D_{SN_2}^{68} - D_{SA}^{67} = bg_{30}(a/b)$, where

$$g_{30}(x) = \frac{4}{5}f_{SN_2}^{68}(x) - f_{SA}^{67}(x) = \frac{1}{10}\big[5x+5 - \sqrt{2x^2+2} - 2\sqrt{2x+2}(\sqrt{x}+1)\big].$$

Now we shall show that $g_{30}(x) \geq 0$, $\forall x > 0$. Let us consider

$$v_{30}(x) = [5x+5]^2 - \big[\sqrt{2x^2+2} + 2\sqrt{2x+2}(\sqrt{x}+1)\big]^2.$$

After simplifications, we have

$$v_{30}(x) = \big[2(x+1) + (\sqrt{x}-1)^2\big]\big[4(x+1) + (\sqrt{x}-1)^2\big] - \\ - 4\sqrt{2x^2+2}\sqrt{2x+2}(\sqrt{x}+1).$$

Let us apply again the procedure (1), we get

$$p_{30}(x) = \big[\big(2(x+1) + (\sqrt{x}-1)^2\big)\big(4(x+1) + (\sqrt{x}-1)^2\big)\big]^2 - \\ - \big[4\sqrt{2x^2+2}\sqrt{2x+2}(\sqrt{x}+1)\big]^2 \\ = (\sqrt{x}-1)^4\big(161x^2 + 36x^{3/2} + 326x + 36\sqrt{x} + 161\big) \geq 0, \quad \forall x > 0.$$

31. For $D_{SA}^{67} \leq \frac{3}{4}D_{SN_3}^{69}$: We have, $\beta_{31} = f_{SA}''(1)/f_{SN_3}''(1) = \frac{3}{4}$. In order to show $\frac{3}{4}D_{SN_3}^{69} - D_{SA}^{67} \geq 0$, let us consider, $\frac{3}{4}D_{SN_3}^{69} - D_{SA}^{67} = bg_{31}(a/b)$, where

$$g_{31}(x) = \frac{3}{4}f_{SN_3}^{69}(x) - f_{SA}^{67}(x) = \frac{1}{8}\big[2(x-\sqrt{x}+1) - \sqrt{2x^2+2}\big].$$

In order to prove $g_{31}(x) \geq 0$, $\forall x > 0$, let us consider

$$v_{31}(x) = \left[2(x - \sqrt{x} + 1)\right]^2 - \left[\sqrt{2x^2 + 2}\right]^2 = 2(\sqrt{x} - 1)^4 \geq 0, \quad \forall x > 0.$$

32. For $D_{SH}^{73} \leq \frac{3}{2}D_{AH}^{42}$: We have, $\beta_{32} = f_{SH}''(1)/f_{AH}''(1) = \frac{3}{2}$. In order to show $\frac{3}{2}D_{AH}^{42} - D_{SH}^{73} \geq 0$, let us consider, $\frac{3}{2}D_{AH}^{42} - D_{SH}^{73} = bg_{32}(a/b)$, where

$$g_{32}(x) = \frac{3}{2}f_{AH}^{42}(x) - f_{SH}^{73}(x) = \frac{3x^2 + 2x + 3 - 2(x+1)\sqrt{2x^2 + 2}}{4(x+1)}.$$

In order to prove $g_{32}(x) \geq 0$, $\forall x > 0$, let us consider

$$v_{32}(x) = \left(3x^2 + 2x + 3\right)^2 - \left[2(x+1)\sqrt{2x^2 + 2}\right]^2 = (x - 1)^4 \geq 0, \quad \forall x > 0.$$

33. For $D_{SN_2}^{68} \leq \frac{5}{6}D_{SN_1}^{70}$: We have, $\beta_{33} = f_{SN_2}''(1)/f_{SN_1}''(1) = \frac{5}{6}$. In order to show $\frac{5}{6}D_{SN_1}^{70} - D_{SN_2}^{68} \geq 0$, let us consider, $\frac{5}{6}D_{SN_1}^{70} - D_{SN_2}^{68} = bg_{33}(a/b)$, where

$$g_{33}(x) = \frac{5}{6}f_{SN_1}^{70}(x) - f_{SN_2}^{68}(x)$$

$$= \frac{1}{24}\left[6\sqrt{2x+2}(\sqrt{x}+1) - 2\sqrt{2x^2+2} - 5(\sqrt{x}+1)^2\right].$$

Now we shall show that $g_{33}(x) \geq 0$, $\forall x > 0$. Let us consider

$$v_{33}(x) = \left[6\sqrt{2x+2}(\sqrt{x}+1)\right]^2 - \left[2\sqrt{2x^2+2} + 5(\sqrt{x}+1)^2\right]^2.$$

After simplifications, we have

$$v_{33}(x) = 39x^2 + 41x^{3/2} + 3\sqrt{x}(\sqrt{x}-1)^2 + 41\sqrt{x} + 39 -$$
$$- 20\sqrt{2x^2+2}(\sqrt{x}+1)^2.$$

Let us apply Note 2 again, we get

$$p_{33}(x) = \left(39x^2 + 41x^{3/2} + 3\sqrt{x}(\sqrt{x}-1)^2 + 41\sqrt{x} + 39\right)^2 -$$
$$- \left[20\sqrt{2x^2+2}(\sqrt{x}+1)^2\right]^2$$
$$= (\sqrt{x}-1)^4\left(721x^2 + 3116x^{3/2} + 4806x + 3116\sqrt{x} + 721\right) \geq 0,$$
$$\forall x > 0.$$

34. For $D_{SN_3}^{69} \leq \frac{8}{9}D_{SN_1}^{70}$: We have, $\beta_{34} = f_{SN_3}''(1)/f_{SN_1}''(1) = \frac{8}{9}$. In order to show $\frac{8}{9}D_{SN_1}^{70} - D_{SN_3}^{69} \geq 0$, let us consider, $\frac{8}{9}D_{SN_1}^{70} - D_{SN_3}^{69} = bg_{34}(a/b)$, where

$$g_{34}(x) = \frac{8}{9}f_{SN_1}^{70}(x) - f_{SN_3}^{69}(x) = \frac{2(x - \sqrt{x} + 1) - \sqrt{2x^2 + 2}}{18}.$$

In order to prove $g_{34}(x) \geq 0$, $\forall x > 0$, let us consider

$$v_{34}(x) = \left[2(x - \sqrt{x} + 1)\right]^2 - \left[\sqrt{2x^2 + 2}\right]^2 = 2(\sqrt{x} - 1)^4 \geq 0, \quad \forall x > 0.$$

35. For $D_{SN_1}^{70} \leq \frac{3}{4}D_{SG}^{71}$: We have, $\beta_{35} = f_{SN_1}''(1)/f_{SG}''(1) = \frac{3}{4}$. In order to show $\frac{3}{4}D_{SG}^{71} - D_{SN_1}^{70} \geq 0$, let us consider, $\frac{3}{4}D_{SG}^{71} - D_{SN_1}^{70} = bg_{35}(a/b)$, where

$$g_{35}(x) = \frac{3}{4}f_{SG}^{71}(x) - f_{SN_1}^{70}(x) = \frac{2(x - \sqrt{x} + 1) - \sqrt{2x^2 + 2}}{8}.$$

In order to prove $g_{35}(x) \geq 0$, $\forall x > 0$, let us consider

$$v_{35}(x) = \left[2(x - \sqrt{x} + 1)\right]^2 - \left[\sqrt{2x^2 + 2}\right]^2 = 2(\sqrt{x} - 1)^4 \geq 0, \quad \forall x > 0.$$

36. For $D_{SN_1}^{70} \leq \frac{1}{2}D_{CG}^{84}$: We have, $\beta_{36} = f_{SN_1}''(1)/f_{CG}''(1) = \frac{1}{2}$. In order to show $\frac{1}{2}D_{CG}^{84} - D_{SN_1}^{70} \geq 0$, let us consider, $\frac{1}{2}D_{CG}^{84} - D_{SN_1}^{70} = bg_{36}(a/b)$, where

$$g_{36}(x) = \frac{1}{2}f_{CG}^{84}(x) - f_{SN_1}^{70}(x) = \frac{3x^2 + 2x + 3 - 2(x + 1)\sqrt{2x^2 + 2}}{4(x + 1)}.$$

In order to prove $g_{36}(x) \geq 0$, $\forall x > 0$, let us consider

$$v_{36}(x) = \left(3x^2 + 2x + 3\right)^2 - \left[2(x + 1)\sqrt{2x^2 + 2}\right]^2 = (x - 1)^4 \geq 0, \quad \forall x > 0.$$

37. For $D_{CG}^{84} \leq \frac{36}{11}D_{RN_2}^{57}$: We have, $\beta_{37} = f_{CG}''(1)/f_{RN_2}''(1) = \frac{36}{11}$. In order to show $\frac{36}{11}D_{RN_2}^{57} - D_{CG}^{84} \geq 0$, let us consider, $\frac{36}{11}D_{RN_2}^{57} - D_{CG}^{84} = bg_{37}(a/b)$, where

$$\begin{aligned}
g_{37}(x) &= \frac{36}{11}f_{RN_2}^{57}(x) - f_{CG}^{84}(x) \\
&= \frac{1}{11(x + 1)}\left[(x + \sqrt{x} + 1)(12(x + 1) + (\sqrt{x} - 1)^2) - \right. \\
&\quad \left. - 9\sqrt{2x + 2}(\sqrt{x} + 1)(x + 1)\right].
\end{aligned}$$

In order to prove $g_{37}(x) \geq 0$, $\forall x > 0$, let us consider

$$\begin{aligned}
v_{37}(x) &= \left[(x + \sqrt{x} + 1)(12(x + 1) + (\sqrt{x} - 1)^2)\right]^2 - \\
&\quad - \left[9\sqrt{2x + 2}(\sqrt{x} + 1)(x + 1)\right]^2 \\
&= (\sqrt{x} - 1)^4\left(7x^2 - 10x^{3/2} + 15x - 10\sqrt{x} + 7\right) \geq 0, \quad \forall x > 0.
\end{aligned}$$

38. For $D_{AH}^{42} \leq \frac{4}{5} D_{SP_4}^{72}$: We have, $\beta_{38} = f''_{AH}(1)/f''_{SP_4}(1) = \frac{4}{5}$. In order to show $\frac{4}{5} D_{SP_4}^{72} - D_{AH}^{42} \geq 0$, let us consider, $\frac{4}{5} D_{SP_4}^{72} - D_{AH}^{42} = b g_{38}(a/b)$, where

$$g_{38}(x) = \frac{4}{5} f_{SP_4}^{72}(x) - f_{AH}^{42}(x) = \frac{u_{38}(x)}{10(\sqrt{x}+1)^2(x+1)}$$

with

$$u_{38}(x) = 4\sqrt{2x^2+2}(\sqrt{x}+1)^2(x+1) - \\ - \left(27x^2 + 27x + 5x^3 + 10\sqrt{x}(x-1)^2 + 5\right).$$

In order to prove $u_{38}(x) \geq 0$, $\forall x > 0$, let us consider

$$v_{38}(x) = \left[4\sqrt{2x^2+2}(\sqrt{x}+1)^2(x+1)\right]^2 - \\ - \left[27x^2 + 27x + 5x^3 + 10\sqrt{x}(x-1)^2 + 5\right]^2.$$

After simplifications, we get

$$v_{38}(x) = (\sqrt{x}-1)^6\left(7x^3 + 70x^{5/2} + \\ + 201x^2 + 340x^{3/2} + 201x + 70\sqrt{x} + 7\right) \geq 0, \quad \forall x > 0.$$

39. For $D_{AH}^{42} \leq \frac{6}{7} D_{CN_3}^{82}$: We have, $\beta_{39} = f''_{AH}(1)/f''_{CN_3}(1) = \frac{6}{7}$. In order to show $\frac{6}{7} D_{CN_3}^{82} - D_{AH}^{42} \geq 0$, let us consider, $\frac{6}{7} D_{CN_3}^{82} - D_{AH}^{42} = b g_{39}(a/b)$, where

$$g_{39}(x) = \frac{6}{7} f_{CN_3}^{82}(x) - f_{AH}^{42}(x) \\ = \frac{x^2 - 4x^{3/2} + 6x - 4\sqrt{x} + 1}{14(x+1)} = \frac{(\sqrt{x}-1)^4}{14(x+1)} \geq 0, \quad \forall x > 0.$$

40. For $D_{SP_4}^{72} \leq D_{P_6N_1}^{83}$: We have, $\beta_{40} = f''_{SP_4}(1)/f''_{P_6N_1}(1) = 1$. In order to show $D_{P_6N_1}^{83} - D_{SP_4}^{72} \geq 0$, let us consider, $D_{P_6N_1}^{83} - D_{SP_4}^{72} = b g_{40}(a/b)$, where

$$g_{40}(x) = f_{P_6N_1}^{83}(x) - f_{SP_4}^{72}(x) = \frac{u_{40}(x)}{4(\sqrt{x}+1)^2(x+1)}$$

with

$$u_{40}(x) = 3x^3 + 4x^{5/2} + 9x^2 + 4x(\sqrt{x}-1)^2 + 9x + 4\sqrt{x} + 3 - \\ - 2(\sqrt{x}+1)^2\sqrt{2x^2+2}(x+1).$$

In order to prove $u_{40}(x) \geq 0$, $\forall x > 0$, let us consider

$$v_{40}(x) = \left[3x^3 + 4x^{5/2} + 9x^2 + 4x(\sqrt{x} - 1)^2 + 9x + 4\sqrt{x} + 3\right]^2 - \left[2(\sqrt{x} + 1)^2\sqrt{2x^2 + 2}(x + 1)\right]^2.$$

After simplifications, we get

$$v_{40}(x) = \left(\sqrt{x} - 1\right)^4 \left[x^3\left(\sqrt{x} - 2\right)^2 + 4x^3 + \right.$$
$$\left. + 20x^{5/2} + 78x^2 + 20x^{3/2} + 4x + \left(2\sqrt{x} - 1\right)^2\right] \geq 0, \ \forall x > 0.$$

41. For $D_{SP_4}^{72} \leq \frac{5}{7}D_{CP_4}^{85}$: We have, $\beta_{41} = f''_{SP_4}(1)/f''_{CP_4}(1) = \frac{5}{7}$. In order to show $\frac{5}{7}D_{CP_4}^{85} - D_{SP_4}^{72} \geq 0$, let us consider, $\frac{5}{7}D_{CP_4}^{85} - D_{SP_4}^{72} = bg_{41}(a/b)$, where

$$g_{41}(x) = \frac{5}{7}f_{CP_4}^{85}(x) - f_{SP_4}^{72}(x) = \frac{u_{41}(x)}{4(\sqrt{x} + 1)^2(x + 1)}$$

with

$$u_{41}(x) = 10x^3 + 20x^{5/2} + 26x^2 + 26x + 20\sqrt{x} + 10 - \\ - 7\sqrt{2x^2 + 2}(\sqrt{x} + 1)^2(x + 1).$$

In order to prove $u_{39}(x) \geq 0$, $\forall x > 0$, let us consider

$$v_{41}(x) = \left(10x^3 + 20x^{5/2} + 26x^2 + 26x + 20\sqrt{x} + 10\right)^2 - \\ - \left[7(\sqrt{x} + 1)^2\sqrt{2x^2 + 2}(x + 1)\right]^2.$$

After simplifications, we get

$$v_{41}(x) = (\sqrt{x} - 1)^4\left(2x^4 + 16x^{7/2} + 188x^3 + 528x^{5/2} + \\ + 772x^2 + 528x^{3/2} + 188x + 16\sqrt{x} + 2\right) \geq 0, \quad \forall x > 0.$$

42. For $D_{CN_3}^{82} \leq \frac{14}{15}D_{CN_1}^{83}$: We have, $\beta_{42} = f''_{CN_3}(1)/f''_{CN_1}(1) = \frac{14}{15}$. In order to show $\frac{14}{15}D_{CN_1}^{83} - D_{CN_3}^{82} \geq 0$, let us consider, $\frac{14}{15}D_{CN_1}^{83} - D_{CN_3}^{82} = bg_{42}(a/b)$, where

$$g_{42}(x) = \frac{14}{15}f_{CN_1}^{83}(x) - f_{CN_3}^{82}(x)$$
$$= \frac{x^2 - 4x^{3/2} + 6x - 4\sqrt{x} + 1}{30(x + 1)} = \frac{(\sqrt{x} - 1)^4}{30(x + 1)} \geq 0, \quad \forall x > 0.$$

43. For $D_{CN_3}^{82} \leq \frac{2}{3} D_{CP_4}^{85}$: We have, $\beta_{43} = f''_{CN_3}(1)/f''_{CP_4}(1) = \frac{2}{3}$. In order to show $\frac{2}{3} D_{CP_4}^{85} - D_{CN_3}^{82} \geq 0$, let us consider, $\frac{2}{3} D_{CP_4}^{85} - D_{CN_3}^{82} = b g_{43}(a/b)$, where

$$g_{43}(x) = \frac{2}{3} f_{CP_4}^{85}(x) - f_{CN_3}^{82}(x)$$

$$= \frac{\sqrt{x}(x^2 - 4x^{3/2} + 6x - 4\sqrt{x} + 1)}{3(\sqrt{x} + 1)^2(x + 1)}$$

$$= \frac{\sqrt{x}(\sqrt{x} - 1)^4}{3(\sqrt{x} + 1)^2(x + 1)} \geq 0, \quad \forall x > 0.$$

44. For $D_{CN_1}^{83} \leq \frac{5}{6} D_{CG}^{84}$: We have, $\beta_{44} = f''_{CN_1}(1)/f''_{CG}(1) = \frac{5}{6}$. In order to show $\frac{5}{6} D_{CG}^{84} - D_{CN_1}^{83} \geq 0$, let us consider, $\frac{5}{6} D_{CG}^{84} - D_{CN_1}^{83} = b g_{44}(a/b)$, where

$$g_{44}(x) = \frac{5}{6} f_{CG}^{84}(x) - f_{CN_1}^{83}(x)$$

$$= \frac{(x^2 - 4x^{3/2} + 6x - 4\sqrt{x} + 1)}{12(x + 1)} = \frac{(\sqrt{x} - 1)^4}{12(x + 1)} \geq 0, \quad \forall x > 0.$$

45. For $D_{CP_4}^{85} \leq \frac{7}{6} D_{CG}^{84}$: We have, $\beta_{45} = f''_{CP_4}(1)/f''_{CG}(1) = \frac{5}{6}$. In order to show $\frac{7}{6} D_{CG}^{84} - D_{CP_4}^{85} \geq 0$, let us consider, $\frac{7}{6} D_{CG}^{84} - D_{CP_4}^{85} = b g_{45}(a/b)$, where

$$g_{45}(x) = \frac{7}{6} f_{CG}^{84}(x) - f_{CP_4}^{85}(x)$$

$$= \frac{x^3 - 5x^{5/2} + 11x^2 - 14x^{3/2} + 11x - 5\sqrt{x} + 1}{6(\sqrt{x} + 1)^4(x + 1)}$$

$$= \frac{(x^{3/2} + 1)(\sqrt{x} - 1)^4}{6(\sqrt{x} + 1)^5(x + 1)} \geq 0, \quad \forall x > 0.$$

46. For $D_{SG}^{71} \leq \frac{12}{7} D_{RN_1}^{59}$: We have, $\beta_{46} = f''_{SG}(1)/f''_{RN_1}(1) = \frac{12}{7}$. In order to show $\frac{12}{7} D_{RN_1}^{59} - D_{SG}^{71} \geq 0$, let us consider, $\frac{12}{7} D_{RN_1}^{59} - D_{SG}^{71} = b g_{46}(a/b)$, where

$$g_{46}(x) = \frac{12}{7} D_{RN_1}^{59} - D_{SG}^{71} = \frac{u_{46}(x)}{14(x + 1)}$$

with

$$u_{46}(x) = 2(5x^2 + x^{3/2} + 2x + \sqrt{x} + 5) - 7(x + 1)\sqrt{2x^2 + 2}.$$

In order to prove $u_{46}(x) \geq 0$, $\forall x > 0$, let us consider

$$v_{46}(x) = \left[2(5x^2 + x^{3/2} + 2x + \sqrt{x} + 5)\right]^2 - \left[7(x + 1)\sqrt{2x^2 + 2}\right]^2.$$

After simplifications, we get

$$v_{46}(x) = 2(\sqrt{x} - 1)^4(x^2 + 24x^{3/2} + 34x + 24\sqrt{x} + 1) \geq 0, \quad \forall x > 0.$$

47. For $D_{SG}^{71} \leq \frac{4}{5}D_{P_5H}^{52}$: We have, $\beta_{47} = f_{SG}''(1)/f_{P_5H}''(1) = \frac{4}{5}$. In order to show $\frac{4}{5}D_{P_5H}^{52} - D_{SG}^{71} \geq 0$, let us consider, $\frac{4}{5}D_{P_5H}^{52} - D_{SG}^{71} = bg_{47}(a/b)$, where

$$g_{47}(x) = \frac{4}{5}f_{P_5H}^{52}(x) - f_{SG}^{71}(x) = \frac{u_{47}(x)}{10(\sqrt{x} + 1)^2(x + 1)}$$

with

$$u_{47}(x) = 2\big[4x^3 + 5x^{5/2} + 11x^2 + 3\sqrt{x}(\sqrt{x} - 1)^2 + 11x + 5\sqrt{x} + 4\big] - \\ - 5(x + 1)(\sqrt{x} + 1)^2\sqrt{2x^2 + 2}.$$

In order to prove $u_{47}(x) \geq 0$, $\forall x > 0$, let us consider

$$v_{47}(x) = 4\big[4x^3 + 5x^{5/2} + 11x^2 + 3\sqrt{x}(\sqrt{x} - 1)^2 + 11x + 5\sqrt{x} + 4\big]^2 - \\ - \big[5(x + 1)(\sqrt{x} + 1)^2\sqrt{2x^2 + 2}\big]^2.$$

After simplifications, we get

$$v_{47}(x) = 2(\sqrt{x} - 1)^4(7x^4 + 8x^{7/2} + 64x^3 + 120x^{5/2} + \\ + 242x^2 + 120x^{3/2} + 64x + 8\sqrt{x} + 7) \geq 0, \quad \forall x > 0.$$

48. For $D_{RN_2}^{57} \leq \frac{11}{14}D_{RN_1}^{59}$: We have, $\beta_{48} = f_{RN_2}''(1)/f_{RN_1}''(1) = \frac{11}{14}$. In order to show $\frac{11}{14}D_{RN_1}^{59} - D_{RN_2}^{57} \geq 0$, let us consider, $\frac{11}{14}D_{RN_1}^{59} - D_{RN_2}^{57} = bg_{48}(a/b)$, where

$$g_{48}(x) = \frac{11}{14}f_{RN_1}^{59}(x) - f_{RN_2}^{57}(x) = \frac{u_{49}(x)}{56(x + 1)}$$

with

$$u_{48}(x) = 14(x + 1)(\sqrt{x} + 1)\sqrt{2x^2 + 2} - (19x^2 + 22x^{3/2} + 30x + 22\sqrt{x} + 19).$$

In order to prove $u_{48}(x) \geq 0$, $\forall x > 0$, let us consider

$$v_{48}(x) = \big[14(x + 1)(\sqrt{x} + 1)\sqrt{2x^2 + 2}\big]^2 \\ - (19x^2 + 22x^{3/2} + 30x + 22\sqrt{x} + 19)^2.$$

After simplifications, we get

$$v_{48}(x) = (\sqrt{x} - 1)^4(31x^2 + 72x^{3/2} + 46x + 72\sqrt{x} + 31) \geq 0, \quad \forall x > 0.$$

49. For $D^{57}_{RN_2} \leq \frac{11}{30} D^{52}_{P_5 H}$: We have, $\beta_{49} = f''_{RN_2}(1)/f''_{P_5 H}(1) = \frac{11}{30}$. In order to show $\frac{11}{30} D^{52}_{P_5 H} - D^{57}_{RN_2} \geq 0$, let us consider, $\frac{11}{30} D^{52}_{P_5 H} - D^{57}_{RN_2} = b g_{49}(a/b)$, where

$$g_{49}(x) = \frac{11}{30} f^{52}_{P_5 H}(x) - f^{57}_{RN_2}(x) = \frac{u_{49}(x)}{60(\sqrt{x}+1)^2(x+1)}$$

with

$$u_{49}(x) = 15\sqrt{2x+2}(\sqrt{x}+1)^3(x+1) - \\ - (8x^3 + 10x^{5/2} + 22x^2 + 6x(\sqrt{x}-1)^2 + 22x + 10\sqrt{x} + 8).$$

In order to prove $u_{49}(x) \geq 0$, $\forall x > 0$, let us consider

$$u_{49}(x) = \left[15\sqrt{2x+2}(\sqrt{x}+1)^3(x+1) \right] - \\ - \left[8x^3 + 10x^{5/2} + 22x^2 + 6x(\sqrt{x}-1)^2 + 22x + 10\sqrt{x} + 8 \right]^2.$$

After simplifications, we get

$$v_{49}(x) = 2(\sqrt{x}-1)^4 \big(63x^4 + 162x^{7/2} + 76x^3 + 470x^{5/2} + \\ + 18x^2 + 470x^{3/2} + 76x + 162\sqrt{x} + 63 \big) \geq 0, \quad \forall x > 0.$$

50. For $D^{52}_{P_5 H} \leq \frac{3}{2} D^{60}_{RG}$: We have, $\beta_{50} = f''_{P_5 H}(1)/f''_{RG}(1) = \frac{3}{2}$. In order to show $\frac{3}{2} D^{60}_{RG} - D^{52}_{P_5 H} \geq 0$, let us consider, $\frac{3}{2} D^{60}_{RG} - D^{52}_{P_5 H} = b g_{50}(a/b)$, where

$$\begin{aligned} g_{50}(x) &= \frac{3}{2} f^{60}_{RG}(x) - f^{52}_{P_5 H}(x) \\ &= \frac{\sqrt{x}(x^2 - 4x^{3/2} + 6x - 4\sqrt{x} + 1)}{2(\sqrt{x}+1)^2(x+1)} \\ &= \frac{\sqrt{x}(\sqrt{x}-1)^4}{2(\sqrt{x}+1)^2(x+1)} \geq 0, \quad \forall x > 0. \end{aligned}$$

51. For $D^{60}_{RG} \leq \frac{20}{3} D^{30}_{N_2 N_1}$: We have, $\beta_{51} = f''_{RG}(1)/f''_{N_2 N}(1) = \frac{20}{3}$. In order to show $\frac{20}{3} D^{30}_{N_2 N_1} - D^{60}_{RG} \geq 0$, let us consider, $\frac{20}{3} D^{30}_{N_2 N_1} - D^{60}_{RG} = b g_{51}(a/b)$, where

$$g_{51}(x) = \frac{20}{3} f^{30}_{N_2 N_1}(x) - f^{60}_{RG}(x) = \frac{u_{51}(x)}{3(x+1)}$$

with

$$u_{51}(x) = 5\sqrt{2x+2}(\sqrt{x}+1)(x+1) - \big(7x^2 + 7x^{3/2} + 12x + 7\sqrt{x} + 7 \big).$$

Let us consider

$$v_{51}(x) = \left[5\sqrt{2x+2}(\sqrt{x}+1)(x+1)\right]^2 - \left(7x^2 + 7x^{3/2} + 12x + 7\sqrt{x}+7\right)^2$$
$$= (\sqrt{x}-1)^4\left(x^2 + 6x^{3/2} + x + 6\sqrt{x}+1\right) \geq 0, \quad \forall x > 0.$$

52. For $D_{RN_1}^{59} \leq \frac{14}{3}D_{N_2N_1}^{30}$: We have, $\beta_{52} = f_{RN_1}''(1)/f_{N_2N_1}''(1) = \frac{14}{3}$. In order to show $\frac{14}{3}D_{N_2N_1}^{30} - D_{RN_1}^{59} \geq 0$, let us consider, $\frac{14}{3}D_{N_2N_1}^{30} - D_{RN_1}^{59} = bg_{52}(a/b)$, where

$$g_{52}(x) = \frac{14}{3}f_{N_2N_1}^{30}(x) - f_{RN_1}^{59}(x) = \frac{u_{52}(x)}{12(x+1)}$$

with

$$u_{52}(x) = 14\sqrt{2x+2}(\sqrt{x}+1)(x+1) - $$
$$- \left(19x^2 + 22x^{3/2} + 30x + 22\sqrt{x}+19\right).$$

Let us consider

$$v_{52}(x) = \left[14\sqrt{2x+2}(\sqrt{x}+1)(x+1)\right]^2 - $$
$$- \left(19x^2 + 22x^{3/2} + 30x + 22\sqrt{x}+19\right)^2$$
$$= (\sqrt{x}-1)^4\left(31x^2 + 72x^{3/2} + 46x + 72\sqrt{x}+31\right) \geq 0, \quad \forall x > 0.$$

53. For $D_{N_2N_1}^{30} \leq \frac{1}{3}D_{N_2G}^{31}$: We have, $\beta_{53} = f_{N_2N_1}''(1)/f_{N_2G}''(1) = \frac{1}{3}$. In order to show $\frac{1}{3}D_{N_2G}^{31} - D_{N_2N_1}^{30} \geq 0$, let us consider, $\frac{14}{3}D_{N_2N_1}^{30} - D_{RN_1}^{59} = bg_{53}(a/b)$, where

$$g_{53}(x) = \frac{1}{3}f_{N_2G}^{31}(x) - f_{N_2N_1}^{30}(x) = \frac{1}{12}u_{53}(x)$$

with

$$u_{53}(x) = 2\sqrt{x} + 3x + 3 - 2(\sqrt{x}+1)\sqrt{2x+2}.$$

Let us consider

$$v_{53}(x) = (2\sqrt{x}+3x+3)^2 - \left[2(\sqrt{x}+1)\sqrt{2x+2}\right]^2 = (\sqrt{x}-1)^4 \geq 0, \quad \forall x > 0.$$

54. For $D_{N_2G}^{31} \leq \frac{3}{8}D_{P_5P_4}^{51}$: We have, $\beta_{54} = f_{N_2N_1}''(1)/f_{N_2G}''(1) = \frac{1}{3}$. In order to show $\frac{3}{8}D_{P_5P_4}^{51} - D_{N_2G}^{31} \geq 0$, let us consider, $\frac{3}{8}D_{P_5P_4}^{51} - D_{N_2G}^{31} = bg_{54}(a/b)$, where

$$g_{54}(x) = \frac{3}{8}f_{P_5P_4}^{51}(x) - f_{N_2G}^{31}(x) = \frac{u_{54}(x)}{8(\sqrt{x}+1)^2}$$

with

$$u_{54}(x) = (\sqrt{x}+1)^2 \left[(3x+2\sqrt{x}+3) - 2(\sqrt{x}+1)\sqrt{2x+2}\right].$$

Let us consider

$$v_{54}(x) = (2\sqrt{x}+3x+3)^2 - \left[2(\sqrt{x}+1)\sqrt{2x+2}\right]^2 = (\sqrt{x}-1)^4 \geq 0, \quad \forall x > 0.$$

55. For $D_{P_5 P_4}^{51} \leq 8 D_{AN_2}^{37}$: We have, $\beta_{55} = f''_{P_5 P_4}(1)/f''_{AN_2}(1) = 8$. In order to show $8D_{AN_2}^{37} - D_{P_5 P_4}^{51} \geq 0$, let us consider, $8D_{AN_2}^{37} - D_{P_5 P_4}^{51} = bg_{55}(a/b)$, where

$$g_{55}(x) = \frac{3}{8} f_{P_5 P_4}^{51}(x) - f_{N_2 G}^{31}(x) = \frac{u_{55}(x)}{(\sqrt{x}+1)^2}$$

with

$$u_{55}(x) = (\sqrt{x}+1)^2 \left[(3x+2\sqrt{x}+3) - 2(\sqrt{x}+1)\sqrt{2x+2}\right].$$

Let us consider

$$v_{55}(x) = (2\sqrt{x}+3x+3)^2 - \left[2(\sqrt{x}+1)\sqrt{2x+2}\right]^2 = (\sqrt{x}-1)^4 \geq 0, \quad \forall x > 0.$$

56. For $D_{AN_2}^{37} \leq \frac{1}{6} D_{P_5 G}^{50}$: We have, $\beta_{56} = f''_{AN_2}(1)/f''_{P_5 G}(1) = \frac{1}{6}$. In order to show $\frac{1}{6} D_{P_5 G}^{50} - D_{AN_2}^{37} \geq 0$, let us consider, $\frac{1}{6} D_{P_5 G}^{50} - D_{AN_2}^{37} = bg_{56}(a/b)$, where

$$g_{56}(x) = \frac{1}{6} f_{P_5 G}^{50}(x) - f_{AN_2}^{37}(x) = \frac{u_{56}(x)}{12(\sqrt{x}+1)^2}$$

with

$$u_{56}(x) = 3(\sqrt{x}+1)^3 \sqrt{2x+2} - 2(2x^2 + 7x^{3/2} + 6x + 7\sqrt{x} + 2).$$

Let us consider

$$v_{56}(x) = \left[3(\sqrt{x}+1)^3\sqrt{2x+2}\right]^2 - \left[2(2x^2 + 7x^{3/2} + 6x + 7\sqrt{x} + 2)\right]^2$$
$$= 2(\sqrt{x}-1)^4(x^2 + 2x^{3/2} + 2\sqrt{x} + 1) \geq 0, \quad \forall x > 0.$$

57. For $D_{P_5 G}^{50} \leq \frac{3}{2} D_{P_5 N_1}^{49}$: We have, $\beta_{57} = f''_{P_5 G}(1)/f''_{P_5 N_1}(1) = \frac{3}{2}$. In order to show $\frac{3}{2} D_{P_5 N_1}^{49} - D_{P_5 G}^{50} \geq 0$, let us consider, $\frac{3}{2} D_{P_5 N_1}^{49} - D_{P_5 G}^{50} = bg_{57}(a/b)$, where

$$g_{57}(x) = \frac{3}{2} f_{P_5 N_1}^{49}(x) - f_{P_5 G}^{50}(x)$$
$$= \frac{(x^2 - 4x^{3/2} + 6x - 4\sqrt{x} + 1)}{8(\sqrt{x}+1)^2} = \frac{(\sqrt{x}-1)^4}{8(\sqrt{x}+1)^2} \geq 0, \quad \forall x > 0.$$

58. For $D^{49}_{P_5 N_1} \leq \frac{6}{5} D^{48}_{P_5 N_3}$: We have, $\beta_{58} = f''_{P_5 N_1}(1)/f''_{P_5 N_3}(1) = \frac{6}{5}$. In order to show $\frac{6}{5} D^{48}_{P_5 N_3} - D^{49}_{P_5 N_1} \geq 0$, let us consider, $\frac{6}{5} D^{48}_{P_5 N_3} - D^{49}_{P_5 N_1} = b g_{58}(a/b)$, where

$$
\begin{aligned}
g_{58}(x) &= \frac{6}{5} f^{48}_{P_5 N_3}(x) - f^{49}_{P_5 N_1}(x) \\
&= \frac{(x^2 - 4x^{3/2} + 6x - 4\sqrt{x} + 1)}{20(\sqrt{x} + 1)^2} = \frac{(\sqrt{x} - 1)^4}{20(\sqrt{x} + 1)^2} \geq 0, \quad \forall x > 0.
\end{aligned}
$$

59. For $D^{48}_{P_5 N_3} \leq \frac{10}{9} D^{47}_{P_5 N_2}$: We have, $\beta_{59} = f''_{P_5 N_3}(1)/f''_{P_5 N_2}(1) = \frac{10}{9}$. In order to show $\frac{10}{9} D^{47}_{P_5 N_2} - D^{48}_{P_5 N_3} \geq 0$, let us consider, $\frac{10}{9} D^{47}_{P_5 N_2} - D^{48}_{P_5 N_3} = b g_{59}(a/b)$, where

$$
g_{59}(x) = \frac{10}{9} f^{47}_{P_5 N_2}(x) - D^{48}_{P_5 N_3}(x) = \frac{u_{59}(x)}{18(\sqrt{x} + 1)^2}
$$

with

$$
u_{59}(x) = 2\big(4x^2 + 9x^{3/2} + 14x + 9\sqrt{x} + 4\big) - 5(\sqrt{x} + 1)^3 \sqrt{2x + 2}.
$$

Let us consider

$$
\begin{aligned}
v_{59}(x) &= \left[2\big(4x^2 + 9x^{3/2} + 14x + 9\sqrt{x} + 4\big) \right]^2 - \left[5(\sqrt{x} + 1)^3 \sqrt{2x + 2} \right]^2 \\
&= 2(\sqrt{x} - 1)^4 \big(7x^2 + 22x^{3/2} + 32x + 22\sqrt{x} + 7\big) \geq 0, \quad \forall x > 0.
\end{aligned}
$$

60. For $D^{47}_{P_5 N_2} \leq \frac{3}{2} D^{46}_{P_5 A}$: We have, $\beta_{60} = f''_{P_5 N_2}(1)/f''_{P_5 A}(1) = \frac{3}{2}$. In order to show $\frac{3}{2} D^{46}_{P_5 A} - D^{47}_{P_5 N_2} \geq 0$, let us consider, $\frac{3}{2} D^{46}_{P_5 A} - D^{47}_{P_5 N_2} = b g_{60}(a/b)$, where

$$
g_{60}(x) = \frac{3}{2} f^{46}_{P_5 A}(x) - D^{47}_{P_5 N_2}(x) = \frac{u_{60}(x)}{4(\sqrt{x} + 1)^2}
$$

with

$$
u_{60}(x) = (\sqrt{x} + 1)^3 \sqrt{2x + 2} - (x + 1)(x + 6\sqrt{x} + 1).
$$

Let us consider

$$
\begin{aligned}
v_{60}(x) &= \left[(\sqrt{x} + 1)^3 \sqrt{2x + 2} \right]^2 - \left[(x + 1)(x + 6\sqrt{x} + 1) \right]^2 \\
&= (\sqrt{x} - 1)^4 (x + 1)(x + 4\sqrt{x} + 1) \geq 0, \quad \forall x > 0.
\end{aligned}
$$

61. For $D^1_{P_2 P_1} \leq \frac{2}{3} D^3_{P_3 P_1}$: We have, $\beta_{61} = f''_{P_2 P_1}(1)/f''_{P_3 P_1}(1) = \frac{2}{3}$. In order to show $\frac{2}{3} D^3_{P_3 P_1} - D^1_{P_2 P_1} \geq 0$, let us consider, $\frac{2}{3} D^3_{P_3 P_1} - D^1_{P_2 P_1} = b g_{61}(a/b)$,

where

$$g_{61}(x) = \frac{2}{3}f^3_{P_3P_1}(x) - D^1_{P_2P_1}(x) = \frac{x(\sqrt{x}+1)u_{61}(x)}{3(x^{3/2}+1)(x^3+1)(x^2+1)}$$

with

$$u_{61}(x) = \sqrt{x}\left(2x^4 - 5x^{7/2} + 3x^3 + x^{5/2} - 2x^2 + x^{3/2} + 3x - 5\sqrt{x} + 2\right)$$
$$= (\sqrt{x}-1)^4\sqrt{x}\left(2x^2 + 3x^{3/2} + 3x + 3\sqrt{x} + 2\right) \geq 0, \quad \forall x > 0.$$

62. For $D^3_{P_3P_1} \leq 3D^2_{P_3P_2}$: We have, $\beta_{62} = f''_{P_3P_1}(1)/f''_{P_3P_2}(1) = 3$. In order to show $3D^2_{P_3P_2} - D^3_{P_3P_1} \geq 0$, let us consider, $3D^2_{P_3P_2} - D^3_{P_3P_1} = bg_{62}(a/b)$, where

$$g_{62}(x) = 3f^2_{P_3P_2}(x) - f^3_{P_3P_1}(x) = \frac{x(\sqrt{x}+1)u_{62}(x)}{(x^{3/2}+1)(x^3+1)(x^2+1)}$$

with

$$u_{62}(x) = \sqrt{x}\left(2x^4 - 5x^{7/2} + 3x^3 + x^{5/2} - 2x^2 + x^{3/2} + 3x - 5\sqrt{x} + 2\right)$$
$$= (\sqrt{x}-1)^4\sqrt{x}\left(2x^2 + 3x^{3/2} + 3x + 3\sqrt{x} + 2\right) \geq 0, \quad \forall x > 0.$$

63. For $D^3_{P_3P_1} \leq \frac{3}{4}D^6_{HP_1}$: We have, $\beta_{63} = f''_{P_3P_1}(1)/f''_{HP_1}(1) = \frac{3}{4}$. In order to show $\frac{3}{4}D^6_{HP_1} - D^3_{P_3P_1} \geq 0$, let us consider, $\frac{3}{4}D^6_{HP_1} - D^3_{P_3P_1} = bg_3(a/b)$, where

$$g_{63}(x) = \frac{3}{4}f^6_{HP_1}(x) - f^3_{P_3P_1}(x)$$
$$= \frac{3x^3 - 7x^{5/2} + x^2 + 6x^{3/2} + x - 7\sqrt{x} + 3}{4(x^3+1)(x-\sqrt{x}+1)}$$
$$= \frac{(3x + 5\sqrt{x} + 3)(\sqrt{x}-1)^4}{4(x^3+1)(x-\sqrt{x}+1)} \geq 0, \quad \forall x > 0.$$

64. For $D^2_{P_3P_2} \leq \frac{1}{2}D^5_{HP_2}$: We have, $\beta_{64} = f''_{P_3P_2}(1)/f''_{HP_2}(1) = \frac{1}{2}$. In order to show $\frac{1}{2}D^5_{HP_2} - D^2_{P_3P_2} \geq 0$, let us consider, $\frac{1}{2}D^5_{HP_2} - D^2_{P_3P_2} = bg_{64}(a/b)$, where

$$g_{64}(x) = \frac{1}{2}f^5_{HP_2}(x) - f^2_{P_3P_2}(x)$$
$$= \frac{x^3 - 3x^{5/2} + 3x^2 - 2x^{3/2} + 3x - 3\sqrt{x} + 1}{2(x-\sqrt{x}+1)(x^2+1)(x+1)}$$
$$= \frac{(x + \sqrt{x} + 1)(\sqrt{x}-1)^4}{2(x-\sqrt{x}+1)(x^2+1)(x+1)} \geq 0, \quad \forall x > 0.$$

65. For $D^6_{HP_1} \leq 2D^5_{HP_2}$: We have, $\beta_{65} = f''_{HP_1}(1)/f''_{HP_2}(1) = 2$. In order to show $2D^5_{HP_2} - D^6_{HP_1} \geq 0$, let us consider, $2D^5_{HP_2} - D^6_{HP_1} = bg_{65}(a/b)$, where

$$g_{65}(x) = 2f^5_{HP_2}(x) - f^6_{HP_1}(x) = \frac{x(x-1)^2}{(x^2+1)(x^3+1)} \geq 0, \quad \forall x > 0.$$

66. For $D^3_{P_3P_1} \leq \frac{3}{4}D^6_{HP_1}$: We have, $\beta_{66} = f''_{P_3P_1}(1)/f''_{HP_1}(1) = \frac{3}{4}$. In order to show $\frac{3}{4}D^6_{HP_1} - D^3_{P_3P_1} \geq 0$, let us consider, $\frac{3}{4}D^6_{HP_1} - D^3_{P_3P_1} = bg_{66}(a/b)$, where

$$\begin{aligned}
g_{66}(x) &= \frac{3}{4}f^6_{HP_1}(x) - D^3_{P_3P_1}(x) \\
&= \frac{3x^3 - 7x^{5/2} + x^2 + 6x^{3/2} - 7\sqrt{x} + x + 3}{4(x^3+1)(x - \sqrt{x}+1)} \\
&= \frac{(3x+5\sqrt{x}+3)(\sqrt{x}-1)^4}{4(x^3+1)(x-\sqrt{x}+1)} \geq 0, \quad \forall x > 0.
\end{aligned}$$

67. For $D^6_{HP_1} \leq \frac{8}{9}D^{10}_{P_4P_1}$: We have, $\beta_{67} = f''_{HP_1}(1)/f''_{P_4P_1}(1) = \frac{8}{9}$. In order to show $\frac{8}{9}D^{10}_{P_4P_1} - D^6_{HP_1} \geq 0$, let us consider, $\frac{8}{9}D^{10}_{P_4P_1} - D^6_{HP_1} = bg_{67}(a/b)$, where

$$\begin{aligned}
g_{67}(x) &= \frac{8}{9}f^{10}_{P_4P_1}(x) - f^6_{HP_1}(x) \\
&= \frac{15x^3 - 34x^{5/2} + x^2 + 36x^{3/2} + x - 34\sqrt{x} + 15}{9(\sqrt{x}+1)^2(x^3+1)} \\
&= \frac{(15x+26\sqrt{x}+15)(\sqrt{x}-1)^4}{9(\sqrt{x}+1)^2(x^3+1)} \geq 0, \quad \forall x > 0.
\end{aligned}$$

68. For $D^{10}_{P_4P_1} \leq \frac{9}{10}D^{15}_{GP_1}$: We have, $\beta_{68} = f''_{P_4P_1}(1)/f''_{GP_1}(1) = \frac{9}{10}$. In order to show $\frac{9}{10}D^{15}_{GP_1} - D^{10}_{P_4P_1} \geq 0$, let us consider, $\frac{9}{10}D^{15}_{GP_1} - D^{10}_{P_4P_1} = bg_{68}(a/b)$, where

$$\begin{aligned}
g_{68}(x) &= \frac{9}{10}f^{15}_{GP_1}(x) - f^{10}_{P_4P_1}(x) \\
&= \frac{\sqrt{x}(9x^4 - 21x^{7/2} + 11x^3 + x^{5/2} + 11x^2 + x^{3/2} - 21\sqrt{x} + 9)}{10(\sqrt{x}+1)^2(x^3+1)} \\
&= \frac{\sqrt{x}(\sqrt{x}-1)^4(9x^2 + 15x^{3/2} + 17x + 15\sqrt{x} + 9)}{10(\sqrt{x}+1)^2(x^3+1)} \geq 0, \quad \forall x > 0.
\end{aligned}$$

69. For $D_{GP_1}^{15} \leq \frac{5}{3}D_{GP_2}^{14}$: We have, $\beta_{69} = f_{GP_1}''(1)/f_{GP_2}''(1) = \frac{5}{3}$. In order to show $\frac{5}{3}D_{GP_2}^{14} - D_{GP_1}^{15} \geq 0$, let us consider, $\frac{5}{3}D_{GP_2}^{14} - D_{GP_1}^{15} = bg_{69}(a/b)$, where

$$g_{69}(x) = \frac{5}{3}f_{GP_2}^{14}(x) - f_{GP_1}^{15}(x) = \frac{u_{69}(x)}{3(x^3+1)(x^2+1)},$$

with

$$u_{69}(x) = \sqrt{x}\left(2x^5 - 2x^{9/2} - 5x^{7/2} + 2x^3 + 6x^{5/2} + 2x^2 - 5x^{3/2} - 2\sqrt{x} + 2\right)$$
$$= \sqrt{x}(\sqrt{x}-1)^4\left(2x^3 + 6x^{5/2} + 12x^2 + 15x^{3/2} + 12x + 6\sqrt{x} + 2\right) \geq 0,$$
$$\forall x > 0.$$

70. For $D_{GP_2}^{14} \leq \frac{6}{7}D_{N_1P_2}^{20}$: We have, $\beta_{70} = f_{GP_2}''(1)/f_{N_1P_2}''(1) = \frac{6}{7}$. In order to show $\frac{6}{7}D_{N_1P_2}^{20} - D_{GP_2}^{14} \geq 0$, let us consider, $\frac{6}{7}D_{N_1P_2}^{20} - D_{GP_2}^{14} = bg_{70}(a/b)$, where

$$g_{70}(x) = \frac{6}{7}f_{N_1P_2}^{20}(x) - f_{GP_2}^{14}(x)$$
$$= \frac{3x^3 - 8x^{5/2} + 5x^2 + 5x - 8\sqrt{x} + 3}{14(x^2+1)}$$
$$= \frac{(3x + 4\sqrt{x} + 3)(\sqrt{x}-1)^4}{14(x^2+1)} \geq 0, \quad \forall x > 0.$$

71. For $D_{GP_1}^{15} \leq \frac{10}{11}D_{N_1P_1}^{21}$: We have, $\beta_{71} = f_{GP_1}''(1)''/f_{N_1P_1}''(1) = \frac{10}{11}$. In order to show $\frac{10}{11}D_{N_1P_1}^{21} - D_{GP_1}^{15} \geq 0$, let us consider, $\frac{10}{11}D_{N_1P_1}^{21} - D_{GP_1}^{15} = bg_{71}(a/b)$, where

$$g_{71}(x) = \frac{10}{11}f_{N_1P_1}^{21}(x) - f_{GP_1}^{15}(x)$$
$$= \frac{5x^4 - 12x^{7/2} + 7x^3 + 7x - 12\sqrt{x} + 5}{22(x^3+1)}$$
$$= \frac{(5x^2 + 8x^{3/2} + 9x + 8\sqrt{x} + 5)(\sqrt{x}-1)^4}{22(x^3+1)} \geq 0, \quad \forall x > 0.$$

72. For $D_{N_1P_1}^{21} \leq \frac{33}{34}D_{NP_1}^{28}$: We have, $\beta_{72} = g_{N_1P_1}''(1)/g_{NP_1}''(1) = \frac{33}{34}$. In order to show $\frac{33}{34}D_{N_3P_1}^{28} - D_{N_1P_1}^{21} \geq 0$, let us consider, $\frac{33}{34}D_{N_3P_1}^{28} - D_{N_1P_1}^{21} =$

$bg_{72}(a/b)$, where

$$g_{72}(x) = \frac{33}{34} f_{N_3 P_1}^{28}(x) - f_{N_1 P_1}^{21}(x)$$
$$= \frac{5x^4 - 12x^{7/2} + 7x^3 + 7x - 12\sqrt{x} + 5}{68(x^3 + 1)}$$
$$= \frac{(5x^2 + 8x^{3/2} + 9x + 8\sqrt{x} + 5)(\sqrt{x} - 1)^4}{68(x^3 + 1)} \geq 0, \quad \forall x > 0.$$

73. For $D_{N_3 P_1}^{28} \leq \frac{68}{69} D_{N_2 P_1}^{36}$: We have, $\beta_{73} = f_{N_3 P_1}''(1)/f_{N_2 P_1}''(1) = \frac{68}{69}$. In order to show $\frac{68}{69} D_{N_2 P_1}^{36} - D_{N_3 P_1}^{28} \geq 0$, let us consider, $\frac{68}{69} D_{N_2 P_1}^{36} - D_{N_3 P_1}^{28} = bg_{73}(a/b)$, where

$$g_{73}(x) = \frac{68}{69} f_{N_2 P_1}^{36}(x) - f_{N_3 P_1}^{28}(x) = \frac{u_{73}(x)}{69(x^3 + 1)}$$

with

$$u_{73}(x) = 17\sqrt{2x + 2}(\sqrt{x} + 1)(x^3 + 1) - \\ - (23x^4 + 23x^{7/2} + 22x^3 + 22x + 23\sqrt{x} + 23).$$

Now we shall show that $u_{73}(x) \geq 0$, $\forall x > 0$. Let us consider

$$v_{73}(x) = \left[17\sqrt{2x + 2}(\sqrt{x} + 1)(x^3 + 1)\right]^2 - \\ - (23x^4 + 23x^{7/2} + 22x^3 + 22x + 23\sqrt{x} + 23)^2.$$

After simplifications, we get

$$v_{73}(x) = (\sqrt{x} - 1)^4 (49 + 294\sqrt{x} + 93x^3 + 564x^{3/2} + \\ + 495x^2 + 49x^6 + 497x^5 + 294x^{11/2} + 564x^{9/2} + \\ + 290x^{5/2} + 495x^4 + 290x^{7/2} + 497x) \geq 0, \quad \forall x > 0.$$

74. For $D_{N_2 P_1}^{36} \leq \frac{23}{24} D_{A P_1}^{45}$: We have, $\beta_{74} = f_{N_2 P_1}''(1)/f_{A P_1}''(1) = \frac{23}{24}$. In order to show $\frac{23}{24} D_{A P_1}^{45} - D_{N_2 P_1}^{36} \geq 0$, let us consider, $\frac{23}{24} D_{A P_1}^{45} - D_{N_2 P_1}^{36} = bg_{74}(a/b)$, where

$$g_{74}(x) = \frac{23}{24} f_{A P_1}^{45}(x) - f_{N_2 P_1}^{36}(x) = \frac{u_{74}(x)}{48(x^3 + 1)}$$

with

$$u_{74}(x) = 23x^4 + 25x^3 + 25x + 23 - 12(x^3 + 1)\sqrt{2x + 2}.$$

Now we shall show that $u_{73}(x) \geq 0$, $\forall x > 0$. Let us consider

$$v_{74}(x) = \left(23x^4 + 25x^3 + 25x + 23\right)^2 - \left[12(x^3 + 1)\sqrt{2x+2}\right]^2$$
$$= (\sqrt{x} - 1)^4 \left(241 + 388\sqrt{x} + 688x^4 + 404x^{7/2} + \right.$$
$$+ 780x^{3/2} + 241x^6 + 680x^5 + 388x^{11/2} +$$
$$+ 780x^{9/2} + 502x^3 + 688x^2 + 404x^{5/2} + 680x\left.\right) \geq 0, \quad \forall x > 0.$$

75. For $D^{36}_{N_2 P_1} \leq \frac{23}{14} D^{20}_{N_1 P_2}$: We have, $\beta_{75} = f''_{N_2 P_1}(1)/f''_{N_1 P_2}(1) = \frac{23}{14}$. In order to show $\frac{23}{14} D^{20}_{N_1 P_2} - D^{36}_{N_2 P_1} \geq 0$, let us consider, $\frac{23}{14} D^{20}_{N_1 P_2} - D^{36}_{N_2 P_1} = bg_{75}(a/b)$, where

$$g_{75}(x) = \frac{23}{14} f^{20}_{N_1 P_2}(x) - f^{36}_{N_2 P_1}(x) = \frac{u_{75}(x)}{56(x^2 + 1)(x^3 + 1)}$$

with

$$u_{75}(x) = 23x^6 + 46x^{11/2} - 13x^5 - 69x^4 + 46x^{7/2} + 158x^3 + 46x^{5/2} -$$
$$- 69x^2 - 13x + 46\sqrt{x} + 23 -$$
$$- 14\sqrt{2x+2}(\sqrt{x} + 1)(x^3 + 1)(x^2 + 1).$$

Now we shall show that $u_{75}(x) \geq 0$, $\forall x > 0$. Let us consider

$$v_{75}(x) = \left(23x^6 + 46x^{11/2} - 13x^5 - 69x^4 + 46x^{7/2} + 158x^3 + 46x^{5/2} - \right.$$
$$- 69x^2 - 13x + 46\sqrt{x} + 23\left.\right)^2 -$$
$$- \left[14\sqrt{2x+2}(\sqrt{x} + 1)(x^3 + 1)(x^2 + 1)\right]^2.$$

After simplifications, we get

$$v_{75}(x) = (\sqrt{x} - 1)^4 \left(137 + 1880\sqrt{x} + 17016x^{3/2} + 28553x^4 + 18122x^3 + \right.$$
$$+ 14600x^{7/2} + 26648x^{5/2} + 26674x^8 + 28553x^6 + 18122x^7 +$$
$$+ 72164x^5 + 26648x^{15/2} + 56208x^{11/2} + 14600x^{13/2} +$$
$$+ 56208x^{9/2} + 137x^{10} + 7432x^9 + 1880x^{19/2} + 17016x^{17/2} +$$
$$+ 26674x^2 + 7432x\left.\right).$$

Since $v_{75}(x) \geq 0$, giving $u_{75}(x) \geq 0$, $\forall x > 0$, hence proving the required result.

76. For $D^{45}_{AP_1} \leq \frac{18}{11} D^{27}_{N_3 P_2}$: We have, $\beta_{76} = f''_{AP_1}(1)/f''_{N_3 P_2}(1) = \frac{18}{11}$. In order to show $\frac{18}{11} D^{27}_{N_3 P_2} - D^{45}_{AP_1} \geq 0$, let us consider, $\frac{18}{11} D^{27}_{N_3 P_2} - D^{45}_{AP_1} =$

$bg_{76}(a/b)$, where

$$g_{76}(x) = \frac{18}{11}D_{N_3P_2}^{27} - D_{AP_1}^{45} = \frac{u_{76}(x)}{22(x^2+1)(x^3+1)},$$

with

$$
\begin{aligned}
u_{76}(x) &= x^6 + 46x^3 - 35x^4 - 13x + 12x^{11/2} + 12x^{5/2} + \\
&\quad + 12x^{7/2} + 12\sqrt{x} - 13x^5 - 35x^2 + 1 \\
&= (\sqrt{x} - 1)^4\big(x^4 + 16x^{7/2} + 45x^3 + 88x^{5/2} + \\
&\quad + 110x^2 + 88x^{3/2} + 45x + 16\sqrt{x} + 1\big) \geq 0, \quad \forall x > 0.
\end{aligned}
$$

77. For $D_{N_1P_2}^{20} \leq \frac{21}{22}D_{N_3P_2}^{27}$: We have, $\beta_{77} = f_{N_1P_2}''(1)/f_{N_3P_2}''(1) = \frac{21}{22}$. In order to show $\frac{21}{22}D_{N_3P_2}^{27} - D_{N_1P_2}^{20} \geq 0$, let us consider, $\frac{21}{22}D_{N_3P_2}^{27} - D_{N_1P_2}^{20} = bg_{77}(a/b)$, where

$$
\begin{aligned}
g_{77}(x) &= \frac{21}{22}f_{N_3P_2}^{27}(x) - D_{N_1P_2}^{20}(x) \\
&= \frac{3x^3 - 8x^{5/2} + 5x^2 + 5x - 8\sqrt{x} + 3}{44(x^2+1)} \\
&= \frac{(3x + 4\sqrt{x} + 3)(\sqrt{x} - 1)^4}{44(x^2+1)} \geq 0, \quad \forall x > 0.
\end{aligned}
$$

78. For $D_{HP_2}^{5} \leq \frac{6}{11}D_{N_3P_2}^{27}$: We have, $\beta_{78} = f_{HP_2}''(1)/f_{N_3P_2}''(1) = \frac{6}{11}$. In order to show $\frac{6}{11}D_{N_3P_2}^{27} - D_{HP_2}^{5} \geq 0$, let us consider, $\frac{6}{11}D_{N_3P_2}^{27} - D_{HP_2}^{5} = bg_{78}(a/b)$, where

$$
\begin{aligned}
g_{78}(x) &= \frac{6}{11}f_{N_3P_2}^{27}(x) - f_{HP_2}^{5}(x) \\
&= \frac{1}{11(x+1)(x^2+1)}\big(2x^4 + 14x^2 - 13x - 13x^3 + \\
&\quad + 2x^{7/2} + 2x^{3/2} + 2\sqrt{x} + 2x^{5/2} + 2\big) \\
&= \frac{(\sqrt{x} - 1)^2(2x^2 + 10x^{3/2} + 15x + 10\sqrt{x} + 2)}{11(x+1)(x^2+1)} \geq 0, \quad \forall x > 0.
\end{aligned}
$$

79. For $D_{N_3P_2}^{27} \leq \frac{44}{45}D_{N_2P_2}^{35}$: We have, $\beta_{79} = f_{N_3P_2}''(1)/f_{N_2P_2}''(1) = \frac{44}{45}$. In order to show $\frac{44}{45}D_{N_2P_2}^{35} - D_{N_3P_2}^{27} \geq 0$, let us consider, $\frac{44}{45}D_{N_2P_2}^{35} - D_{N_3P_2}^{27} = bg_{79}(a/b)$, where

$$g_{79}(x) = \frac{44}{45}D_{N_2P_2}^{35} - D_{N_3P_2}^{27} = \frac{u_{79}(x)}{45(x^2+1)},$$

with

$$u_{79}(x) = 11\sqrt{2x+2}(\sqrt{x}+1)(x^2+1) -$$
$$- \left(15x^3 + 15x^{5/2} + 14x^2 + 14x + 15\sqrt{x} + 15\right).$$

Let us consider

$$v_{79}(x) = \left[11\sqrt{2x+2}(\sqrt{x}+1)(x^2+1)\right]^2 -$$
$$- \left(15x^3 + 15x^{5/2} + 14x^2 + 14x + 15\sqrt{x} + 15\right)^2.$$

After simplifications, we get

$$v_{79}(x) = (\sqrt{x}-1)^4\left(17x^4 + 102x^{7/2} + 145x^3 + 100x^{5/2} +\right.$$
$$\left. + 31x^2 + 100x^{3/2} + 145x + 102\sqrt{x} + 17\right) \geq 0, \quad \forall x > 0.$$

80. For $D_{N_2P_2}^{35} \leq \frac{45}{4}D_{SR}^{66}$: We have, $\beta_{80} = f_{N_2P_2}''(1)/f_{SR}''(1) = \frac{45}{4}$. In order to show $\frac{45}{4}D_{SR}^{66} - D_{N_2P_2}^{35} \geq 0$, let us consider, $\frac{45}{4}D_{SR}^{66} - D_{N_2P_2}^{35} = bg_{80}(a/b)$, where

$$g_{80}(x) = \frac{45}{4}f_{SR}^{66}(x) - f_{N_2P_2}^{35}(x) = \frac{u_{80}(x)}{8(x+1)(x^2+1)}$$

with

$$u_{80}(x) = 45\sqrt{2x^2+2}(x+1)(x^2+1) -$$
$$- \left[2\sqrt{2x+2}(\sqrt{x}+1)(x+1)(x^2+1) +\right.$$
$$\left. + 15x^4 + 13x^3 + 26x^2 + 13x + 15\right].$$

Let us consider

$$v_{80}(x) = \left[45\sqrt{2x^2+2}(x+1)(x^2+1)\right]^2 -$$
$$- \left[2\sqrt{2x+2}(\sqrt{x}+1)(x+1)(x^2+1) +\right.$$
$$\left. + 15x^4 + 13x^3 + 26x^2 + 13x + 15\right]^2.$$

After simplifications, we get

$$v_{80}(x) = 442 - 16\sqrt{x} - 48x^{3/2} - 80x^{5/2} + 442x^8 + 952x^6 +$$
$$+ 1828x^7 + 7148x^5 - 16x^{15/2} - 80x^{11/2} - 48x^{13/2} -$$
$$- 112x^{9/2} + 764x^4 + 7148x^3 + 952x^2 + 1828x - 112x^{7/2} -$$
$$- 16\sqrt{2x+2}(\sqrt{x}+1)(x+1)(x^2+1) \times$$
$$\times \left(15x^4 + 13x^3 + 26x^2 + 13x + 15\right).$$

Again, after simplifications, we have

$$
\begin{aligned}
v_{80}(x) = {}& 2\big(217x^8 + 898x^7 + 444x^6 + 3526x^5 + 326x^4 + 3526x^3 + 444x^2 + \\
& + 898x + 217\big) + 8(\sqrt{x} - 1)^2 \times \\
& \times \big(x^7 + 3x^6 + 5x^5 + 7x^4 + 7x^3 + 5x^2 + 3x + 1\big) \\
& - 16\sqrt{2x+2}(\sqrt{x} + 1)(x + 1)\big(x^2 + 1\big) \times \\
& \times \big(15x^4 + 13x^3 + 26x^2 + 13x + 15\big).
\end{aligned}
$$

From the above expression, we are unable to say about the positivity of the expression. Let us apply again Note 2. This gives

$$
\begin{aligned}
p_{80}(x) = {}& \big[2\big(217x^8 + 898x^7 + 444x^6 + 3526x^5 + 326x^4 + 3526x^3 + \\
& + 444x^2 + 898x + 217\big) + 8(\sqrt{x} - 1)^2 \times \\
& \times \big(x^7 + 3x^6 + 5x^5 + 7x^4 + 7x^3 + 5x^2 + 3x + 1\big)\big]^2 - \\
& - \big[16\sqrt{2x+2}(\sqrt{x} + 1)(x + 1)\big(x^2 + 1\big) \times \\
& \times \big(15x^4 + 13x^3 + 26x^2 + 13x + 15\big)\big]^2.
\end{aligned}
$$

After simplifications, we get

$$
\begin{aligned}
p_{80}(x) = {}& 4(\sqrt{x} - 1)^4 \big(20041 + 194798x + 1171167x^2 + 3096436x^3 + \\
& + 19028\sqrt{x} + 447316x^{3/2} + 12955920x^{15/2} + 12955920x^{13/2} + \\
& + 10231739x^8 + 17135064x^7 + 5412701x^{10} + 11456882x^9 + \\
& + 8082668x^{19/2} + 9625068x^{17/2} + 1861304x^{23/2} + \\
& + 3373240x^{21/2} + 3096436x^{11} + 20041x^{14} + 194798x^{13} + \\
& + 19028x^{27/2} + 447316x^{25/2} + 1171167x^{12} + 5412701x^4 + \\
& + 3373240x^{7/2} + 1861304x^{5/2} + 10231739x^6 + 9625068x^{11/2} + \\
& + 8082668x^{9/2} + 11456882x^5\big).
\end{aligned}
$$

Since $p_{80}(x) \geq 0$, giving $v_{80}(x) \geq 0$ and $u_{80}(x) \geq 0$, $\forall x > 0$, hence proving the required result.

81. For $D^{35}_{N_2 P_2} \leq \frac{3}{2} D^{19}_{N_1 P_3}$: We have, $\beta_{81} = f''_{N_2 P_2}(1)/f''_{N_1 P_3}(1) = \frac{3}{2}$. In order to show $\frac{3}{2} D^{19}_{N_1 P_3} - D^{35}_{N_2 P_2} \geq 0$, let us consider, $\frac{3}{2} D^{19}_{N_1 P_3} - D^{35}_{N_2 P_2} = bg_{81}(a/b)$, where

$$
g_{81}(x) = \frac{3}{2} f^{19}_{N_1 P_3}(x) - f^{35}_{N_2 P_2}(x) = \frac{u_{81}(x)}{8(x - \sqrt{x} + 1)(x^2 + 1)}
$$

with

$$u_{81}(x) = 3x^4 + 3x^{7/2} - 4x^3 - 5x^{5/2} + 22x^2 - 5x^{3/2} - 4x + 3\sqrt{x} + 3 - \\ - 2\sqrt{2x+2}(\sqrt{x}+1)(x - \sqrt{x}+1)(x^2+1).$$

After simplifications, we have

$$u_{81}(x) = 3x^4 + 3x^{7/2} - 4x^3 - 5x^{5/2} + 22x^2 - 5x^{3/2} - 4x + 3\sqrt{x} + 3 - \\ - 2\sqrt{2x+2}(\sqrt{x}+1)(x - \sqrt{x}+1)(x^2+1).$$

Now we shall show that $u_{81}(x) \geq 0$, $\forall x > 0$. Let us consider

$$v_{81}(x) = \left[(3x^2+3)(x-1)^2 + 2\sqrt{x}(x^3+1) + 10x^2 + \\ + \sqrt{x}(x^2+1)((\sqrt{x}-1)^2 + \sqrt{x}) + 3x(x+1)(\sqrt{x}-1)^2 \right]^2 - \\ - \left[2\sqrt{2x+2}(\sqrt{x}+1)(x - \sqrt{x}+1)(x^2+1) \right]^2.$$

After simplifications, we get

$$v_{81}(x) = (\sqrt{x}-1)^4 \left[x^6 + 22x^{11/2} + 59x^5 + 31x^{9/2} + x^4 + \\ + 7x^{3/2}(x^2+1)(\sqrt{x}-1)^2 + 53x^{7/2} + 182x^3 + \\ + 53x^{5/2} + x^2 + 31x^{3/2} + 59x + 22\sqrt{x} + 1 \right] \geq 0, \quad \forall x > 0.$$

82. For $D^{35}_{N_2 P_2} \leq \frac{15}{16} D^{44}_{AP_2}$: We have, $\beta_{82} = f''_{N_2 P_2}(1)/f''_{AP_2}(1) = \frac{15}{16}$. In order to show $\frac{15}{16}D^{44}_{AP_2} - D^{35}_{N_2 P_2} \geq 0$, let us consider, $\frac{15}{16}D^{44}_{AP_2} - D^{35}_{N_2 P_2} = bg_{82}(a/b)$, where

$$g_{82}(x) = \frac{15}{16}D^{44}_{AP_2} - D^{35}_{N_2 P_2} = \frac{u_{82}(x)}{32(x^2+1)},$$

with

$$u_{82}(x) = (x+1)(15x^2+2x+15) - 8\sqrt{2x+2}(\sqrt{x}+1)(x^2+1).$$

Now we shall show that $u_{82}(x) \geq 0$, $\forall x > 0$. Let us consider

$$v_{82}(x) = \left[(x+1)(15x^2+2x+15) \right]^2 - \left[8\sqrt{2x+2}(\sqrt{x}+1)(x^2+1) \right]^2 \\ = (\sqrt{x}-1)^4(x+1)(97x^3 + 132x^{5/2} + 103x^2 + \\ + 8x^{3/2} + 103x + 132\sqrt{x} + 97) \geq 0, \quad \forall x > 0.$$

83. For $D_{AP_2}^{44} \leq \frac{4}{5} D_{SP_2}^{75}$: We have, $\beta_{83} = f_{AP_2}''(1)/f_{SP_2}''(1) = \frac{4}{5}$. In order to show $\frac{4}{5} D_{SP_2}^{75} - D_{AP_2}^{44} \geq 0$, let us consider, $\frac{4}{5} D_{SP_2}^{75} - D_{AP_2}^{44} = b g_{83}(a/b)$, where

$$g_{83}(x) = \frac{4}{5} f_{SP_2}^{75}(x) - f_{AP_2}^{44}(x) = \frac{u_{83}(x)}{10(x^2 + 1)},$$

with

$$u_{83}(x) = 4(x^2 + 1)\sqrt{2x^2 + 2} - (5x^3 + 3x^2 + 3x + 5).$$

Now we shall show that $u_{83}(x) \geq 0$, $\forall x > 0$. Let us consider

$$
\begin{aligned}
v_{83}(x) &= \left[4(x^2 + 1)\sqrt{2x^2 + 2}\right]^2 - (5x^3 + 3x^2 + 3x + 5)^2 \\
&= (x - 1)^4 \left[6(x^2 + 1) + (x - 1)^2\right] \geq 0, \quad \forall x > 0.
\end{aligned}
$$

84. For $D_{AP_2}^{44} \leq \frac{6}{7} D_{RP_2}^{64}$: We have, $\beta_{84} = f_{AP_2}''(1)/f_{RP_2}''(1) = \frac{6}{7}$. In order to show $\frac{6}{7} D_{RP_2}^{64} - D_{AP_2}^{44} \geq 0$, let us consider, $\frac{6}{7} D_{RP_2}^{64} - D_{AP_2}^{44} = b g_{84}(a/b)$, where

$$g_{84}(x) = \frac{6}{7} f_{RP_2}^{64}(x) - f_{AP_2}^{44}(x) = \frac{(x - 1)^2}{14(x + 1)(x^2 + 1)} \geq 0, \quad \forall x > 0.$$

85. For $D_{SR}^{66} \leq \frac{1}{8} D_{N_3 P_3}^{26}$: We have, $\beta_{85} = f_{SR}''(1)/f_{N_3 P_3}''(1) = \frac{1}{8}$. In order to show $\frac{1}{8} D_{N_3 P_3}^{26} - D_{SR}^{66} \geq 0$, let us consider, $\frac{1}{8} D_{N_3 P_3}^{26} - D_{SR}^{66} = b g_{85}(a/b)$, where

$$g_{85}(x) = \frac{1}{8} f_{N_3 P_3}^{26}(x) - f_{SR}^{66}(x) = \frac{u_{85}(x)}{24(x - \sqrt{x} + 1)(x + 1)}$$

with

$$
\begin{aligned}
u_{85}(x) &= 17x^3 - 16x^{5/2} + 31x^2 - 16x^{3/2} + 31x - 16\sqrt{x} + 17 - \\
&\quad - 12\sqrt{2x^2 + 2}(x^3 + 1) \\
&= 9(x^3 + 1) + 8(x^2 + x + 1)(\sqrt{x} - 1)^2 + 15x(x + 1) - \\
&\quad - 12\sqrt{2x^2 + 2}(x^3 + 1).
\end{aligned}
$$

Now we shall show that $u_{85}(x) \geq 0$, $\forall x > 0$. Let us consider

$$
\begin{aligned}
v_{85}(x) &= \left[9(x^3 + 1) + 8(x^2 + x + 1)(\sqrt{x} - 1)^2 + 15x(x + 1)\right]^2 - \\
&\quad - \left[12\sqrt{2x^2 + 2}(x^3 + 1)\right]^2.
\end{aligned}
$$

After simplifications, we have

$$v_{85}(x) = (\sqrt{x} - 1)^4(\sqrt{x} + 1)^2(x^3 + 34x^{5/2} -$$
$$- 61x^2 + 100x^{3/2} - 61x + 34\sqrt{x} + 1)$$
$$= (\sqrt{x} - 1)^4(\sqrt{x} + 1)^2[31\sqrt{x}(x + 1)(\sqrt{x} - 1)^2 +$$
$$+ x^3 + 3x^{5/2} + x^2 + 38x^{3/2} + x + 3\sqrt{x} + 1] \geq 0, \quad \forall x > 0.$$

86. For $D_{N_1 P_3}^{19} \leq \frac{15}{16} D_{N_3 P_3}^{26}$: We have, $\beta_{86} = f''_{N_1 P_3}(1)/f''_{N_3 P_3}(1) = \frac{15}{16}$. In order to show $\frac{15}{16} D_{N_3 P_3}^{26} - D_{N_1 P_3}^{19} \geq 0$, let us consider, $\frac{15}{16} D_{N_3 P_3}^{26} - D_{N_1 P_3}^{19} = bg_{86}(a/b)$, where

$$g_{86}(x) = \frac{15}{16} f_{N_3 P_3}^{26}(x) - f_{N_1 P_3}^{19}(x)$$
$$= \frac{x^2 - 4x^{3/2} - 4\sqrt{x} + 6x + 1}{16(x - \sqrt{x} + 1)} = \frac{(\sqrt{x} - 1)^4}{16(x - \sqrt{x} + 1)} \geq 0, \quad \forall x > 0.$$

87. For $D_{N_3 P_3}^{26} \leq \frac{32}{33} D_{N_2 P_3}^{34}$: We have, $\beta_{87} = f''_{N_3 P_3}(1)/f''_{N_2 P_3}(1) = \frac{32}{33}$. In order to show $\frac{32}{33} D_{N_2 P_3}^{34} - D_{N_3 P_3}^{26} \geq 0$, let us consider, $\frac{32}{33} D_{N_2 P_3}^{34} - D_{N_3 P_3}^{26} = bg_{87}(a/b)$, where

$$g_{87}(x) = \frac{32}{33} f_{N_2 P_3}^{34}(x) - f_{N_3 P_3}^{26}(x) = \frac{u_{87}(x)}{33(x - \sqrt{x} + 1)}$$

with

$$u_{87}(x) = 8\sqrt{2x + 2}(\sqrt{x} + 1)(x^3 + 1) - (11x^2 + 10x + 11).$$

Now we shall show that $u_{87}(x) \geq 0$, $\forall x > 0$. Let us consider

$$u_{87}(x) = [8\sqrt{2x + 2}(\sqrt{x} + 1)(x^3 + 1)]^2 - (11x^2 + 10x + 11)^2$$
$$= (\sqrt{x} - 1)^4(7x^2 + 28x^{3/2} - 22x + 28\sqrt{x} + 7)$$
$$= (\sqrt{x} - 1)^4[7x^2 + 17x^{3/2} + 11\sqrt{x}(\sqrt{x} - 1)^2 + 17\sqrt{x} + 7] \geq 0,$$
$$\forall x > 0.$$

88. For $D_{N_2 P_3}^{34} \leq \frac{11}{12} D_{AP_3}^{43}$: We have, $\beta_{88} = f''_{N_2 P_3}(1)/f''_{AP_3}(1) = \frac{11}{12}$. In order to show $\frac{11}{12} D_{AP_3}^{43} - D_{N_2 P_3}^{34} \geq 0$, let us consider, $\frac{11}{12} D_{AP_3}^{43} - D_{N_2 P_3}^{34} = bg_{88}(a/b)$, where

$$g_{88}(x) = \frac{11}{12} f_{AP_3}^{43}(x) - f_{N_2 P_3}^{34}(x) = \frac{u_{88}(x)}{24(x - \sqrt{x} + 1)}$$

with

$$u_{88}(x) = 11x^2 - 11x^{3/2} + 24x - 11\sqrt{x} + 11 - \sqrt{2x+2}\left(x^{3/2} + 1\right)$$
$$= 6(x+1)(\sqrt{x}-1)^2 + 5x^2 + x^{3/2} + 12x + \sqrt{x} + 5.$$

Let us consider

$$v_{88}(x) = \left[6(x+1)(\sqrt{x}-1)^2 + 5x^2 + x^{3/2} + 12x + \sqrt{x} + 5\right]^2 -$$
$$- \left[\sqrt{2x+2}\left(x^{3/2}+1\right)\right]^2.$$

After simplifications, we get

$$v_{88}(x) = (\sqrt{x}-1)^4\left(49x^2 - 46x^{3/2} + 99x - 46\sqrt{x} + 49\right)$$
$$= (\sqrt{x}-1)^4\left[23(x+1)(\sqrt{x}-1)^2 + 26(x+1)^2 + 11\right] \ge 0, \quad \forall x > 0.$$

89. For $D_{SP_2}^{75} \le \frac{5}{3}D_{AP_3}^{43}$: We have, $\beta_{89} = f_{SP_2}''(1)/f_{AP_3}''(1) = \frac{5}{3}$. In order to show $\frac{5}{3}D_{AP_3}^{43} - D_{SP_2}^{75} \ge 0$, let us consider, $\frac{5}{3}D_{AP_3}^{43} - D_{SP_2}^{75} = bg_{89}(a/b)$, where

$$g_{89}(x) = \frac{5}{3}f_{AP_3}^{43}(x) - f_{SP_2}^{75}(x) = \frac{u_{89}(x)}{6(x^2+1)(x-\sqrt{x}+1)},$$

with

$$u_{89}(x) = 5x^4 - 5x^{7/2} + 6x^3 - 11x^{5/2} + 22x^2 - 11x^{3/2} +$$
$$+ 6x - 5\sqrt{x} + 5 - 3\sqrt{2x^2+2}(x - \sqrt{x} + 1)\left(x^2+1\right)$$
$$= \left(3\left(x^3+1\right) + x^{3/2}\right)(\sqrt{x}-1)^2 + 3x^2(\sqrt{x}-2)^2 + 3x(2\sqrt{x}-1)^2 +$$
$$+ 2x^4 + x^{7/2} + \sqrt{x} + 2 - 3\sqrt{2x^2+2}(x - \sqrt{x} + 1)\left(x^2+1\right).$$

Let us consider

$$v_{89}(x) = \left[\left(3\left(x^3+1\right) + x^{3/2}\right)(\sqrt{x}-1)^2 + 3x^2(\sqrt{x}-2)^2 +$$
$$+ 3x(2\sqrt{x}-1)^2 + 2x^4 + x^{7/2} + \sqrt{x} + 2\right]^2 -$$
$$- \left[3\sqrt{2x^2+2}(x - \sqrt{x} + 1)\left(x^2+1\right)\right]^2.$$

After simplifications, we have

$$v_{89}(x) = (\sqrt{x}-1)^2\left[5x\left(x^3+1\right)(\sqrt{x}-1)^2 + 7x^6 +$$
$$+ 14x^{11/2} + 40x^5 + 27x^4 + 4x^{7/2} +$$
$$+ 126x^3 + 4x^{5/2} + 27x^2 + 40x + 14\sqrt{x} + 7\right] \ge 0, \quad \forall x > 0.$$

90. For $D_{RP_2}^{64} \leq \frac{14}{9} D_{AP_3}^{43}$: We have, $\beta_{90} = f_{RP_2}''(1)/f_{AP_3}''(1) = \frac{14}{9}$. In order to show $\frac{14}{9} D_{AP_3}^{43} - D_{RP_2}^{64} \geq 0$, let us consider, $\frac{14}{9} D_{AP_3}^{43} - D_{RP_2}^{64} = bg_{90}(a/b)$, where

$$g_{90}(x) = \frac{14}{9} f_{AP_3}^{43}(x) - f_{RP_2}^{64}(x)$$
$$= \frac{(\sqrt{x}-1)^4(x^3 + 3x^{5/2} + 10x^2 + 9x^{3/2} + 10x + 3\sqrt{x} + 1)}{9(x^2+1)(x+1)(x-\sqrt{x}+1)} \geq 0,$$
$$\forall x > 0.$$

91. For $D_{AP_3}^{43} \leq 2D_{N_1 H}^{18}$: We have, $\beta_{91} = f_{AP_3}''(1)/f_{N_1 H}''(1) = 2$. In order to show $2D_{N_1 H}^{18} - D_{AP_3}^{43} \geq 0$, let us consider, $2D_{N_1 H}^{18} - D_{AP_3}^{43} = bg_{91}(a/b)$, where

$$g_{91}(x) = 2f_{N_1 H}^{18}(x) - f_{AP_3}^{43}(x) = \frac{\sqrt{x}(\sqrt{x}-1)^4}{(x+1)(x-\sqrt{x}+1)} \geq 0, \quad \forall x > 0.$$

92. For $D_{AP_3}^{43} \leq \frac{3}{4} D_{SP_3}^{74}$: We have, $\beta_{92} = f_{AP_3}''(1)/f_{SP_3}''(1) = \frac{3}{4}$. In order to show $\frac{3}{4} D_{SP_3}^{74} - D_{AP_3}^{43} \geq 0$, let us consider, $\frac{3}{4} D_{SP_3}^{74} - D_{AP_3}^{43} = bg_{92}(a/b)$, where

$$g_{92}(x) = \frac{3}{4} f_{SP_3}^{74}(x) - f_{AP_3}^{43}(x) = \frac{u_{92}(x)}{8(x - \sqrt{x}+1)},$$

with

$$u_{92}(x) = 3\sqrt{2x^2 + 2}(x - \sqrt{x} + 1) - 2[x^2 + (x+1)(\sqrt{x}-1)^2 + x^2 + x + 1].$$

Let us consider

$$v_{92}(x) = [3\sqrt{2x^2 + 2}(x - \sqrt{x} + 1)]^2 -$$
$$- 2[x^2 + (x+1)(\sqrt{x}-1)^2 + x^2 + x + 1]^2$$
$$= 2(\sqrt{x}-1)^4[2\sqrt{x}(\sqrt{x}-1)^2 + x^2 + x + 1] \geq 0, \quad \forall x > 0.$$

93. For $D_{SP_3}^{74} \leq \frac{12}{11} D_{RP_3}^{63}$: We have, $\beta_{93} = f_{SP_3}''(1)/f_{RP_3}''(1) = \frac{12}{11}$. In order to show $\frac{12}{11} D_{RP_3}^{63} - D_{SP_3}^{74} \geq 0$, let us consider, $\frac{12}{11} D_{RP_3}^{63} - D_{SP_3}^{74} = bg_{93}(a/b)$, where

$$g_{93}(x) = \frac{12}{11} f_{RP_3}^{63}(x) - f_{SP_3}^{74}(x) = \frac{u_{93}(x)}{(x+1)(x - \sqrt{x}+1)},$$

with

$$u_{93}(x) = 8x^3 - 8x^{5/2} + 15x^2 - 8x^{3/2} + 15x - 8\sqrt{x} + 8 -$$
$$- 11\sqrt{2x^2 + 2}(x - \sqrt{x} + 1)(x + 1)$$
$$= 4(x^2 + x + 1)(\sqrt{x} - 1)^2 + 4x^3 + 7x^2 + 7x + 4 -$$
$$- 11\sqrt{2x^2 + 2}(x - \sqrt{x} + 1)(x + 1).$$

Let us consider

$$v_{93}(x) = \left[4(x^2 + x + 1)(\sqrt{x} - 1)^2 + 4x^3 + 7x^2 + 7x + 4\right]^2 -$$
$$- \left[11\sqrt{2x^2 + 2}(x - \sqrt{x} + 1)(x + 1)\right]^2$$
$$= 2(\sqrt{x} - 1)^4 \left(7x^4 + 14x^{7/2} + 17x^3 + 2x^{5/2} +\right.$$
$$\left. + 52x^2 + 2x^{3/2} + 17x + 14\sqrt{x} + 7\right) \geq 0, \quad \forall x > 0.$$

94. For $D_{N_1H}^{18} \leq \frac{9}{10} D_{NH}^{25}$: We have, $\beta_{94} = f_{N_1H}''(1)/f_{NH}''(1) = \frac{9}{10}$. In order to show $\frac{9}{10} D_{NH}^{25} - D_{N_1H}^{18} \geq 0$, let us consider, $\frac{9}{10} D_{NH}^{25} - D_{N_1H}^{18} = bg_{94}(a/b)$, where

$$g_{94}(x) = \frac{9}{10} f_{N_3H}^{25}(x) - f_{N_1H}^{18}(x)$$
$$= \frac{x^2 - 4x^{3/2} - 4\sqrt{x} + 6x + 1}{20(x + 1)} = \frac{(\sqrt{x} - 1)^4}{20(x + 1)} \geq 0, \quad \forall x > 0.$$

95. For $D_{RP_3}^{63} \leq \frac{11}{5} D_{NH}^{25}$: We have, $\beta_{95} = f_{RP_3}''(1)/f_{NH}''(1) = \frac{11}{5}$. In order to show $\frac{11}{5} D_{NH}^{25} - D_{RP_3}^{63} \geq 0$, let us consider, $\frac{11}{5} D_{NH}^{25} - D_{RP_3}^{63} = bg_{95}(a/b)$, where

$$g_{95}(x) = \frac{11}{5} f_{NH}^{25}(x) - f_{RP_3}^{63}(x)$$
$$= \frac{x^3 + 10x^{5/2} - 49x^2 + 76x^{3/2} - 49x + 10\sqrt{x} + 1}{15(x + 1)(x - \sqrt{x} + 1)}$$
$$= \frac{(\sqrt{x} - 1)^4(x + 14\sqrt{x} + 1)}{15(x + 1)(x - \sqrt{x} + 1)} \geq 0, \quad \forall x > 0.$$

96. For $D_{NH}^{25} \leq \frac{20}{21} D_{N_2H}^{33}$: We have, $\beta_{96} = f_{NH}''(1)/f_{N_2H}''(1) = \frac{20}{21}$. In order to show $\frac{20}{21} D_{N_2H}^{33} - D_{NH}^{25} \geq 0$, let us consider, $\frac{20}{21} D_{N_2H}^{33} - D_{NH}^{25} = bg_{96}(a/b)$, where

$$g_{96}(x) = \frac{20}{21} f_{N_2H}^{33}(x) - f_{NH}^{25}(x) = \frac{u_{96}(x)}{21(x + 1)},$$

with

$$u_{96}(x) = 5\sqrt{2x+2}(\sqrt{x}+1)(x+1) - (7x^2 + 7x^{3/2} + 12x + 7\sqrt{x}+7).$$

Let us consider

$$v_{96}(x) = \left[5\sqrt{2x+2}(\sqrt{x}+1)(x+1)\right]^2 - (7x^2 + 7x^{3/2} + 12x + 7\sqrt{x}+7)^2$$
$$= (\sqrt{x}-1)^4(x^2 + 6x^{3/2} + x + 6\sqrt{x}+1) \geq 0, \quad \forall x > 0.$$

97. For $D^{33}_{N_2H} \leq 21D^{29}_{N_2N}$: We have, $\beta_{97} = f''_{N_2H}(1)/f''_{N_2N}(1) = 21$. In order to show $21D^{29}_{N_2N} - D^{33}_{N_2H} \geq 0$, let us consider, $21D^{29}_{N_2N} - D^{33}_{N_2H} = bg_{96}(a/b)$, where

$$g_{97}(x) = 21f^{29}_{N_2N}(x) - f^{33}_{N_2H}(x) = \frac{u_{97}(x)}{(x+1)}$$

with

$$u_{97}(x) = 5\sqrt{2x+2}(\sqrt{x}+1)(x+1) - (12x + 7x^2 + 7x^{3/2} + 7\sqrt{x}+7).$$

Let us consider

$$v_{97}(x) = \left[5\sqrt{2x+2}(\sqrt{x}+1)(x+1)\right]^2 - (12x + 7x^2 + 7x^{3/2} + 7\sqrt{x}+7)^2$$
$$= (\sqrt{x}-1)^4(x^2 + 6x^{3/2} + x + 6\sqrt{x}+1) \geq 0, \quad \forall x > 0.$$

98. For $D^{33}_{N_2H} \leq \frac{7}{8}D^{42}_{AH}$: We have, $\beta_{98} = f''_{N_2H}(1)/f''_{AH}(1) = \frac{7}{8}$. In order to show $\frac{7}{8}D^{42}_{AH} - D^{33}_{N_2H} \geq 0$, let us consider, $\frac{7}{8}D^{42}_{AH} - D^{33}_{N_2H} = bg_{98}(a/b)$, where

$$g_{98}(x) = \frac{7}{8}f^{42}_{AH}(x) - f^{33}_{N_2H}(x) = \frac{u_{98}(x)}{16(x+1)}$$

with

$$u_{98}(x) = 7x^2 + 18x + 7 - 4\sqrt{2x+2}(\sqrt{x}+1)(x+1).$$

Let us consider

$$v_{98}(x) = (7x^2 + 18x + 7)^2 - \left[4\sqrt{2x+2}(\sqrt{x}+1)(x+1)\right]^2$$
$$= (\sqrt{x}-1)^4(17x^2 + 4x^{3/2} + 38x + 4\sqrt{x}+17) \geq 0, \quad \forall x > 0.$$

99. For $D^{42}_{AH} \leq \frac{8}{9}D^{81}_{CN_2}$: We have, $\beta_{99} = f''_{AH}(1)/f''_{CN_2}(1) = \frac{8}{9}$. In order to show $\frac{8}{9}D^{81}_{CN_2} - D^{42}_{AH} \geq 0$, let us consider, $\frac{8}{9}D^{81}_{CN_2} - D^{42}_{AH} = bg_{99}(a/b)$, where

$$g_{99}(x) = \frac{8}{9}f^{81}_{CN_2}(x) - f^{42}_{AH}(x) = \frac{u_{99}(x)}{8(x+1)}$$

with

$$u_{99}(x) = 7x^2 + 18x + 7 - 4\sqrt{2x+2}(\sqrt{x}+1)(x+1).$$

Let us consider

$$v_{99}(x) = \left(7x^2 + 18x + 7\right)^2 - \left[4\sqrt{2x+2}(\sqrt{x}+1)(x+1)\right]^2$$
$$= (\sqrt{x}-1)^4\left(17x^2 + 4x^{3/2} + 38x + 4\sqrt{x} + 17\right) \geq 0, \quad \forall x > 0.$$

100. For $D_{CN_2}^{81} \leq \frac{3}{4}D_{CG}^{84}$: We have, $\beta_{100} = f_{CN_2}''(1)/f_{CG}''(1) = \frac{3}{4}$. In order to show $\frac{3}{4}D_{CG}^{84} - D_{CN_2}^{81} \geq 0$, let us consider, $\frac{3}{4}D_{CG}^{84} - D_{CN_2}^{81} = bg_{100}(a/b)$, where

$$g_{100}(x) = \frac{3}{4}f_{CG}^{84}(x) - f_{CN_2}^{81}(x) = \frac{u_{100}(x)}{4(x+1)}$$

with

$$u_{100}(x) = \sqrt{2x+2}(\sqrt{x}+1)(x+1) - \left(x^2 + 3x^{3/2} + 3\sqrt{x} + 1\right).$$

Let us consider

$$v_{100}(x) = \left[\sqrt{2x+2}(\sqrt{x}+1)(x+1)\right]^2 - \left(x^2 + 3x^{3/2} + 3\sqrt{x} + 1\right)^2$$
$$= (\sqrt{x}-1)^2\left(x^2 + 2x^{3/2} + x + 2\sqrt{x} + 1\right) \geq 0, \quad \forall x > 0.$$

101. For $D_{CN_2}^{81} \leq \frac{27}{26}D_{RP_4}^{61}$: We have, $\beta_{101} = g_{CN_2}''(1)/g_{RP_4}''(1) = \frac{27}{26}$. In order to show $\frac{27}{26}D_{RP_4}^{61} - D_{CN_2}^{81} \geq 0$, let us consider, $\frac{27}{26}D_{RP_4}^{61} - D_{CN_2}^{81} = bg_{101}(a/b)$, where

$$g_{101}(x) = \frac{27}{26}f_{RP_4}^{61}(x) - f_{CN_2}^{81} = \frac{u_{101}(x)}{52(\sqrt{x}+1)^2(x+1)}$$

with

$$u_{101}(x) = 13\sqrt{2x+2}(\sqrt{x}+1)^3(x+1) - \left[16x^3 + 32x^{5/2} + 160x^2 + 36x(\sqrt{x}-1)^2 + 160x + 32\sqrt{x} + 16\right].$$

Let us consider

$$v_{101}(x) = \left[13\sqrt{2x+2}(\sqrt{x}+1)^3(x+1)\right]^2 - \left[16x^3 + 32x^{5/2} + 160x^2 + 36x(\sqrt{x}-1)^2 + 160x + 32\sqrt{x} + 16\right]^2.$$

After simplifications, we get

$$v_{101}(x) = (\sqrt{x} - 1)^4\big(82x^4 + 1332x^{7/2} + 3624x^3 + 9436x^{5/2} + $$
$$+ 2460x^2 + 9436x^{3/2} + 3624x + 1332\sqrt{x} + 82\big) \geq 0, \quad \forall x > 0.$$

102. For $D_{N_2N}^{29} \leq \frac{1}{12}D_{N_1P_4}^{17}$: We have, $\beta_{102} = f_{N_2N}''(1)/f_{N_1P_4}''(1) = \frac{1}{12}$. In order to show $\frac{1}{12}D_{N_1P_4}^{17} - D_{N_2N_3}^{29} \geq 0$, let us consider, $\frac{1}{12}D_{N_1P_4}^{17} - D_{N_2N_3}^{29} = bg_{102}(a/b)$, where

$$g_{102}(x) = \frac{1}{12}f_{N_1P_4}^{17}(x) - f_{N_2N_3}^{29}(x) = \frac{u_{102}(x)}{48(\sqrt{x} + 1)^2},$$

with

$$u_{102}(x) = 17x^2 + 52x^{3/2} + 54x + 52\sqrt{x} + 17 - 12\sqrt{2x+2}(\sqrt{x} + 1)^3.$$

Let us consider

$$v_{102}(x) = \big(17x^2 + 52x^{3/2} + 54x + 52\sqrt{x} + 17\big)^2 - \big[12\sqrt{2x+2}(\sqrt{x} + 1)^3\big]^2$$
$$= (\sqrt{x} - 1)^4\big(x^2 + 44x^{3/2} + 102x + 44\sqrt{x} + 1\big) \geq 0, \quad \forall x > 0.$$

103. For $D_{N_2N_3}^{29} \leq \frac{1}{42}D_{CP_4}^{85}$: We have, $\beta_{103} = f_{N_2N_3}''(1)/f_{CP_4}''(1) = \frac{1}{42}$. In order to show $\frac{1}{42}D_{CP_4}^{85} - D_{N_2N_3}^{29} \geq 0$, let us consider, $\frac{1}{42}D_{CP_4}^{85} - D_{N_2N_3}^{29} = bg_{103}(a/b)$, where

$$g_{103}(x) = \frac{1}{42}f_{CP_4}^{85}(x) - f_{N_2N_3}^{29}(x) = \frac{u_{103}(x)}{84(\sqrt{x} + 1)^2(x + 1)}$$

with

$$u_{103}(x) = 2\big(15x^3 + 44x^{5/2} + 67x^2 + 84x^{3/2} + 67x + 44\sqrt{x} + 15\big) - $$
$$- 21\sqrt{2x+2}(\sqrt{x} + 1)^3(x + 1).$$

Let us consider

$$u_{103}(x) = 4\big(15x^3 + 44x^{5/2} + 67x^2 + 84x^{3/2} + 67x + 44\sqrt{x} + 15\big)^2 - $$
$$- \big[21\sqrt{2x+2}(\sqrt{x} + 1)^3(x + 1)\big]^2$$
$$= 2(\sqrt{x} - 1)^4\big(9x^4 + 30x^{7/2} + 20x^3 + 10x^{5/2} + $$
$$+ 30x^2 + 10x^{3/2} + 20x + 30\sqrt{x} + 9\big) \geq 0, \quad \forall x > 0.$$

104. For $D_{AH}^{42} \leq 2D_{N_1P_4}^{17}$: We have, $\beta_{104} = f_{AH}''(1)/f_{N_1P_4}''(1) = 2$. In order to show $2D_{N_1P_4}^{17} - D_{AH}^{42} \geq 0$, let us consider, $2D_{N_1P_4}^{17} - D_{AH}^{42} = bg_{104}(a/b)$, where

$$g_{104}(x) = 2f_{N_1P_4}^{17}(x) - f_{AH}^{42}(x)$$
$$= \frac{\sqrt{x}(x^2 - 4x^{3/2} + 6x - 4\sqrt{x} + 1)}{(x+1)(\sqrt{x}+1)^2} = \frac{\sqrt{x}(\sqrt{x}-1)^4}{(x+1)(\sqrt{x}+1)^2} \geq 0,$$
$$\forall x > 0.$$

105. For $D_{AH}^{42} \leq \frac{6}{7}D_{CN}^{82}$: We have, $\beta_{105} = f_{AH}''(1)/f_{CN}''(1) = \frac{6}{7}$. In order to show $\frac{6}{7}D_{CN}^{82} - D_{AH}^{42} \geq 0$, let us consider, $\frac{6}{7}D_{CN}^{82} - D_{AH}^{42} = bg_{105}(a/b)$, where

$$g_{105}(x) = \frac{6}{7}f_{CN}^{82}(x) - D_{AH}^{42}(x)$$
$$= \frac{x^2 - 4x^{3/2} + 6x - 4\sqrt{x} + 1}{14(x+1)} = \frac{(\sqrt{x}-1)^4}{14(x+1)} \geq 0, \quad \forall x > 0.$$

106. For $D_{N_1P_4}^{17} \leq \frac{4}{5}D_{N_2P_4}^{32}$: We have, $\beta_{106} = f_{N_1P_4}''(1)/f_{N_2P_4}''(1) = \frac{4}{5}$. In order to show $\frac{4}{5}D_{N_2P_4}^{32} - D_{N_1P_4}^{17} \geq 0$, let us consider, $\frac{4}{5}D_{N_2P_4}^{32} - D_{N_1P_4}^{17} = bg_{106}(a/b)$, where

$$g_{106}(x) = \frac{4}{5}f_{N_2P_4}^{32}(x) - f_{N_1P_4}^{17}(x) = \frac{u_{106}(x)}{20(\sqrt{x}+1)^2}$$

with

$$u_{106}(x) = 4(\sqrt{x}+1)^3\sqrt{2x+2} - (14x + 5x^2 + 20x^{3/2} + 20\sqrt{x} + 5).$$

Let us consider

$$v_{106}(x) = \left[4(\sqrt{x}+1)^3\sqrt{2x+2}\right]^2 - (14x + 5x^2 + 20x^{3/2} + 20\sqrt{x} + 5)^2$$
$$= 2(\sqrt{x}-1)^4(7x^2 + 20x^{3/2} + 10x + 20\sqrt{x} + 7) \geq 0, \quad \forall x > 0.$$

107. For $D_{N_2P_4}^{32} \leq \frac{15}{14}D_{N_3P_4}^{24}$: We have, $\beta_{107} = f_{N_2P_4}''(1)/f_{N_3P_4}''(1) = \frac{15}{14}$. In order to show $\frac{15}{14}D_{N_3P_4}^{24} - D_{N_2P_4}^{32} \geq 0$, let us consider, $\frac{15}{14}D_{N_3P_4}^{24} - D_{N_2P_4}^{32} = bg_{107}(a/b)$, where

$$g_{107}(x) = \frac{15}{14}f_{N_3P_4}^{24}(x) - f_{N_2P_4}^{32}(x) = \frac{u_{107}(x)}{28(\sqrt{x}+1)^2}$$

with

$$u_{107}(x) = 2\big(5x^2 + 15x^{3/2} + 16x + 15\sqrt{x} + 5\big) - 7(\sqrt{x}+1)^3\sqrt{2x+2}.$$

Let us consider

$$\begin{aligned}v_{107}(x) &= \big[2\big(5x^2 + 15x^{3/2} + 16x + 15\sqrt{x} + 5\big)\big]^2 - \big[7(\sqrt{x}+1)^3\sqrt{2x+2}\big]^2 \\ &= 2(\sqrt{x}-1)^4\big(x^2 + 10x^{3/2} + 20x + 10\sqrt{x} + 1\big) \geq 0, \quad \forall x > 0.\end{aligned}$$

108. For $D^{85}_{CP_4} \leq 3D^{24}_{N_3P_4}$: We have, $\beta_{108} = f''_{CP_4}(1)/f''_{N_3P_4}(1) = 3$. In order to show $3D^{24}_{N_3P_4} - D^{85}_{CP_4} \geq 0$, let us consider, $3D^{24}_{N_3P_4} - D^{85}_{CP_4} = bg_{108}(a/b)$, where

$$\begin{aligned}g_{108}(x) &= 3f^{24}_{N_3P_4}(x) - f^{85}_{CP_4}(x) \\ &= \frac{\sqrt{x}(x^2 - 4x^{3/2} + 6x - 4\sqrt{x} + 1)}{(\sqrt{x}+1)^2(x+1)} = \frac{\sqrt{x}(\sqrt{x}-1)^2}{(\sqrt{x}+1)^2(x+1)} \geq 0, \\ &\forall x > 0.\end{aligned}$$

109. For $D^{24}_{NP_4} \leq \frac{7}{9}D^{41}_{AP_4}$: We have, $\beta_{109} = f''_{NP_4}(1)/f''_{AP_4}(1) = \frac{7}{9}$. In order to show $\frac{7}{9}D^{41}_{AP_4} - D^{24}_{NP_4} \geq 0$, let us consider, $\frac{7}{9}D^{41}_{AP_4} - D^{24}_{NP_4} = bg_{109}(a/b)$, where

$$\begin{aligned}g_{109}(x) &= \frac{7}{9}f^{41}_{AP_4}(x) - f^{24}_{NP_4}(x) \\ &= \frac{x^2 - 4x^{3/2} + 6x - 4\sqrt{x} + 1}{18(\sqrt{x}+1)^2} = \frac{(\sqrt{x}-1)^4}{18(\sqrt{x}+1)^2} \geq 0, \quad \forall x > 0.\end{aligned}$$

110. For $D^{84}_{CG} \leq 2D^{41}_{AP_4}$: We have, $\beta_{110} = f''_{CG}(1)/f''_{AP_4}(1) = 2$. In order to show $2D^{41}_{AP_4} - D^{84}_{CG} \geq 0$, let us consider, $2D^{41}_{AP_4} - D^{84}_{CG} = bg_{110}(a/b)$, where

$$\begin{aligned}g_{110}(x) &= 2f^{41}_{AP_4}(x) - f^{84}_{CG}(x) \\ &= \frac{\sqrt{x}(x^2 - 4x^{3/2} + 6x - 4\sqrt{x} + 1)}{(\sqrt{x}+1)^2(x+1)} = \frac{\sqrt{x}(\sqrt{x}-1)^2}{(\sqrt{x}+1)^2(x+1)} \geq 0, \\ &\forall x > 0.\end{aligned}$$

111. For $D^{61}_{RP_4} \leq \frac{13}{9}D^{41}_{AP_4}$: We have, $\beta_{111} = f''_{RP_4}(1)/f''_{AP_4}(1) = \frac{13}{9}$. In order to show $\frac{13}{9}D^{41}_{AP_4} - D^{61}_{RP_4} \geq 0$, let us consider, $\frac{13}{9}D^{41}_{AP_4} - D^{61}_{RP_4} = bg_{111}(a/b)$,

where

$$g_{111}(x) = \frac{13}{9} f_{AP_4}^{41}(x) - f_{RP_4}^{61}(x)$$

$$= \frac{x^3 + 2x^{5/2} - 17x^2 + 28x^{3/2} - 17x + 2\sqrt{x} + 1}{18(\sqrt{x} + 1)^2 (x + 1)}$$

$$= \frac{(\sqrt{x} - 1)^2 (x + 6\sqrt{x} + 1)}{18(\sqrt{x} + 1)^2 (x + 1)} \geq 0, \quad \forall x > 0.$$

112. For $D_{AP_4}^{41} \leq \frac{3}{4} D_{P_5 P_4}^{51}$: We have, $\beta_{112} = f_{AP_4}''(1)/f_{P_5 P_4}''(1) = \frac{3}{4}$. In order to show $\frac{3}{4} D_{P_5 P_4}^{51} - D_{AP_4}^{41} \geq 0$, let us consider, $\frac{3}{4} D_{P_5 P_4}^{51} - D_{AP_4}^{41} = b g_{112}(a/b)$, where

$$g_{112}(x) = \frac{3}{4} f_{P_5 P_4}^{51}(x) - f_{AP_4}^{41}(x)$$

$$= \frac{x^2 - 4x^{3/2} + 6x - 4\sqrt{x} + 1}{4(\sqrt{x} + 1)^2} = \frac{(\sqrt{x} - 1)^4}{4(\sqrt{x} + 1)^2} \geq 0, \quad \forall x > 0.$$

113. For $D_{AP_1}^{45} \leq \frac{12}{13} D_{P_5 P_1}^{55}$: We have, $\beta_{113} = f_{AP_1}''(1)/f_{P_5 P_1}''(1) = \frac{12}{13}$. Now, in order to show $\frac{12}{13} D_{P_5 P_1}^{55} - D_{AP_1}^{45} \geq 0$, let us consider, $\frac{12}{13} D_{P_5 P_1}^{55} - D_{AP_1}^{45} = b g_{113}(a/b)$, where

$$g_{113}(x) = \frac{12}{13} f_{P_5 P_1}^{55}(x) - f_{AP_1}^{45}(x) = \frac{u_{113}(x)}{26(\sqrt{x} + 1)^2 (x^3 + 1)}$$

with

$$u_{113}(x) = 11x^5 - 26x^{9/2} + 24x^4 - 22x^{7/2} + 13x^3 +$$
$$+ 13x^2 - 22x^{3/2} + 24x - 26\sqrt{x} + 11$$
$$= (\sqrt{x} - 1)^4 (11x^3 + 18x^{5/2} + 30x^2 + 34x^{3/2} +$$
$$+ 30x + 18\sqrt{x} + 11) \geq 0, \quad \forall x > 0.$$

114. For $D_{AP_1}^{45} \leq \frac{9}{10} D_{RP_1}^{65}$: We have, $\beta_{114} = f_{AP_1}''(1)/f_{RP_1}''(1) = \frac{9}{10}$. Now, in order to show $\frac{9}{10} D_{RP_1}^{65} - D_{AP_1}^{45} \geq 0$, let us consider, $\frac{9}{10} D_{RP_1}^{65} - D_{AP_1}^{45} = b g_{114}(a/b)$, where

$$g_{114}(x) = \frac{9}{10} f_{RP_1}^{65}(x) - f_{AP_1}^{45}(x) = \frac{(x - 1)^4}{10(x^3 + 1)} \geq 0, \quad \forall x > 0.$$

115. For $D_{RP_1}^{65} \leq \frac{20}{21} D_{SP_1}^{76}$: We have, $\beta_{115} = f_{RP_1}''(1)/f_{SP_1}''(1) = \frac{20}{21}$. Now, in order to show $\frac{20}{21} D_{SP_1}^{76} - D_{RP_1}^{65} \geq 0$, let us consider, $\frac{20}{21} D_{SP_1}^{76} - D_{RP_1}^{65} =$

$bg_{115}(a/b)$, where

$$g_{115}(x) = \frac{20}{21}f_{SP_1}^{76}(x) - f_{RP_1}^{65}(x) = \frac{u_{115}(x)}{21(x^3+1)}$$

with

$$u_{115}(x) = 10\sqrt{2x^2+2}\,(x^3+1) - \left(14x^4 - x^3 + 14x^2 - x + 14\right)$$
$$= 10\sqrt{2x^2+2}\,(x^3+1) -$$
$$- \left[(x^2+1)(x-1)^2 + 13x^4 + x^3 + 12x^2 + x + 13\right].$$

Let us consider

$$v_{116}(x) = \left[10\sqrt{2x^2+2}\,(x^3+1)\right]^2 -$$
$$- \left[(x^2+1)(x-1)^2 + 13x^4 + x^3 + 12x^2 + x + 13\right]^2$$
$$= (x-1)^4\left(4x^4 + 23x^3 + 21x(x-1)^2 + 23x + x^2 + 4\right) \geq 0, \quad \forall x > 0.$$

116. For $D_{SP_1}^{76} \leq \frac{7}{4}D_{AP_2}^{44}$: We have, $\beta_{116} = f_{SP_1}''(1)/f_{AP_2}''(1) = \frac{7}{4}$. Now, in order to show $\frac{7}{4}D_{AP_2}^{44} - D_{SP_1}^{76} \geq 0$, let us consider, $\frac{7}{4}D_{AP_2}^{44} - D_{SP_1}^{76} = bg_{116}(a/b)$, where

$$g_{116}(x) = \frac{7}{4}f_{AP_2}^{44}(x) - f_{SP_1}^{76}(x) = \frac{u_{116}(x)}{8(x^2+1)(x^3+1)}$$

with

$$u_{116}(x) = \left(7x^6 + x^5 - 7x^4 + 30x^3 - 7x^2 + x + 7\right) -$$
$$- 10\sqrt{2x^2+2}\,(x^3+1)(x^2+1)$$
$$= 7x^6 + x^3(x-4)^2 + x^2(x-1)^2 + x(4x-1)^2 + 7 -$$
$$- 10\sqrt{2x^2+2}\,(x^3+1)(x^2+1).$$

Let us consider

$$v_{116}(x) = \left[7x^6 + x^3(x-4)^2 + x^2(x-1)^2 + x(4x-1)^2 + 7\right]^2 -$$
$$- \left[10\sqrt{2x^2+2}\,(x^3+1)(x^2+1)\right]^2$$
$$= (x-1)^4\left(17x^8 + 82x^7 + 33x^6 + 50x^5 +\right.$$
$$\left. + 228x^4 + 50x^3 + 33x^2 + 82x + 17\right) \geq 0, \quad \forall x > 0.$$

117. For $D_{AP_2}^{44} \leq \frac{4}{3}D_{AP_3}^{43}$: We have, $\beta_{117} = f_{AP_2}''(1)/f_{AP_3}''(1) = \frac{4}{3}$. Now, in order to show $\frac{4}{3}D_{AP_3}^{43} - D_{AP_2}^{44} \geq 0$, let us consider, $\frac{4}{3}D_{AP_3}^{43} - D_{AP_2}^{44} =$

$bg_{117}(a/b)$, where

$$g_{117}(x) = \frac{4}{3}f_{AP_3}^{43}(x) - f_{AP_2}^{44}(x)$$

$$= \frac{x^4 - x^{7/2} - 7x^{5/2} + 14x^2 - 7x^{3/2} - \sqrt{x} + 1}{6(x^2+1)(x-\sqrt{x}+1)}$$

$$= \frac{(\sqrt{x}-1)^4(x^2 + 3x^{3/2} + 6x + 3\sqrt{x} + 1)}{6(x^2+1)(x-\sqrt{x}+1)} \geq 0, \quad \forall x > 0.$$

118. For $D_{P_5P_1}^{55} \leq \frac{13}{6}D_{AP_3}^{43}$: We have, $\beta_{118} = f_{P_5P_1}''(1)/f_{AP_3}''(1) = \frac{13}{6}$. Now, in order to show $\frac{13}{6}D_{AP_3}^{43} - D_{P_5P_1}^{55} \geq 0$, let us consider, $\frac{13}{6}D_{AP_3}^{43} - D_{P_5P_1}^{55} = bg_{118}(a/b)$, where

$$g_{118}(x) = \frac{13}{6}f_{AP_3}^{43}(x) - f_{P_5P_1}^{55}(x) = \frac{u_{119}(x)}{4(x^3+1)(\sqrt{x}+1)(x^{3/2}+1)}$$

with

$$u_{118}(x) = \left(x^6 + 25x^{11/2} - 37x^5 + 10x^{9/2} - 49x^4 + 37x^{7/2} + \right.$$
$$+ 26x^3 + 37x^{5/2} - 49x^2 + 10x^{3/2} - 37x + 25\sqrt{x} + 1\big)$$
$$= (\sqrt{x}-1)^4\big(x^4 + 29x^{7/2} + 73x^3 + 132x^{5/2} + $$
$$+ 156x^2 + 132x^{3/2} + 73x + 29\sqrt{x} + 1\big) \geq 0, \quad \forall x > 0.$$

119. For $D_{AP_3}^{43} \leq \frac{9}{11}D_{RP_3}^{63}$: We have, $\beta_{119} = f_{AP_3}''(1)/f_{RP_3}''(1) = \frac{9}{11}$. Now, in order to show $\frac{9}{11}D_{RP_3}^{63} - D_{AP_3}^{43} \geq 0$, let us consider, $\frac{9}{11}D_{RP_3}^{63} - D_{AP_3}^{43} = bg_{119}(a/b)$, where

$$g_{119}(x) = \frac{9}{11}f_{RP_3}^{63}(x) - D_{AP_3}^{43}(x)$$

$$= \frac{x^3 - x^{5/2} - 5x^2 + 10x^{3/2} - 5x - \sqrt{x} + 1}{22(x-\sqrt{x}+1)(x+1)}$$

$$= \frac{(\sqrt{x}-1)^4(x + 3\sqrt{x} + 1)}{22(x-\sqrt{x}+1)(x+1)} \geq 0, \quad \forall x > 0.$$

120. For $D_{RP_3}^{63} \leq \frac{22}{21}D_{P_5P_3}^{53}$: We have, $\beta_{120} = f_{RP_3}''(1)/f_{P_5P_3}''(1) = \frac{22}{21}$. Now, in order to show $\frac{22}{21}D_{P_5P_3}^{53} - D_{RP_3}^{63} \geq 0$, let us consider, $\frac{22}{21}D_{P_5P_3}^{53} - D_{RP_3}^{63} = bg_{120}(a/b)$, where

$$g_{120}(x) = \frac{22}{21}f_{P_5P_3}^{53}(x) - f_{RP_3}^{63}(x) = \frac{u_{120}(x)}{21(\sqrt{x}+1)(x+1)(x^{3/2}+1)}$$

with

$$u_{120}(x) = 8x^4 - 36x^{7/2} + 73x^3 - 96x^{5/2} +$$
$$+ 102x^2 - 96x^{3/2} + 73x - 36\sqrt{x} + 8$$
$$= (\sqrt{x} - 1)^4 [6(x^2 + 1) + 2(x+1)(\sqrt{x} - 1)^2 + 5x] \geq 0, \quad \forall x > 0.$$

121. **For $D_{P_5 P_3}^{53} \leq \frac{3}{2} D_{CN}^{82}$:** We have, $\beta_{121} = f_{P_5 P_3}''(1)/f_{CN}''(1) = \frac{3}{2}$. Now, in order to show $\frac{3}{2} D_{CN}^{82} - D_{P_5 P_3}^{53} \geq 0$, let us consider, $\frac{3}{2} D_{CN}^{82} - D_{P_5 P_3}^{53} = bg_{121}(a/b)$, where

$$g_{121}(x) = \frac{3}{2} f_{CN}^{82}(x) - f_{P_5 P_3}^{53}(x)$$
$$= \frac{3(x^{7/2} + 3x^{5/2} - 3x^3 - 2x^2 + 3x^{3/2} - 3x + \sqrt{x})}{2(\sqrt{x} + 1)(x^{3/2} + 1)(x+1)}$$
$$= \frac{3\sqrt{x}(\sqrt{x} - 1)^4(x + \sqrt{x} + 1)}{2(\sqrt{x} + 1)(x^{3/2} + 1)(x+1)} \geq 0, \quad \forall x > 0.$$

122. **For $D_{CN}^{82} \leq \frac{2}{3} D_{CP_4}^{85}$:** We have, $\beta_{122} = f_{CN}''(1)/f_{CP_4}''(1) = \frac{2}{3}$. Now, in order to show $\frac{2}{3} D_{CP_4}^{85} - D_{CN}^{82} \geq 0$, let us consider, $\frac{2}{3} D_{CP_4}^{85} - D_{CN}^{82} = bg_{122}(a/b)$, where

$$g_{122}(x) = \frac{2}{3} f_{CP_4}^{85}(x) - f_{CN}^{82}(x)$$
$$= \frac{\sqrt{x}(x^2 - 4x^{3/2} + 6x - 4\sqrt{x} + 1)}{3(x+1)(\sqrt{x} + 1)^2} = \frac{\sqrt{x}(\sqrt{x} - 1)^4}{3(x+1)(\sqrt{x} + 1)^2} \geq 0,$$
$$\forall x > 0.$$

123. **For $D_{P_5 P_3}^{53} \leq \frac{14}{9} D_{CN_2}^{81}$:** We have, $\beta_{123} = f_{P_5 P_3}''(1)/f_{CN_2}''(1) = \frac{14}{9}$. Now, in order to show $\frac{14}{9} D_{CN_2}^{81} - D_{P_5 P_3}^{53} \geq 0$, let us consider, $\frac{14}{9} D_{CN_2}^{81} - D_{P_5 P_3}^{53} = bg_{123}(a/b)$, where

$$g_{123}(x) = \frac{14}{9} f_{CN_2}^{81}(x) - f_{P_5 P_3}^{53}(x) = \frac{u_{123}(x)}{18(x+1)(\sqrt{x} + 1)^2(x - \sqrt{x} + 1)}$$

with

$$u_{123}(x) = (10x^4 + 46x^{7/2} - 54x^3 + 118x^{5/2} - 16x^2 + 118x^{3/2} - 54x +$$
$$+ 46\sqrt{x} + 10) - 7\sqrt{2x + 2}(x^{3/2} + 1)(\sqrt{x} + 1)^2(x+1)$$
$$= \sqrt{x}(27x^2 + 8x + 27)(\sqrt{x} - 1)^2 + 10(x^4 + 1) + 19\sqrt{x}(x^3 + 1) +$$
$$+ 83x^{3/2}(x+1)46\sqrt{x} + 10 -$$
$$- 7\sqrt{2x + 2}(x^{3/2} + 1)(\sqrt{x} + 1)^2(x+1).$$

Now we shall show that $u_{123}(x) \geq 0$, $\forall x > 0$. Let us consider

$$v_{123}(x) = \left[\sqrt{x}\left(27x^2 + 8x + 27\right)(\sqrt{x} - 1)^2 + 10\left(x^4 + 1\right) + \right.$$
$$\left. + 19\sqrt{x}\left(x^3 + 1\right) + 83x^{3/2}(x+1)46\sqrt{x} + 10\right]^2 -$$
$$- \left[7\sqrt{2x+2}\left(x^{3/2} + 1\right)(\sqrt{x} + 1)^2(x+1)\right]^2.$$

After simplification, we get

$$v_{123}(x) = (\sqrt{x} - 1)^2\left(x^6 + 268x^{11/2} + 1143x^5 + 782x^{9/2} + \right.$$
$$+ 2597x^4 + 2014x^{7/2} + 4478x^3 + 2014x^{5/2} +$$
$$+ 2597x^2 + 782x^{3/2} + 1143x + 268\sqrt{x} + 1\right) \geq 0, \quad \forall x > 0.$$

124. For $D_{GP_2}^{14} \leq 2D_{CP_5}^{79}$: We have, $\beta_{124} = f_{GP_2}''(1)/f_{CP_5}''(1) = 2$. Now, in order to show $2D_{CP_5}^{79} - D_{GP_2}^{14} \geq 0$, let us consider, $2D_{CP_5}^{79} - D_{GP_2}^{14} = bg_{124}(a/b)$, where

$$g_{124}(x) = 2f_{CP_5}^{79}(x) - f_{GP_2}^{14}(x)$$
$$= \frac{\sqrt{x}(3x^4 - 5x^{7/2} + 3x^{5/2} + 10x^2 - 3x^{3/2} - 5\sqrt{x} + 3)}{(x+1)(x^2+1)(\sqrt{x}+1)^2}$$
$$= \frac{\sqrt{x}(x + \sqrt{x} + 1)(3x + 4\sqrt{x} + 3)(\sqrt{x} - 1)^4}{(x+1)(x^2+1)(\sqrt{x}+1)^2} \geq 0, \quad \forall x > 0.$$

125. For $D_{CP_5}^{79} \leq 3D_{GP_4}^{11}$: We have, $\beta_{125} = f_{CP_5}''(1)/f_{GP_4}''(1) = 3$. Now, in order to show $3D_{GP_4}^{11} - D_{CP_5}^{79} \geq 0$, let us consider, $3D_{GP_4}^{11} - D_{CP_5}^{79} = bg_{125}(a/b)$, where

$$g_{125}(x) = \frac{33}{34}f_{N_3P_1}^{28}(x) - f_{N_1P_1}^{21}(x)$$
$$= \frac{\sqrt{x}(x^2 - 4x^{3/2} + 6x - 4\sqrt{x} + 1)}{(x+1)(\sqrt{x}+1)^2} = \frac{\sqrt{x}(\sqrt{x} - 1)^4}{(x+1)(\sqrt{x}+1)^2} \geq 0,$$
$$\forall x > 0.$$

126. For $D_{GP_4}^{11} \leq \frac{1}{4}D_{AH}^{42}$: We have, $\beta_{126} = f_{GP_4}''(1)/f_{AH}''(1) = \frac{1}{4}$. Now, in order to show $\frac{1}{4}D_{AH}^{42} - D_{GP_4}^{11} \geq 0$, let us consider, $\frac{1}{4}D_{AH}^{42} - D_{GP_4}^{11} = bg_{126}(a/b)$,

where

$$g_{126}(x) = \frac{1}{4}f_{AH}^{42}(x) - f_{GP_4}^{11}(x)$$

$$= \frac{x^3 - 6x^{5/2} + 15x^2 - 20x^{3/2} + 15x - 6\sqrt{x} + 1}{8(\sqrt{x}+1)^2(x+1)}$$

$$= \frac{(\sqrt{x}-1)^2}{8(\sqrt{x}+1)^2(x+1)} \geq 0, \quad \forall x > 0.$$

127. For $D_{CP_4}^{85} \leq \frac{7}{2}D_{CS}^{77}$: We have, $\beta_{127} = f_{CP_4}''(1)/f_{CS}''(1) = \frac{7}{2}$. Now, in order to show $\frac{7}{2}D_{CS}^{77} - D_{CP_4}^{85} \geq 0$, let us consider, $\frac{7}{2}D_{CS}^{77} - D_{CP_4}^{85} = bg_{127}(a/b)$, where

$$g_{127}(x) = \frac{7}{2}D_{CS}^{77} - D_{CP_4}^{85} = \frac{u_{127}(x)}{4(\sqrt{x}+1)^2(x+1)}$$

with

$$u_{127}(x) = 10x^3 + 20x^{5/2} + 26x^2 + 26x + 20\sqrt{x} + 10 -$$
$$- 7\sqrt{2x^2 + 2}(\sqrt{x}+1)^2(x+1).$$

Let us consider

$$v_{127}(x) = \left(10x^3 + 20x^{5/2} + 26x^2 + 26x + 20\sqrt{x} + 10\right)^2 -$$
$$- \left[7\sqrt{2x^2 + 2}(\sqrt{x}+1)^2(x+1)\right]^2$$
$$= 2(\sqrt{x}-1)^4\left(x^4 + 8x^{7/2} + 94x^3 + 264x^{5/2} +\right.$$
$$\left. + 386x^2 + 264x^{3/2} + 94x + 8\sqrt{x} + 1\right) \geq 0, \quad \forall x > 0.$$

128. For $D_{CN_2}^{81} \leq \frac{9}{4}D_{CS}^{77}$: We have, $\beta_{128} = f_{CN_2}''(1)/f_{CS}''(1) = \frac{9}{4}$. Now, in order to show $\frac{9}{4}D_{CS}^{77} - D_{CN_2}^{81} \geq 0$, let us consider, $\frac{9}{4}D_{CS}^{77} - D_{CN_2}^{81} = bg_{128}(a/b)$, where

$$g_{128}(x) = \frac{9}{4}f_{CS}^{77}(x) - f_{CN_2}^{81}(x) = \frac{u_{128}(x)}{8(x+1)}$$

with

$$u_{128}(x) = 10\left(x^2 + 1\right) + 2\sqrt{2x+2}(\sqrt{x}+1)(x+1) - 9\sqrt{2x^2 + 2}(x+1).$$

Now we shall show that $u_{130}(x) \geq 0$, $\forall x > 0$. Let us consider

$$v_{128}(x) = \left[10(x^2 + 1) + 2\sqrt{2x + 2}(\sqrt{x} + 1)(x + 1)\right]^2 -$$
$$- \left[9\sqrt{2x^2 + 2}(x + 1)\right]^2$$
$$= 2\left[26(x^4 + 1) + x^3(\sqrt{x} - 4)^2 + 118x(x^2 + 1) +\right.$$
$$\left. + 12x(x + 1)(\sqrt{x} - 1)^2 + (4\sqrt{x} - 1)^2 + 14x^2\right] -$$
$$- 40\sqrt{2x + 2}(\sqrt{x} + 1)(x^2 + 1)(x + 1).$$

Let us consider

$$p_{128}(x) = 4\left[26(x^4 + 1) + x^3(\sqrt{x} - 4)^2 + 118x(x^2 + 1) +\right.$$
$$+ 12x(x + 1)(\sqrt{x} - 1)^2 + (4\sqrt{x} - 1)^2 + 14x^2\Big]^2 -$$
$$- \left[40\sqrt{2x + 2}(\sqrt{x} + 1)(x^2 + 1)(x + 1)\right]^2.$$

After simplifications, we get

$$p_{128}(x) = (\sqrt{x} - 1)^4 \big(284 + 9264\sqrt{x} + 284x^6 + 16360x^5 + 9264x^{11/2} +$$
$$+ 44720x^{9/2} + 48084x^2 + 44720x^{3/2} + 48084x^4 +$$
$$+ 24016x^3 + 47840x^{7/2} + 47840x^{5/2} + 16360\big) \geq 0, \quad \forall x > 0.$$

129. For $D_{CS}^{77} \leq \frac{1}{2} D_{P_5 P_4}^{51}$: We have, $\beta_{129} = f_{CS}''(1)/f_{P_5 P_4}''(1) = \frac{1}{2}$. Now, in order to show $\frac{1}{2} D_{P_5 P_4}^{51} - D_{CS}^{77} \geq 0$, let us consider, $\frac{1}{2} D_{P_5 P_4}^{51} - D_{CS}^{77} = b g_{129}(a/b)$, where

$$g_{129}(x) = \frac{1}{2} D_{P_5 P_4}^{51} - D_{CS}^{77} = \frac{u_{129}(x)}{2(\sqrt{x} + 1)^2(x + 1)}$$

with

$$u_{129}(x) = (\sqrt{x} + 1)^2 \left[\sqrt{2x^2 + 2}(x + 1)(\sqrt{x} + 1)^2 -\right.$$
$$\left. - \left(x^2 + x^{3/2} + \sqrt{x}(\sqrt{x} - 1)^2 + \sqrt{x} + 1\right)\right].$$

Let us consider

$$v_{129}(x) = (\sqrt{x} + 1)^4 \left[\left(\sqrt{2x^2 + 2}(x + 1)(\sqrt{x} + 1)^2\right)^2 -\right.$$
$$\left. - \left(x^2 + x^{3/2} + \sqrt{x}(\sqrt{x} - 1)^2 + \sqrt{x} + 1\right)^2\right]$$
$$= (\sqrt{x} + 1)^6(\sqrt{x} - 1)^6 = (x - 1)^6 \geq 0, \quad \forall x > 0.$$

Combining Results 1–129, we get the proof of (94), (95) and (96).   □

6.4 Particular Cases

Inequalities given in (94), (95) and (96) work with nonnegative differences arising due to 14 means given in (89). Considering only *seven means*, we are left with the following three inequalities:

$$\left\{ \begin{matrix} D_{GH}^{12} \\ 3D_{SR}^{66} \end{matrix} \right\} \le f_{SA}^{67} \le \left. \begin{cases} \left. \begin{matrix} \frac{1}{3}D_{SH}^{73} \le \frac{1}{2}D_{AH}^{42} \\ \frac{3}{4}D_{SN}^{69} \end{matrix} \right\} \le \frac{3}{7}D_{CN}^{82} \le \frac{1}{3}D_{CG}^{84} \\ \frac{3}{4}D_{SN}^{69} \le \begin{cases} \frac{1}{3}D_{CG}^{84} \\ \frac{1}{2}D_{SG}^{71} \end{cases} \end{cases} \right\} \le$$

$$\le \frac{3}{5}D_{RG}^{60} \le D_{AG}^{40} \tag{97}$$

$$3D_{SR}^{66} \le \frac{3}{5}D_{NH}^{25} \le \frac{1}{2}D_{AH}^{42} \le \frac{1}{3}D_{CG}^{84} \le D_{AG}^{40} \tag{98}$$

and

$$\frac{1}{2}D_{AH}^{42} \le \frac{3}{7}D_{CN}^{82} \le D_{CS}^{77} \le D_{AG}^{40}. \tag{99}$$

Combining (97), (98) and (99), we get the same result as given in (46).

6.5 Reverse Order and Corner Inequalities

Theorem 9 is based on the nonnegative differences given in terms of obvious inequalities given in (93). Leaving first line with single measure, we have other 12 lines with inequalities. Each line can be written in *reverse order inequalities*. These are given in following theorem.

Theorem 10. *The 13 lines appearing in (93) satisfy the following inequalities in reverse order:*

1. $D_{P_2P_1}^1.$

2. $\frac{1}{3}D_{P_3P_1}^3 \le D_{P_3P_2}^2.$

3. $\frac{1}{2}D_{HP_1}^6 \le D_{HP_2}^5 \le 2D_{HP_3}^4.$

4. $\frac{1}{9}D_{P_4P_1}^{10} \le \frac{1}{5}D_{P_4P_2}^9 \le \frac{1}{3}D_{P_4P_3}^8 \le 3D_{P_4H}^7.$

5. $\frac{1}{5}D_{GP_1}^{15} \le \frac{1}{3}D_{GP_2}^{14} \le \frac{1}{2}D_{GP_3}^{13} \le D_{GH}^{12} \le 2D_{GP_4}^{11}.$

6. $\frac{1}{11}D_{N_1P_1}^{21} \le \frac{1}{7}D_{N_1P_2}^{20} \le \frac{1}{5}D_{N_1P_3}^{19} \le \frac{1}{3}D_{N_1H}^{18} \le \frac{1}{2}D_{N_1P_4}^{17} \le D_{N_1G}^{16}.$

7. $\frac{1}{17}D_{NP_1}^{28} \le \frac{1}{11}D_{NP_2}^{27} \le \frac{1}{8}D_{NP_3}^{26} \le \frac{1}{5}D_{NH}^{25} \le \frac{2}{7}D_{NP_4}^{24} \le D_{NG}^{23} = 4f_{NN_1}^{22}.$

8. $\frac{1}{23}D_{N_2P_1}^{36} \le \frac{1}{15}D_{N_2P_2}^{35} \le \frac{1}{11}D_{N_2P_3}^{34} \le \frac{1}{7}D_{N_2H}^{33} \le 3D_{N_2N}^{29} \le \frac{1}{5}D_{N_2P_4}^{32} \le$

$$\leq D_{N_2 N_1}^{30} \leq \tfrac{1}{3} D_{N_2 G}^{31}.$$

9. $$\tfrac{1}{3} D_{AP_1}^{45} \leq \tfrac{1}{2} D_{AP_2}^{44} \leq \tfrac{2}{3} D_{AP_3}^{43} \leq D_{AH}^{42} \leq \tfrac{4}{3} D_{AP_4}^{41} \leq 2 D_{AG}^{40} = 4 D_{AN_1}^{39} =$$
$$= 6 D_{AN}^{38} \leq 8 D_{AN_2}^{37}.$$

10. $$\tfrac{1}{13} D_{P_5 P_1}^{55} \leq \tfrac{1}{9} D_{P_5 P_2}^{54} \leq \tfrac{1}{7} D_{P_5 P_3}^{53} \leq \tfrac{1}{5} D_{P_5 H}^{52} \leq \tfrac{1}{4} D_{P_5 P_4}^{51} \leq \tfrac{1}{3} D_{P_5 G}^{50} \leq$$
$$\leq \tfrac{1}{2} D_{P_5 N_1}^{49} \leq \tfrac{3}{5} D_{P_5 N}^{48} \leq \tfrac{2}{3} D_{P_5 N_2}^{47} \leq D_{P_5 A}^{46}.$$

11. $$\tfrac{1}{10} D_{RP_1}^{65} \leq \tfrac{1}{7} D_{RP_2}^{64} \leq \tfrac{2}{11} D_{RP_3}^{63} \leq \tfrac{1}{4} D_{RH}^{62} = D_{RA}^{56} \leq$$
$$\leq \begin{Bmatrix} \tfrac{4}{13} D_{RP_4}^{61} \\ \tfrac{2}{3} D_{RN}^{58} \leq \tfrac{8}{11} D_{RN_2}^{57} \leq \tfrac{4}{7} D_{RN_1}^{59} \end{Bmatrix} \leq \tfrac{2}{5} D_{RG}^{60}.$$

12. $$\tfrac{1}{7} D_{SP_1}^{76} \leq \begin{Bmatrix} \tfrac{1}{5} D_{SP_2}^{75} \\ 3 D_{SR}^{66} \end{Bmatrix} \leq \tfrac{1}{4} D_{SP_3}^{74} \leq D_{SA}^{67} \leq \begin{Bmatrix} \tfrac{1}{3} D_{SH}^{73} \\ \tfrac{3}{4} D_{SN}^{69} \\ \tfrac{4}{5} D_{SN_2}^{68} \end{Bmatrix} \leq \begin{Bmatrix} \tfrac{2}{5} D_{SP_4}^{72} \\ \tfrac{2}{3} D_{SN_1}^{70} \end{Bmatrix} \leq$$
$$\leq \tfrac{1}{2} D_{SG}^{71}.$$

13. $$\begin{Bmatrix} \tfrac{1}{2} D_{CP_1}^{89} \leq \tfrac{2}{3} D_{CP_2}^{88} \leq \tfrac{4}{5} D_{CP_3}^{87} \\ \tfrac{8}{3} D_{CP_5}^{79} \end{Bmatrix} \leq D_{CH}^{86} =$$

$$= 2 D_{CA}^{80} = 3 D_{CR}^{78} \leq \begin{cases} \tfrac{12}{5} D_{CN}^{82} \leq \tfrac{8}{5} D_{CP_4}^{85} \leq \begin{Bmatrix} \tfrac{28}{15} D_{CG}^{84} \\ \tfrac{28}{5} D_{CS}^{77} \end{Bmatrix} \\ \begin{Bmatrix} \tfrac{12}{5} D_{CN}^{82} \\ \tfrac{16}{9} D_{CN_2}^{81} \end{Bmatrix} \leq \tfrac{8}{5} D_{CN_1}^{83} \leq \tfrac{4}{3} D_{CG}^{84} \end{cases}.$$

$$\dots \quad (100)$$

Proof is based on the similar lines of Theorem 9. The first line with single measure does not have any effect.

Theorem 11. *The corners and initial members of each line of (93) satisfy the following inequalities:*

1. Corner members

$$\tfrac{1}{2} D_{P_2 P_1}^{1} \leq \tfrac{1}{3} D_{P_3 P_1}^{3} \leq \tfrac{1}{4} D_{HP_1}^{6} \leq \tfrac{2}{9} D_{P_4 P_1}^{10} \leq \tfrac{1}{5} D_{GP_1}^{15} \leq \tfrac{2}{11} D_{N_1 P_1}^{21} \leq$$
$$\leq \tfrac{3}{17} D_{NP_1}^{28} \leq \tfrac{4}{23} D_{N_2 P_1}^{36} \leq$$
$$\leq \tfrac{1}{6} D_{AP_1}^{45} \leq \begin{Bmatrix} \tfrac{2}{13} D_{P_5 P_1}^{55} \\ \tfrac{3}{20} D_{RP_1}^{65} \leq \tfrac{1}{7} D_{SP_1}^{76} \leq \tfrac{1}{8} D_{CP_1}^{89} \end{Bmatrix}. \quad (101)$$

2. Initial members

$$\frac{1}{4}D^1_{P_2P_1} \leq \frac{1}{2}D^2_{P_3P_2} \leq \frac{1}{2}D^4_{HP_3} \leq \begin{Bmatrix} D^7_{P_4H} \leq \frac{1}{3}D^{79}_{P_6P_5} \leq D^{11}_{GP_4} \\ \frac{3}{2}D^{66}_{SR} \end{Bmatrix} \leq$$

$$\leq \begin{Bmatrix} 6D^{29}_{N_2N} \\ \frac{3}{4}D^{56}_{RA} \end{Bmatrix} \leq$$

$$\leq \frac{1}{2}D^{77}_{CS} \leq D^{16}_{N_1G} = 3D^{22}_{NN_1} \leq 2D^{37}_{AN_2} \leq D^{46}_{P_5A}. \tag{102}$$

Proof is again based on the similar lines of Theorem 9.

6.6 Individual-Type Inequalities

The results appearing in Theorems 9, 10 and 11 can be separated to write in terms of means. These separations give many individual-type inequalities given in two groups.

Group 6

1. $P_1 \leq \frac{7P_2+3R}{10}$.

2. $P_2 \leq \frac{P_1+2P_3}{3}$.

3. $P_2 \leq \frac{H+P_1}{2}$.

4. $P_2 \leq \frac{3P_1+2G}{5}$.

5. $P_2 \leq \frac{3P_1+C}{4}$.

6. $P_2 \leq \frac{5P_1+4P_4}{9}$.

7. $P_2 \leq \frac{15P_1+8N_2}{23}$.

8. $P_2 \leq \frac{7P_1+4N_1}{11}$.

9. $P_2 \leq \frac{9P_1+4P_5}{13}$.

10. $P_2 \leq \frac{7P_3+2P_5}{9}$.

11. $P_2 \leq \frac{5P_1+2S}{7}$.

12. $P_2 \leq \frac{P_1+2P_3}{3}$.

13. $P_2 \leq \frac{A+2P_1}{3}$.

14. $P_2 \leq \frac{P_1+2P_3}{3}$.

15. $P_3 \leq \frac{P_1+3H}{4}$.

16. $P_3 \leq \frac{H+P_2}{2}$.

17. $P_3 \leq \frac{3H+P_1}{4}$.

18. $P_3 \leq \frac{C+5P_2}{6}$.

19. $P_3 \leq \frac{2P_4+3P_2}{5}$.

20. $P_3 \leq \frac{A+3P_2}{4}$.

21. $P_3 \leq \frac{4P_2+S}{5}$.

22. $P_3 \leq \frac{11P_2+3R}{14}$.

23. $P_3 \leq \frac{5P_2+2N_1}{7}$.

24. $P_3 \leq \frac{H+P_2}{2}$.

25. $P_3 \leq \frac{2P_2+G}{3}$.

26. $P_3 \leq \frac{11P_2+4N_2}{15}$.

27. $H \leq \frac{4P_4+P_2}{5}$.

28. $H \leq \frac{2G+P_2}{3}$.

29. $H \leq \frac{2P_4+P_3}{3}$.

30. $H \leq \frac{8P_4+P_1}{9}$.

31. $H \leq \frac{5P_2+6N}{11}$.

32. $H \leq \frac{P_3+G}{2}$.

33. $H \leq \frac{A+2P_3}{3}$.

34. $H \leq \frac{P_3+2P_4}{3}$.

35. $H \leq \frac{4P_3+C}{5}$.

36. $H \leq \frac{3P_3+2N_1}{5}$.

37. $H \leq \frac{7P_3+4N_2}{11}$.

38. $H \leq \frac{5P_3+2P_5}{7}$.

39. $H \leq \frac{8P_3+3R}{11}$.

40. $P_4 \leq \frac{P_1+9G}{10}$.

41. $P_4 \leq \frac{G+H}{2}$.

42. $P_4 \leq \frac{3G+P_3}{4}$.

43. $P_4 \leq \frac{5H+S}{6}$.

44. $P_4 \leq \frac{4H+P_5}{5}$.

45. $P_4 \leq \frac{2H+N_1}{3}$.

46. $P_4 \leq \frac{A+3H}{4}$.

47. $G \leq \frac{6N_1+P_2}{7}$.

48. $G \leq \frac{C+6P_4}{7}$.

49. $G \leq \frac{N_1+6P_2}{7}$.

50. $G \leq \frac{P_1+10N_1}{11}$.

51. $G \leq \frac{A+2P_4}{3}$.

52. $G \leq \frac{4P_4+S}{5}$.

53. $G \leq \frac{3P_4+P_5}{4}$.

54. $G \leq \frac{10P_4+3R}{13}$.

55. $\frac{S+3G}{4} \leq N_1$.

56. $\frac{C+5G}{6} \leq N_1$.

57. $\frac{G+2N_2}{3} \leq N_1$.

58. $N_1 \leq \frac{P_2+21N}{22}$.

59. $N_1 \leq \frac{P_1+33N}{34}$.

60. $N_1 \leq \frac{21N_3+P_2}{22}$.

61. $N_1 \leq \frac{4N_2+P_4}{5}$.

62. $N_1 \leq \frac{9N_3+H}{10}$.

63. $N_1 \leq \frac{P_5+2G}{3}$.

64. $N_1 \leq \frac{P_3+15N_3}{16}$.

65. $\frac{5G+C}{6} \leq N_1$.

66. $\frac{7G+3R}{10} \leq N_1$.

67. $N_1 \leq \frac{H+S}{2}$.

68. $\frac{C+14N_1}{15} \leq N$.

69. $\frac{C+2P_4}{3} \leq N$.

70. $\frac{P_4+14N_2}{15} \leq N$.

71. $\frac{C+2P_4}{3} \leq N$.

72. $\frac{S+8N_1}{9} \leq N$.

73. $N \leq \frac{11C+7P_2}{18}$.

74. $N \leq \frac{22P_5+5P_2}{27}$.

75. $N \leq \frac{P_1+68N_2}{69}$.

76. $N \leq \frac{P_2+44N_2}{45}$.

77. $N \leq \frac{P_5+5N_1}{6}$.

78. $N \leq \frac{H+20N_2}{21}$.

79. $N \leq \frac{P_3+32N_2}{33}$.

80. $N \leq \frac{7A+2P_4}{9}$.

81. $\frac{8P_4+7S}{15} \leq N$.

82. $\frac{2P_4+C}{3} \leq N$.

83. $\frac{3R+11N_1}{14} \leq N_2$.

84. $\frac{C+3G}{4} \leq N_2$.

85. $\frac{S+5N_1}{6} \leq N_2$.

86. $N_2 \leq \frac{7A+H}{8}$.

87. $N_2 \leq \frac{P_1+23A}{24}$.

88. $N_2 \leq \frac{P_2+15A}{16}$.

89. $N_2 \leq \frac{P_5+9N_3}{10}$.

90. $N_2 \leq \frac{P_3+11A}{12}$.

91. $N_2 \leq \frac{A+3N}{4}$.

92. $N_2 \leq \frac{11N_3+R}{12}$.

93. $\frac{9N_1+C}{10} \leq N_2$.

94. $\frac{P_4+S}{2} \leq N_2$.

95. $\frac{9R+4P_4}{13} \leq A$.

96. $\frac{2S+H}{3} \leq A$.

97. $\frac{S+4N_2}{5} \leq A$.

98. $\frac{S+3N}{4} \leq A$.

99. $A \leq \frac{3S+P_3}{4}$.

100. $A \leq \frac{P_5+2N_2}{3}$.

101. $A \leq \frac{4S+P_2}{5}$.

102. $A \leq \frac{6R+P_2}{7}$.

103. $A \leq \frac{P_1+12P_5}{13}$.

104. $A \leq \frac{P_1+9R}{10}$.

105. $A \leq \frac{P_4+3P_5}{4}$.

106. $A \leq \frac{9R+2P_3}{11}$.

107. $A \leq \frac{H+3R}{4}$.

108. $A \leq \frac{H+C}{2}$.

109. $\frac{2N+R}{3} \leq A$.

110. $S \leq \frac{5C+2P_4}{7}$.

111. $S \le \frac{5C+2P_4}{7}$.

112. $S \le \frac{5C+4N_2}{9}$.

113. $S \le \frac{P_1+7C}{8}$.

114. $\frac{21R+P_3}{22} \le P_5$.

115. $\frac{3H+5C}{8} \le P_5$.

116. $\frac{11S+P_3}{12} \le R$.

117. $R \le \frac{P_1+20S}{21}$.

118. $\frac{A+2S}{3} \le R$.

119. $R \le \frac{2A+C}{3}$.

120. $R \le \frac{P_1+20S}{21}$.

121. $\frac{4N_3+3C}{7} \le R$.

122. $\frac{16N_2+11C}{27} \le R$.

Group 7

1. $P_1 + 4H \le C + 4P_2$.

2. $P_2 + 3P_4 \le P_5 + 3P_3$.

3. $23P_2 + 14N_2 \le 23N_1 + 14P_1$.

4. $18P_2 + 11A \le 18N + 11P_1$.

5. $3P_2 + 2S \le 3P_5 + 2H$.

6. $7P_2 + 4S \le 7A + 4P_1$.

7. $P_2 + 18S \le C + 18R$.

8. $16P_2 + 9C \le 16P_5 + 9P_1$.

9. $P_3 + 5G \le C + 5H$.

10. $3P_3 + 2N_2 \le 3N_1 + 2P_2$.

11. $5P_3 + 3S \le 5A + 3P_2$.

12. $P_3 + 8S \le N_3 + 8R$.

13. $13P_3 + 6P_5 \le 13A + 6P_1$.

14. $14P_3 + 9R \le 14A + 9P_2$.

15. $H + 4G \le A + 4P_4$.

16. $2H + A \le 2N_1 + P_3$.

17. $4H + 5S \le 4P_5 + 5G$.

18. $11H + 30R \le 11P_5 + 30N_2$.

19. $11H + 5R \le 11N_3 + 5P_3$.

20. $H + 3R \le P_3 + 3S$.

21. $H + 2C \le S + 2P_5$.

22. $P_4 + 12N_2 \le N_1 + 12N_3$.

23. $P_4 + 42N_2 \le C + 42N_3$.

24. $3P_4 + 8N_2 \le 3P_5 + 8G$.

25. $2P_4 + A \le 2N_1 + H$.

26. $4P_4 + 5A \le 4S + 5H$.

27. $3P_4 + P_5 \le 3H + C$.

28. $2P_4 + C \le 2A + G$.

29. $3P_4 + C \le 3G + P_5$.

30. $P_4 + 2C \le P_5 + 2S$.

31. $27P_4 + 26C \le 27R + 26N_2$.

32. $G + 6N_3 \le P_4 + 6N_2$.

33. $3G + 2P_5 \le 3H + 2C$.

34. $3G + 2P_5 \le 3R + 2H$.

35. $G + 2P_5 \le P_2 + 2C$.

36. $G + 6A \le P_5 + 6N_2$.

37. $4G + 3A \le 4P_4 + 3R$.

38. $G + 2S \le C + 2N_1$.

39. $G + 6R \le P_3 + 6S$.

40. $2G + C \le 2N_1 + S$.

41. $N_1 + S \le C + P_4$.

42. $12N_1 + 7S \le 12R + 7G$.

43. $20N_1 + 3R \le 20N_2 + 3G$.

44. $3N + 2N_2 \le 3N_1 + 2A$.

45. $6N + 7A \le 6C + 7H$.

46. $27N + 8P_5 \le 27S + 8P_2$.

47. $3N + 2P_5 \le 3C + 2P_3$.

48. $4N + S \le 4N_2 + R$.

49. $9N + 8C \le 9S + 8P_5$.

50. $8N_2 + 9A \le 8C + 9H$.

51. $8N_2 + P_5 \le 8A + P_4$.

52. $14N_2 + 9P_5 \le 14C + 9P_3$.

53. $18N_2 + 5P_5 \le 18S + 5P_2$.

54. $4N_2 + 45R \le 4P_2 + 45S$.

55. $12N_2 + S \le 12N_3 + C$.

56. $6N_2 + 5C \le 6S + 5P_5$.

57. $36N_2 + 11C \le 36R + 11G$.

58. $5A + C \le 5S + P_3$.

59. $3R + 2S \le 3A + 2C$.

Remarks 9. Similar to Theorem 3 (Eq. (39)) or Theorem 5 (Eq. (47)), the results appearing in Groups 6 and 7 can be used to improve the inequalities appearing in (89). This shall be studied elsewhere.

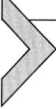

7. CONNECTIONS WITH DIVERGENCE MEASURES

Let us rewrite again the expression (28):

$$\frac{1}{4}\Delta \leq I \leq 4M_1 \leq \frac{4}{3}M_2 \leq h \leq 4M_3 \leq \frac{1}{8}J \leq T \leq \frac{1}{8}K \leq \frac{1}{16}\Psi \leq \frac{1}{16}F \leq \frac{1}{64}L.$$

We shall connect the measures given in (28) with the one appearing in (94), (95) and (96). It is understood that the measures appearing in (94), (95) and (96) are with a pair of probability distributions as in Sections 1 and 2.

Theorem 12. *The following inequalities hold:*

$$\frac{1}{2}D_{AH} \leq \frac{1}{3}D_{CG}^{84} \leq \begin{cases} \frac{12}{11}D_{RN_2}^{57} \leq \begin{cases} \frac{2}{5}D_{P_5H}^{52} \leq \frac{3}{5}D_{RG}^{60} \\ \frac{6}{7}D_{RN_1}^{59} \end{cases} \\ \frac{2}{3}D_{AP_4}^{41} \leq I \end{cases} \leq$$

$$\leq 4D_{N_2N_1}^{30} \leq \frac{4}{3}D_{N_2G}^{31} \leq D_{AG}^{40} \leq 4D_{AN_2}^{37} \leq \frac{2}{3}D_{P_5G}^{50} \leq$$

$$\leq \begin{cases} D_{P_5N_1}^{49} \leq \frac{6}{5}D_{P_5N_3}^{48} \leq \frac{1}{3}D_{P_5N_2}^{47} \leq 2D_{P_5A}^{46} \\ \frac{1}{8}J \end{cases} \leq T \qquad (103)$$

Proof. Comparing with previous results, the following equalities hold:

$$D_{AH} = 2\Delta, \quad D_{N_2N_1}^{30} = M_1, \quad D_{N_2G}^{31} = M_2, \quad D_{AG}^{40} = h \quad \text{and}$$
$$D_{AN_2}^{37} = M_3. \qquad (104)$$

In view of (28), (94), (95), (96) and (104), it is sufficient to show the following:
1. $\frac{1}{3}D_{CG}^{84} \leq \frac{2}{3}D_{AP_4}^{41} \leq I.$
2. $D_{P_5G}^{50} \leq \frac{3}{16}J.$
3. $D_{P_5A}^{46} \leq \frac{1}{2}T.$
Proof is on similar lines of Theorem 1. It is given in different parts:

    1. For $D_{CG}^{84} \leq 2D_{AP_4}^{41}$: We have, $\beta_{CG\_AP_4} = \frac{f_{CG}''(1)}{f_{AP_4}''(1)} = \frac{f_C''(1)-f_G''(1)}{f_A''(1)-f_{P_4}''(1)} = \frac{\frac{1}{2}+\frac{1}{4}}{0+\frac{3}{8}} = 2$, where the second order derivatives are given in (92). In order to show

$2D^{41}_{AP_4} - D^{84}_{CG} \geq 0$, let us consider,

$$2D^{41}_{AP_4} - D^{84}_{CG} = \sum_{i=1}^{n} q_i g_{CG\_AP_4}\left(\frac{p_i}{q_i}\right),$$

where

$$
\begin{aligned}
g_{CG\_AP_4}(x) &= 2f^{41}_{AP_4}(x) - f^{84}_{CG}(x) \\
&= \frac{\sqrt{x}(x^2 - 4x^{3/2} + 6x - 4\sqrt{x} + 1)}{(\sqrt{x}+1)(x+1)} \\
&= \frac{\sqrt{x}(\sqrt{x}-1)^4}{3(x+1)(x^2+1)} \geq 0, \quad \forall x > 0.
\end{aligned}
$$

Since $g_{CG\_AP_4}(x) \geq 0$, $\forall x > 0$, hence proving the required result.

2. For $D_{AP_4} \leq \frac{3}{2}I$: Let us consider

$$g_{AP_4\_I}(x) = \frac{f''_{AP_4}(x)}{f''_I(x)} = \frac{12x(x+1)}{\sqrt{x}(\sqrt{x}+1)^4}, \quad x \neq 1, \ x > 0.$$

Calculating the first order derivative of the function $g_{AP_4\_I}(x)$ with respect to x, one gets

$$g'_{AP_4\_I}(x) = -\frac{6(x^{3/2} - 3x + 3\sqrt{x} - 1)}{\sqrt{x}(\sqrt{x}+1)^5} = -\frac{6(\sqrt{x}-1)^3}{\sqrt{x}(\sqrt{x}+1)^5} \begin{cases} > 0, & x < 1 \\ < 0, & x > 1 \end{cases}$$
$$(105)$$

In view of (106), we conclude that the function $g_{AP_4\_I}(x)$ is increasing in $x \in (0,1)$ and decreasing in $x \in (1, \infty)$. Also,

$$\beta = \sup_{x \in (0,\infty)} g_{AP_4\_I}(x) = g_{AP_4\_I}(1) = \frac{3}{2}. \quad (106)$$

By the application of Lemma 2 with (105) and (106), we get the required result.

3. For $D_{P_5 G} \leq \frac{3}{16}J$: Let us consider

$$g_{P_5 G\_J}(x) = \frac{f''_{P_5 G}(x)}{f''_J(x)} = \frac{3\sqrt{x}(4x^{3/2} + 4\sqrt{x} + x^2 - 2x + 1)}{4(x+1)(\sqrt{x}+1)^4}.$$

Calculating the first order derivative of the function $g_{P_5 G\_J}(x)$ with respect to x, one gets

$$g'_{P_5 G\_J}(x) = -\frac{3(13x^2 - 1 + 15x - 5\sqrt{x} - 13x^{3/2} + 5x^3 - 15x^{5/2} + x^{7/2})}{8\sqrt{x}(\sqrt{x}+1)^5(x+1)^2}$$

$$= -\frac{3(\sqrt{x}-1)^3(x^2 + 8x^{3/2} + 6x + 8\sqrt{x} + 1)}{8\sqrt{x}(\sqrt{x}+1)^5(x+1)^2} \begin{cases} > 0, & x < 1 \\ < 0, & x > 1 \end{cases}$$

$$(107)$$

In view of (107), we conclude that the function $g_{P_5 G\_J}(x)$ is increasing in $x \in (0, 1)$ and decreasing in $x \in (1, \infty)$. Also,

$$\beta = \sup_{x \in (0,\infty)} g_{P_5 G\_J}(x) = g_{P_5 G\_J}(1) = \frac{3}{16}. \tag{108}$$

By the application of Lemma 2 with (107) and (108), we get the required result.

4. For $D_{P_5 A} \le \frac{1}{2} T$: Let us consider

$$g_{P_5 A\_T}(x) = \frac{f''_{P_5 A}(x)}{f''_T(x)} = \frac{2\sqrt{x}(4x^{3/2} + 4\sqrt{x} + x^2 - 6x + 1)(x+1)}{(x^2+1)(\sqrt{x}+1)^4}.$$

Calculating the first order derivative of the function $g_{P_5 A\_T}(x)$ with respect to x, one gets

$$g'_{P_5 A\_T}(x) = -\frac{(\sqrt{x}-1)^3 \begin{pmatrix} x^4 + 8x^{7/2} - 2x^3 + 8x^{5/2} + \\ +10x^2 + 8x^{3/2} - 2x + 8\sqrt{x} + 1 \end{pmatrix}}{\sqrt{x}(\sqrt{x}+1)^5(x^2+1)^2}$$

$$= -\frac{(\sqrt{x}-1)^3 \begin{pmatrix} 8\sqrt{x}(x^2+1)(\sqrt{x}-1)^2 + \\ +x^4 + 14x^3 + 10x^2 + 14x + 1 \end{pmatrix}}{\sqrt{x}(\sqrt{x}+1)^5(x^2+1)^2} \begin{cases} > 0, & x < 1 \\ < 0, & x > 1 \end{cases}$$

$$(109)$$

In view of (109), we conclude that the function $g_{P_5 A\_T}(x)$ is increasing in $x \in (0, 1)$ and decreasing in $x \in (1, \infty)$. Also,

$$\beta = \sup_{x \in (0,\infty)} g_{P_5 A\_T}(x) = g_{P_5 A\_T}(1) = \frac{1}{2}. \tag{110}$$

By the application of Lemma 2 with (109) and (110) we get the required result.

Parts 1–4 complete the proof of Theorem 12.   □

8. FINAL COMMENTS

The work is based on difference of measures arising due to inequalities satisfied by divergence measures and extended means. The divergence measures considered are of two types one with logarithmic expressions, such as, *J-divergence, Jensen difference divergence* and *arithmetic and geometric mean*, another with non–logarithmic measures, such as *triangular discrimination, Hellinger's distance*, etc. Extended means include particular cases of *Gini mean of order r and s* combined with Eves' (2003) *seven means*, such as, *arithmetic, geometric, harmonic, square-root means*, etc. The whole work contained 12 theorems. Out of them only Theorem 1 is an extension of author's theorem (Taneja, 2013a). Theorem 2 is an extended version of the one given in Taneja (2013c). Theorems 3 and 4 are new and are not done before. Theorems 5–8 are without proof and can be seen in Taneja (2013c). Theorem 9 is a considerable extension of author's work (Taneja, 2012). The previous work contains only 41 parts, while here we have extended it for 129 parts. Theorems 10 and 11 give reverse order, corner and initial members type inequalities. Finally Theorem 12 combines both types of inequalities, i.e., one arisen due mean differences (94), (95) and (96), and another based on classical divergence measures appearing in (28). Individual type results given in Groups 6 and 7 (Section 6.6) are due to inequalities (94), (95) and (96). These individual type inequalities are useful in bringing refinement of the inequalities given in (89). Since, there are 179 results to check, it requires a lot of time and space, therefore it shall be dealt elsewhere. Author's work in this direction has been refereed by many researchers. It can be seen in Chul, Wang, and Gong (2011), Jain and Chhabra (2016), Li and Zheng (2013), Ohlan (2015), Shi et al. (2010), Tomar and Ohlan (2014), Wu and Qi (2012), Wu, Qi, and Shi (2014), etc.

REFERENCES

Burbea, J., & Rao, C. R. (1982). On the convexity of some divergence measures based on entropy functions. *IEEE Transactions on Information Theory, IT-28*, 489–495.

Chen, C.-P. (2008). Asymptotic representations for Stolarsky, Gini and the generalized Muirhead means. *RGMIA Research Report Collection, 11*(4), 1–13.

Chul, Y., Wang, M., & Gong, W. (2011). Two sharp double inequalities for Seiffert mean. *Journal of Inequalities and Applications, 44*, 1–7.

Csiszár, I. (1967). Information type measures of differences of probability distribution and indirect observations. *Studia Scientiarum Mathematicarum Hungarica, 2*, 299–318.

Czinder, P., & Pales, Z. (2005). Local monotonicity properties of two-variable Gini means and the comparison theorem revisited. *Journal of Mathematical Analysis and Applications, 301*, 427–438.

Dragomir, S. S., Sunde, J., & Buse, C. (2000). New inequalities for Jeffreys divergence measure. *Tamsui Oxford University Journal of Mathematical Sciences, 16*, 295–309.

Eves, H. (2003). Means appearing in geometrical figures. *Mathematics Magazine, 76*, 292–294.

Gini, C. (1938). Di una formula compressiva delle medie. *Metron, 13*, 3–22.

Hellinger, E. (1909). Neue Begründung der Theorie der quadratischen Formen von unendlichen vielen Veränderlichen. *Journal für die Reine und Angewandte Mathematik, 136*, 210–271.

Jain, K. C., & Chhabra, P. (2016). New series of information divergence measures and their properties. *Applied Mathematics & Information Sciences, 10*(4), 1433–1446.

Jain, K. C., & Srivastava, A. (2007). On symmetric information divergence measures of Csiszar's f-divergence class. *Journal of Applied Mathematics, Statistics and Informatics, 3*, 85–102.

Jeffreys, H. (1946). An invariant form for the prior probability in estimation problems. *Proceedings of the Royal Society of London, Series A, 186*, 453–461.

Kullback, S., & Leibler, R. A. (1951). On information and sufficiency. *The Annals of Mathematical Statistics, 22*, 79–86.

Kumar, P., & Johnson, A. (2005). On a symmetric divergence measure and information inequalities. *Journal of Inequalities in Pure and Applied Mathematics, 6*, 1–13.

LeCam, L. (1986). Asymptotic methods in statistical decision theory. New York, NY: Springer.

Lehmer, D. H. (1971). On the compounding of certain means. *Journal of Mathematical Analysis and Applications, 36*, 183–200.

Li, W.-H., & Zheng, M.-M. (2013). Some inequalities for bounding Toader mean. *Journal of Function Spaces and Applications, 2013*, 394194, 5 pp.

Ohlan, A. (2015). A new generalized fuzzy divergence measure and applications. *Fuzzy Information and Engineering, 7*, 507–523.

Sánder, J. (2004). A note on Gini-mean. *General Mathematics, 12*(4), 17–21.

Shi, H. N., Zhang, J., & Li, D. (2010). Schur-geometric convexity for difference of means. *Applied Mathematics E-Notes, 10*, 275–284.

Sibson, R. (1969). Information radius. *Zeitschrift für Wahrscheinlichkeitstheorie und Verwandte Gebiete, 14*, 149–160.

Simic, S. (2009a). A simple proof of monotonicity for Stolarsky and Gini means. *Kragujevac Journal of Mathematics, 32*, 75–79.

Simic, S. (2009b). On certain new inequalities in information theory. *Acta Mathematica Hungarica, 124*(4), 353–361.

Taneja, I. J. (1995). New developments in generalized information measures. In P. W. Hawkes (Series Ed.), *Advances in imaging and electron physics: Vol. 91* (pp. 37–135).

Taneja, I. J. (2004). On a difference of Jensen inequality and its applications to mean divergence measures. *RGMIA Research Report Collection, 7*(4), 16, 16 pp.

Taneja, I. J. (2005a). On symmetric and non-symmetric divergence measures and their generalizations. In P. W. Hawkes (Series Ed.), *Advances in imaging and electron physics: Vol. 138* (pp. 177–250).

Taneja, I. J. (2005b). Refinement inequalities among symmetric divergence measures. *The Australian Journal of Mathematical Analysis and Applications*, *2*, 8, 23 pp.

Taneja, I. J. (2006a). Bounds on triangular discrimination, harmonic mean and symmetric chi-square divergences. *Journal of Concrete and Applicable Mathematics*, *4*, 91–111.

Taneja, I. J. (2006b). Refinement of inequalities among means. *Journal of Combinatorics, Information & System Sciences*, *31*, 357–378.

Taneja, I. J. (2012). Sequences of inequalities among differences of Gini means and divergence measures. *Journal of Applied Mathematics, Statistics and Informatics*, *8*(2), 49–65.

Taneja, I. J. (2013a). Nested inequalities among divergence measures. *Applied Mathematics & Information Sciences*, *7*(1), 49–72.

Taneja, I. J. (2013b). Generalized symmetric divergence measures and the probability of error. *Communications in Statistics. Theory and Methods*, *42*, 1654–1672.

Taneja, I. J. (2013c). Seven means, generalized triangular discrimination, and generating divergence measures. *Information*, *4*, 198–239.

Taneja, I. J., & Kumar, P. (2004). Relative information of type s, Csiszar's f-divergence, and information inequalities. *Information Sciences*, *166*, 105–125.

Tomar, V. P., & Ohlan, A. (2014). Sequence of inequalities among fuzzy mean difference divergence measures and their applications. *Springer Open Plus*, *3*, 623, 20 pp.

Topsoe, F. (2000). Some inequalities for information divergence and related measures of discrimination. *IEEE Transactions on Information Theory*, *46*, 1602–1609.

Wu, Y., & Qi, F. (2012). Schur-harmonic convexity for differences of some means. *International Mathematical Journal of Analysis and Its Applications*, *32*(4), 263–270.

Wu, Y., Qi, F., & Shi, H. (2014). Schur-harmonic convexity for differences of some means. *Journal of Mathematical Inequalities*, *8*(2), 321–330.

CHAPTER FOUR

Electron Microscopy at Very High Voltages

Gaston Dupouy✠
CNRS Laboratory of Electron Optics, Toulouse, France

Contents

✠ Deceased. "Reprinted from G. Dupouy (1968). Advances in Optical and Electron Microscopy, vol. 2, pp. 167–250 (Academic Press, London & New York)".

Advances in Imaging and Electron Physics, Volume 201
ISSN 1076-5670
http://dx.doi.org/10.1016/bs.aiep.2017.05.001

1. INTRODUCTION

Modern electron microscopes are so highly perfected that they yield pictures of striking beauty. Their employment in the various branches of science has made possible new advances in the examination of the infinitely small. The electron microscope suffers from serious limitations, however, which restrict its range of application. In particular, (a) the object must be placed in a vacuum and (b) the penetrating power of the electrons is too low, in the majority of cases.

(a) The electrons must necessarily travel through a good vacuum, for they would otherwise be slowed down or scattered by collisions with the gas molecules. Unfortunately, when living material (bacteria or cells, for example) is inserted in the evacuated column of the instrument, the substance will be dehydrated and its structure modified: *only dead objects can be examined.*

(b) Furthermore, the penetrating power of the electrons is in most cases very low, so that if a specimen is to be examined by transmission, its thickness must normally not exceed one thousand Angstrom units. It is absolutely imperative that higher accelerating voltages than the 50 to 100 kV of ordinary microscopes be employed when the thickness is greater than this.

Several authors have already built instruments which work at appreciably higher voltages; we mention, for example, the work of von Ardenne (1941), Müller and Ruska (1941), Hillier, Vance, and Zworykin (1941), Le Poole, Oosterkamp, and Van Dorsten (1947), Coupland (1955), Maruse, Morito, Sakaki, and Tadano (1956), Popov (1959), Popov and Zviagin (1959), and Hori, Iwanaga, Kobayashi, Shimadzu, and Suito (1964). Three Japanese firms have constructed electron microscopes which operate, in some cases, at 500 kV and in others, at 1000 kV. Finally, a 750 kV microscope has been designed by Dr. K. C. A. Smith under the direction of Dr. V. E. Cosslett and is at present in operation at the Cavendish Laboratory, Cambridge (Smith, Considine, & Cosslett, 1966).

For our own part, an electron microscope installation has been built in the CNRS Laboratory of Electron Optics in Toulouse, in collaboration with Professor F. Perrier (Dupouy, Durrieu, & Perrier, 1960; Dupouy, Perrier, & Fabre, 1961), the accelerating voltage of which is considerably higher. For several years, we have been using an accelerating voltage of 1 MV as everyday procedure, and more recently, this has been raised to 1.2 MV.

These working conditions are advantageous from several points of view.

i. In theory, the resolving power of the microscope increases when the voltage is made appreciably higher (Cosslett, 1946; Glaser, 1949; Marton, 1946).

ii. Thick specimens, with which no image would have been obtained at a lower voltage, can now be studied.

iii. The chromatic aberration, due to energy losses suffered by the electrons in the object, decreases when the electron energy is very high.

iv. The same is true of the heating and ionization of the preparation under the action of the beam. An important consequence of this is that the deterioration of the specimen is distinctly less serious (Kobayashi & Sakaoku, 1964).

v. Another important fact has also become apparent: the disturbing effects of spherical aberration upon microdiffraction patterns are about 50 times less at 1000 kV than at 100 kV.

We shall study these different points in turn and present a few results to illustrate the value of electron microscopy at a very high voltage. Finally, we shall discuss image contrast. It is known that for amorphous substances, the contrast falls if the voltage is increased appreciably. We have overcome this difficulty by a very simple method. Its effectiveness is demonstrated by the results that will be presented.

1.1 Some Properties of High Energy Electrons

We recall that with accelerating voltages of several hundred kV, the electrons are strongly relativistic: their velocity may be an appreciable fraction of the velocity of light. Under these circumstances, the potential difference V through which the electrons are accelerated must be replaced by the *relativistic potential* V^*, which is distinctly higher; thus

$$V^* = \left(1 + 0.9785 \times 10^{-6} V\right) V \quad (V \text{ in volts}). \tag{1}$$

The mass of the electron, its velocity and the associated de Broglie wavelength λ all depend strongly upon V. Some information which will later prove useful is to be found in Table 1, and also in Fig. 1. $\beta = v/c$ denotes

Table 1 Variation of the relativistic voltage V^*, the velocity ratio $\beta = v/c$, the electron mass m, and the wavelength λ as a function of the accelerating voltage V

V (kV)	V^* (kV)	$\beta = v/c$	β^2	λ (Å)	m/m_0
50	52.4	0.413	0.170	0.0536	1.098
100	109.8	0.548	0.301	0.0370	1.196
250	311.2	0.741	0.549	0.0220	1.489
500	744.6	0.863	0.745	0.0142	1.979
750	1300.4	0.914	0.836	0.0108	2.468
1000	1978.5	0.941	0.886	0.00872	2.957
2000	5914.0	0.979	0.959	0.00504	4.914
3000	11806.5	0.989	0.979	0.00357	6.870

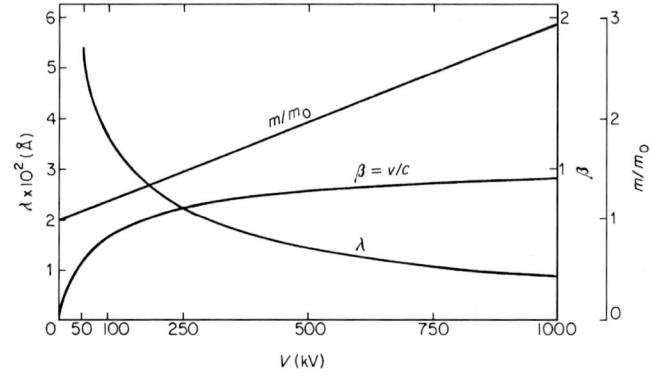

Figure 1 Variation of the velocity ratio $\beta = v/c$, the electron mass m, and the wavelength λ as a function of the accelerating voltage.

the ratio of the electron velocity v to the velocity of light, c; m_0 is the rest mass of the electron.

The following facts are now important.

a. At an operating voltage of 100 kV, the relativistic potential is only about 10% greater than the real voltage; at 1000 kV, however, the relativistic potential increases to 1978 kV, and at 3000 kV it almost reaches 12,000 kV.

b. Likewise, the ratio m/m_0, which is equal to 1.2 at 100 kV, becomes equal to about 3 for $V = 1000$ kV and to 7 for $V = 3000$ kV.

c. The corresponding wavelength decreases very noticeably as the energy of the electrons is raised.

d. The velocity ratio $\beta = v/c$ has, on the contrary, a limiting value of unity; it does not increase very much when accelerating voltages greater than a million volts are employed.

2. THE RESOLVING POWER

2.1 Preliminary Remarks

In electron microscopy the three most important defects affecting the quality of the image are spherical aberration, diffraction, and chromatic aberration. For the moment, we disregard the influence of the chromatic aberration, which *provisionally* we consider to be negligible. We shall, however, return to this point later.

The spherical aberration of an electron lens is corrected by reducing the angular aperture α, which entails placing a suitable diaphragm in the image focal plane. If the angle α is too small, however, diffraction becomes important. There is thus an optimum angular aperture α_M, for which the separation d of two points which can still be resolved passes through a minimum, d_m.

There are at least two aspects of the problem of resolution in electron microscopy.

i. The resolving power of the microscope is a property of the instrument and is particularly dependent upon the characteristics of the objective.

ii. The influence of the object under observation cannot be ignored, however, for it may impose certain limitations on the results that would be expected if allowance were not made for its presence.

Let us consider an object of the simplest possible kind: a point emitting monochromatic radiation, for example. We first assume that the microscope is equipped with a perfect objective, free of spherical aberration. The intensity distribution in the image plane is now given by the Airy curve, possessing a marked central maximum and secondary maxima of much lower intensity (Fig. 2A).

We now proceed to the case in which the objective has spherical aberration. Let C_s be the spherical aberration coefficient. The presence of spherical aberration has several consequences.

(*a*) The best focus no longer occurs at the Gaussian image plane, but at the plane of the circle of least confusion which lies between the marginal and paraxial foci; this "defocusing" is achieved by altering the focal length f of the objective very slightly.

(*b*) The refracted wave emerging from the objective is no longer a sphere but an aspheric surface; the spherical aberration introduces a path difference of $\delta = \frac{1}{4}C_s\alpha^4/\lambda$ wavelengths between the paraxial and the marginal rays.

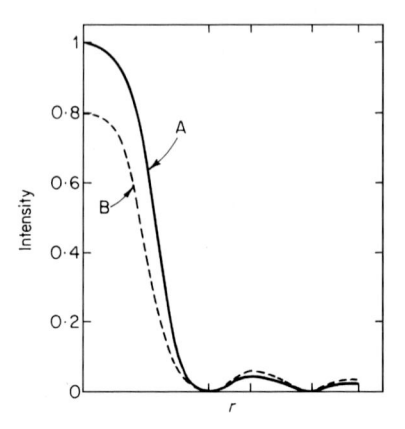

Figure 2 (A) Distribution of luminous intensity in the image plane in the absence of spherical aberration. (B) Distribution of luminous intensity in the image plane in the presence of spherical aberration. The distribution is given along a radius of the diffraction pattern.

(*c*) The curve that represents the intensity distribution in the image plane is also modified: a variable proportion of the intensity in the central maximum is now transferred into the secondary maxima (Fig. 2B).

Various authors (Cosslett, 1951; Haine, 1961; Haine & Mulvey, 1954; Black & Linfoot, 1957) have made a detailed analysis of this problem. They all come to the conclusion that if the effects of spherical aberration and defocussing are simultaneously taken into account, the intensity distribution in the image plane differs little from the Airy curve, provided that the following two conditions are satisfied:

i.   the spherical aberration is not greater than a wavelength; and
ii. the defocussing is equal to half a wavelength.

From this, the optimum value α_M of the angular aperture can be deduced

$$\frac{1}{4} C_s \alpha_M^4 = \lambda$$

so that

$$\alpha_M = 1.41 \lambda^{\frac{1}{4}} C_s^{-\frac{1}{4}}. \tag{2}$$

Let us now consider an object consisting of two points emitting monochromatic radiation of wavelength λ incoherently. According to Rayleigh's criterion, the minimum distance for which they can still be re-

Table 2 Variation of the optimum angular aperture of the objective and the resolving power as a function of the voltage V

V (kV)	λ (Å)	C_s (mm)	$\lambda^{\frac{3}{4}} C_s^{\frac{1}{4}}$ (Å)	α_M (rad)	d_m (Å)
100	0.0370	3.2	6.35	$8.2 \cdot 10^{-3}$	2.7
1000	0.00872	6.5	2.56	$4.8 \cdot 10^{-3}$	1.1

solved, d_m, is given by

$$d_{\mathrm{m}} = 0.61 \frac{\lambda}{\alpha_M} = 0.43 C_s^{\frac{1}{4}} \lambda^{\frac{3}{4}}. \tag{3}$$

Other expressions for α_M and d_m are to be found in the literature, depending upon the hypothesis made in combining the effects of spherical aberration and diffraction. Nevertheless, all the methods lead to expressions of the form:

$$d_m = A C_s^{\frac{1}{4}} \lambda^{\frac{3}{4}} \tag{4}$$

and

$$\alpha_M = B \lambda^{\frac{1}{4}} C_s^{-\frac{1}{4}} \tag{5}$$

The product $R = \lambda^{\frac{3}{4}} C_s^{\frac{1}{4}}$ is commonly referred to as the "resolution parameter".

Henceforward, we shall take $A = 0.43$ and $B = 1.41$.

2.2 The Advantage of Very High Voltages

When we work at very high voltages, the wavelength is appreciably diminished and the resolution is improved. Nevertheless, the fact that the increase in electron energy does in general lead to an increase in the focal length of the objective and in its spherical aberration coefficient C_s must not be overlooked. The formula for d_m shows, however, that even if C_s doubles when we go from 100 to 1000 kV, the gain in resolution remains appreciable; this is illustrated in Table 2. The theoretical resolving power, which is equal to 2.7 Å for an ordinary microscope working at 100 kV, can be reduced to 1.1 Å, if 1 MeV electrons are employed, and this constitutes a gain of about 2.5 times.

For reasons that are more or less well known, the theoretical resolving power has not yet been attained in ordinary microscopes. This is probably a consequence of parasitic electric and magnetic fields, of mechanical defects of various kinds in the microscope column, and of thermal effects at the specimen–holder which cause the object to drift. Thanks to continual improvements in the construction of the instrument, however, the goal is being gradually approached. About 20 years ago, the best resolution attained corresponded to $d_m = 20$ or 30 Å; today, a resolution of 4 Å has been measured (Engel, Koppen, & Wolff, 1962) using an evaporated film of platinum–iridium as the object. Recent work has shown that a resolution of the order of 2 Å can be achieved when periodic structures, the lattice planes of certain gold crystals for example, are under observation (Komoda & Otsuki, 1964). As a means of measuring the resolution of the electron microscope, however, it is well known that such objects should be treated circumspectly.

We cannot as yet draw any conclusions about the gain in resolution achieved by using high energy electrons. Work has not yet proceeded far enough in this direction. Nevertheless, Hori et al. (1964) and R. Uyeda (personal communication) have already succeeded in photographing the lattice planes of copper phthalocyanine (9.61 Å spacing) and platinum phthalocyanine (12.5 Å) at 500 kV. In Toulouse, we are just beginning to study this question. We have obtained good images of the lattice planes of $PtCl_4K_2$ (6.8 Å spacing) and with a preparation containing particles of gold, the observed resolution was 5 Å.

At 1 MV, the theoretical gain in resolution would actually allow us to photograph certain atoms. We are aware that in such an enterprise, we shall encounter many difficulties; however, every effort must be made to overcome these, in consideration of the goal in view.

2.3 The Resolving Power Using the "Condenser-Objective"

Independently of the question of using high voltages, Ruska (1965) and Riecke and Ruska (1966) have recently drawn attention to a new possibility of reducing the spherical aberration of the objective of the electron microscope. This involves the device known as a "condenser-objective", which was originally mentioned by Glaser (1941). In such a lens, the first half of the magnetic field is used as a condenser and the second half as the objective. The specimen is placed in the center of the gap, at the point where the field on the optic axis reaches its maximum value.

The important thing to notice is that the spherical aberration of such a lens may be, in some cases, ten times smaller than that of the usual kinds of objective. As a result, the smallest separation d_m between two points which can just be resolved is divided by 1.8.

This "single-field condenser-objective" had not previously been employed. As Ruska emphasized, there have been two reasons for this:

i. the lens was more difficult to construct and

ii. the resolution of the microscope was limited by other defects to a much higher value than the theoretical limit.

Ruska (1964) has now built a "condenser-objective" capable of operating at 100 kV. We might add that, if this becomes necessary, there is no major obstacle to prevent our combining the advantages of the "condenser-objective" with those of very high voltage.

Measurements made at Toulouse on the objective of our microscope have yielded the following result: when we operate at a voltage of about 700 kV, we are working in the "condenser-objective" mode with a focal length of about 4.9 mm (Dupouy, Perrier, Trinquier, & Fayet, 1967; Dupouy, Perrier, Trinquier, & Murillo, 1967). Dissipation of the heat generated by the lens current is no problem.

One other important advantage of the "condenser-objective" lens should be indicated here; its chromatic aberration coefficient, C_c, is approximately half that of the classical objective. This extremely important result has been confirmed by measurements of C_c, made on our objective lens.

2.4 The Resolving Power in Practice

Throughout the foregoing discussion, we have considered the resolving power of the electron microscope purely as a property of the instrument, disregarding the nature and thickness of the object under observation and the phenomena which result from the interaction between the incident beam of electrons and the specimen. All the operating conditions that must be fulfilled if the theoretical resolution is to be attained are only very exceptionally satisfied, however, and the results obtained in practice are quite different. In this connection, we find it useful to adopt a suggestion made by Beer (1964), that in electron microscopy different "levels" of resolution should be considered, depending upon the nature of the object and the kind of research one is engaged in. Although any classification of this nature is, of course, somewhat arbitrary, we propose to define these levels in the following way:

Level 1: $d_m < 10$ Å;
Level 2: 10 Å $< d_m < 50$ Å;
Level 3: 50 Å $< d_m < 200$ Å.

To work at level 1, it would first of all be necessary to have as good an instrumental resolving power as possible. The thickness of the specimen under observation would also have to be small.

Level 2 includes most of the work regularly performed with ordinary microscopes.

We might mention that at level 3, the resolution is still between 10 and 40 times better than that of the light microscope, and thus covers a wide range of useful research.

At levels 2 and 3, the resolution will not be limited by spherical aberration and diffraction but primarily by chromatic aberration, arising from energy losses suffered by the electrons as they pass through the object. The thickness of the latter can, however, be appreciably increased and in many cases this is indispensable. To make these various points clear, we must first discuss the penetrating power of electrons.

3. THE PENETRATING POWER OF HIGH ENERGY ELECTRONS

Another advantage of using high energy electrons springs to mind immediately, and experiment has indeed confirmed its usefulness: much thicker specimens than usual can now be examined. Observations made in our laboratory on metals and alloys show that good images can still be obtained when the thickness of the specimen is as large as 6μ or even 9μ (Dupouy & Perrier, 1964). By comparison, we recall that with ordinary microscopes, the maximum thickness is of the order of 0.1μ.

This increase of penetrating power can be explained in a first analysis as follows.

(*a*) *The mean free path* of an electron, in a given substance, increases with its energy. The term "mean free path" represents the average value of the distance traveled by an electron before interacting with an atom, either by an elastic or an inelastic scattering event. We know that we may equally well consider not the mean free path but the cross-section for the electron in the two kinds of scattering that the atoms of the object under observation may provoke. The scattering cross-section varies inversely with electron energy.

(*b*) Another important point must be mentioned, although it is only relevant in the case of crystalline substances. For a crystal, the absorption coefficient for electrons decreases as the voltage is increased; alternatively, we may say that the object becomes more transparent as the energy of the incident electrons is raised.

The interaction between the electrons and the specimen can of course be regarded from many points of view. In particular, if the electrons of the incident beam are monokinetic, the same will not be true after they have passed through the object. The *energy dispersion* among the various electrons of the emergent beam is responsible for chromatic aberration in the image. In a later section, we shall study this aberration in the case of very high voltages.

First, however, let us examine in rather more detail the two points mentioned above, with a view to estimating the order of magnitude of the gain in penetration.

3.1 Mean Free Path and Cross-Section

Let us first recapitulate some familiar ideas (Mott & Massey, 1950). The mean free path t_m of an electron which is scattered in the object is given by the equation

$$t_m = \frac{1}{N\sigma}, \tag{6}$$

in which σ is the cross-section for the type of scattering considered and N is the number of atoms/cm^3. We have $N = \rho N_A/A$, in which A denotes the atomic weight of the substance, ρ the density, and N_A is Avogadro's number.

The cross-section that we shall need to consider subsequently is closely related to the angular aperture of the microscope objective, as is the mean free path. We must always bear in mind, therefore, the fact that the cross-section σ corresponds to electron scattering into a well-defined angle α. The condition for the electron to contribute to the image formation, is $\alpha \leq \alpha_0$ where α_0 denotes the angular aperture of the objective.

Statistically, the number of interactions in which an incident electron is involved as it passes through the object depends upon the thickness t of the latter. When the electron is scattered only once – the case of *single scattering* – we have $t \leq t_m$. If $t > t_m$, we shall be concerned with *plural scattering*. Finally, when $t \gg t_m$, the scattering is said to be *multiple*.

According as the scattering occurs with or without exchange of energy, it is called inelastic or elastic scattering, respectively. In each case, a cross-section, σ_i or σ_e, is defined, and the total scattering cross-section is expressed by the equation

$$\sigma_t = \sigma_i + \sigma_e \tag{7}$$

Table 3 Variation of the *elastic* cross-section σ_e (expressed in 10^{-18} cm^2) for carbon as a function of the accelerating voltage V (in kV) at given values of the angle α (radians)

V (kV)	α radians			
	10^{-4}	10^{-3}	$5 \cdot 10^{-3}$	10^{-2}
50	1.320	1.318	1.267	1.129
100	0.748	0.745	0.688	0.550
250	0.409	0.405	0.327	0.196
500	0.302	0.295	0186	0.077
1000	0.254	0.239	0.090	0.025
1500	0.240	0.215	0.050	0.010

Table 4 Variation of the *elastic* cross-section σ_e (in 10^{-18} cm^2) for aluminum as a function of the accelerating voltage V (in kV) at given values of the angle α (radians)

V (kV)	α radians			
	10^{-4}	10^{-3}	$5 \cdot 10^{-3}$	10^{-2}
50	4.250	4.239	3.986	3.354
100	2.408	2.395	2.114	1.548
250	1.318	1.298	0.944	0.543
500	0.972	0.937	0.508	0.252
1000	0.817	0.743	0.266	0.107
1500	0.772	0.651	0.176	0.052

The phenomena of elastic and inelastic scattering have been analyzed by several authors (Lenz, 1954; Burge & Smith, 1962a, 1962b). We shall employ the more recent results of Burge and Smith concerning three elements with quite widely separated atomic numbers: carbon, aluminum and gold. Carbon exists in various forms which have slightly different densities; we shall adopt the values $\rho = 2$ for amorphous carbon and $\rho = 2.25$ for graphite.

In Tables 3, 4, and 5, the various values of the elastic cross-section σ_e, expressed in 10^{-18} cm^2, are listed as a function of the accelerating voltage V of the electrons, for a given scattering angle α.

Fig. 3 shows the dependence of the elastic cross-section of carbon upon the accelerating voltage V for several values of the angular aperture α. The mean free paths, t_m, for the elements considered, are given in Tables 6, 7, 8, and 9. Fig. 4 shows how the mean free path t_m varies with V for three values of the angular aperture, α, in the case of graphite. Dr. Burge has been kind enough to supply the data listed in Tables 10 and 11 which represent

Table 5 Variation of the *elastic* cross-section σ_e (in 10^{-18} cm^2) for gold as a function of the accelerating voltage V (in kV) at the given values of the angle α (radians)

V (kV)	α radians			
	10^{-4}	10^{-3}	$5 \cdot 10^{-3}$	10^{-2}
50	57.18	57.12	55.66	51.63
100	32.40	32.33	30.66	26.58
250	17.73	17.62	15.32	11.33
500	13.08	12.88	9.63	5.98
1000	10.99	10.56	5.98	2.85
1500	10.40	9.67	4.28	1.67

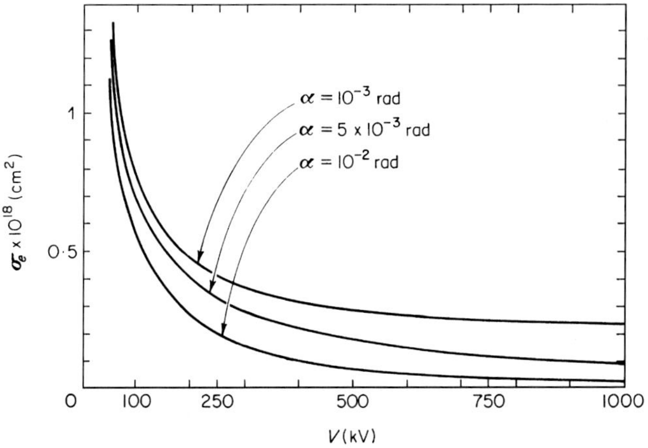

Figure 3 Dependence of the elastic cross-section of carbon upon the voltage V for several values of the angular aperture α of the objective.

the inelastic cross-sections for carbon and gold at various angles when the accelerating voltage varies from 75 to 1000 kV.

3.2 Crystalline Substances

The case of crystalline substances deserves a separate comment; for these, *the coefficient measuring the absorption of electrons by the substance decreases as the accelerating voltage is raised.* In fact, until recently, the data available on this point were essentially of a theoretical nature (Bethe, 1928; Hashimoto, Howie, & Whelan, 1960a, 1960b; Heidenreich, 1949, 1962; Howie & Whelan, 1960, 1961, 1962; McGillavry, 1940; Miyake, 1959); there has

Table 6 Variation of the mean free path t_m (Å) for elastically scattered electrons as a function of the accelerating voltage, for carbon, at the given values of α (radians)

V (kV)	α radians			
	10^{-4}	10^{-3}	$5 \cdot 10^{-3}$	10^{-2}
50	755	757	787	883
100	1333	1338	1450	1813
250	2435	2459	3051	5082
500	3303	3380	5365	13,008
1000	3930	4176	11,117	40,039
1500	4153	4642	19,847	101,720

Table 7 Variation of the mean free path t_m (Å) for elastically scattered electrons as a function of the accelerating voltage, for graphite, at the given values of α (radians)

V (kV)	α radians			
	10^{-4}	10^{-3}	$5 \cdot 10^{-3}$	10^{-2}
50	671	672	700	785
100	1185	1189	1289	1611
250	2165	2186	2712	4517
500	2936	3005	4769	11,563
1000	3493	3712	9881	35,590
1500	3692	4126	17,642	90,418

Table 8 Variation of the mean free path t_m (Å) for elastically scattered electrons as a function of the accelerating voltage, for aluminum, at the given values of α (radians)

V (kV)	α radians			
	10^{-4}	10^{-3}	$5 \cdot 10^{-3}$	10^{-2}
50	390	391	416	494
100	688	692	784	1071
250	1258	1277	1756	3054
500	1706	1769	3264	6571
1000	2030	2232	6226	15,476
1500	2146	2546	9425	31,615

also been some experimental work, however (Hashimoto et al., 1960b; Kamiya, 1963; Kamiya & Uyeda, 1962; Kohra & Watanabe, 1959; Fukuhara, Kohra, & Watanabe, 1962).

Table 9 Variation of the mean free path t_m (Å) for elastically scattered electrons as a function of the accelerating voltage, for gold, at the given values of α (radians)

V (kV)	α radians			
	10^{-4}	10^{-3}	$5 \cdot 10^{-3}$	10^{-2}
50	29.6	29.7	30.5	32.8
100	52.3	52.4	55.3	63.8
250	95.6	96.2	111	150
500	130	132	176	283
1000	154	160	283	595
1500	163	175	396	1017

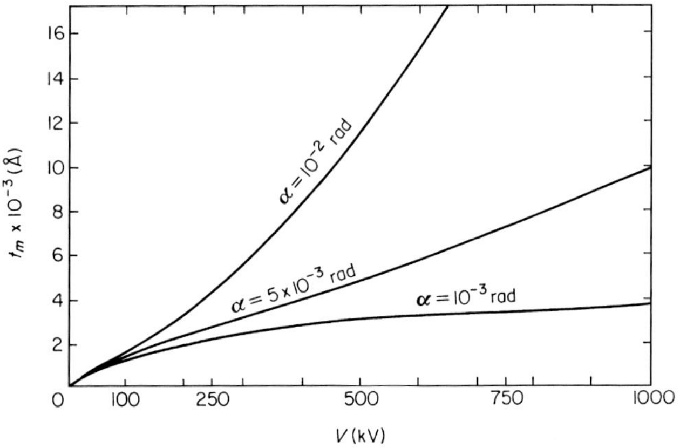

Figure 4 Variation of the elastic mean free path as a function of V for three values of the objective angular aperture in the case of graphite.

An important article on this topic has been published by Hashimoto (1964), who has worked on aluminum, using electron energies up to 300 keV.

Using the Toulouse microscope, we have been able to extend this work to voltages of 1200 kV (Ayroles, Dupouy, Mazel, Perrier, & Uyeda 1965). We have employed magnesium oxide crystals (MgO), following the method outlined by Kohra and Watanabe (1959, 1961). Before describing the results of these measurements, however, this seems a fit opportunity to recapitulate some of the fundamental points of the theory.

According to the dynamical theory, the intensity of the two waves, the transmitted wave and the reflected wave, varies periodically through the crystal. If absorption is neglected, the expression for the intensity I_R of the

Table 10 *Inelastic cross-section σ_i (in 10^{-18} cm^2) for carbon at voltages between 75 and 1000 kV, for given values of α (radians)*

V (kV)	α radians			
	0	10^{-3}	$5 \cdot 10^{-3}$	10^{-2}
75	4.378	2.082	0.848	0.443
300	1.95	0.600	0.157	0.056
500	1.63	0.428	0.083	0.025
1000	1.42	0.280	0.031	0.008

Table 11 *Inelastic cross-section σ_i (in 10^{-18} cm^2) for gold at voltages between 75 and 1000 kV, for given values of α (radians)*

V (kV)	α radians			
	0	10^{-3}	$5 \cdot 10^{-3}$	10^{-2}
75	25.61	12.07	5.955	3.333
300	10.92	3.581	1.195	0.510
500	9.133	2.571	0.668	0.256
1000	8.023	1.804	0.292	0.096

reflected wave, for example, is of the form

$$\frac{I_R}{I_0} = \frac{1}{1+w^2} \sin^2 2\pi p \left(1+w^2\right)^{\frac{1}{2}} t \tag{8}$$

where I_0 denotes the intensity of the incident wave and

$$p = \frac{U_h \lambda_0}{2}, \qquad U_h = \frac{2m_0 e V_h}{h^2}, \qquad w = \frac{2\sin\theta \cdot \Delta\theta}{U_h \lambda_0^2} \tag{9}$$

λ_0 denotes the wavelength of the associated wave, V_h is the hth Fourier coefficient of the internal potential, θ is the Bragg angle, $\Delta\theta$ is the deviation from the exact Bragg position, and t is the distance traveled within the crystal; h is Planck's constant.

The intensity I_R oscillates through the crystal with a spatial period known as the *extinction distance*, which is given by

$$t_w = \frac{t_0}{(1+w^2)^{\frac{1}{2}}} \tag{10}$$

in which

$$t_0 = \frac{h^2}{2m_0 e \lambda_0 V_h}. \tag{11}$$

So far, relativistic effects have not been considered, although the velocity of electrons accelerated through 1000 kV that we regularly use is an appreciable fraction of the velocity of light: $\beta = v/c = 0.941$. To take this relativistic aspect of the electron motion into account, Fujiwara (1961) and Hirsch (1962) have developed a more exact theory. One particular result of their calculations shows that Eqs. (10) and (11) must be modified, and that the relativistic expression for t_0 is

$$t_0 = \frac{h^2}{2m_0 e \Lambda V_h} \cdot \frac{v}{c} \qquad (12)$$

in which v and c denote the velocities of the electrons and of light, respectively, and Λ is the Compton wavelength, $\Lambda = 0.02426$ Å.

Furthermore, the dynamical theory enables us to calculate an expression for the intensity of each of the waves passing through the crystal, provided that we make allowance for absorption. We give here simply the expression corresponding to the reflected wave, in the form employed by Nonoyama and Uyeda (1965):

$$\frac{I_R}{I_0} = \frac{I}{2(1+w^2)} \left\{ \cosh\left[(1+w^2)^{-\frac{1}{2}} \mu_h t\right] + \cos\left[2\pi (1+w^2)^{\frac{1}{2}} t/t_0\right] \right\} e^{-\mu_0 t} \qquad (13)$$

in which μ_0 denotes the mean absorption coefficient, μ_h the Fourier coefficient of the absorption coefficient, and t the thickness of the crystal. I_0 is the intensity of the incident beam and the parameter w is given by

$$w = \frac{h^2}{2me\lambda^2 V_h} 2\sin\theta \cdot \Delta\theta \qquad (14)$$

in which m is the relativistic mass of the electron and λ is the corresponding wavelength.

The curve C, which represents the intensity of the reflected beam (Fig. 5) as a function of the thickness t of the crystal, varies periodically. This curve is enveloped by two others (C_1 and C_2), the difference between the ordinates of which is expressed by

$$I_t = I_0 e^{-\mu_0 t} \qquad (15)$$

for $w = 0$. The results for the beam transmitted through the crystal are analogous. The parameter μ_0 can be determined by measuring the ratio I_t/I_0 for a crystal of thickness t, using the blackening produced at the photographic plate by the electrons.

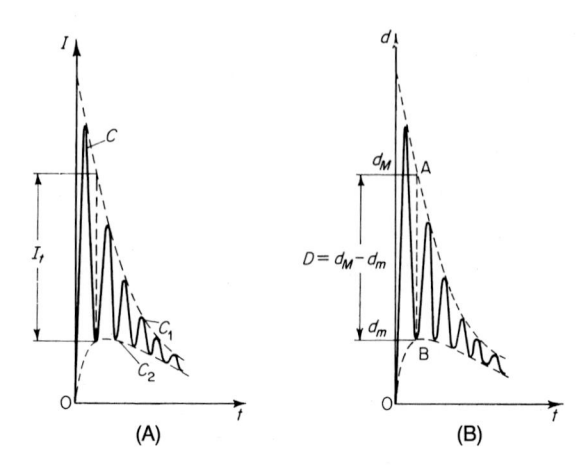

Figure 5 (A) Variation of the intensity of the reflected beam as a function of the crystal thickness t. (B) Curve showing the corresponding optical densities on the photographic plate.

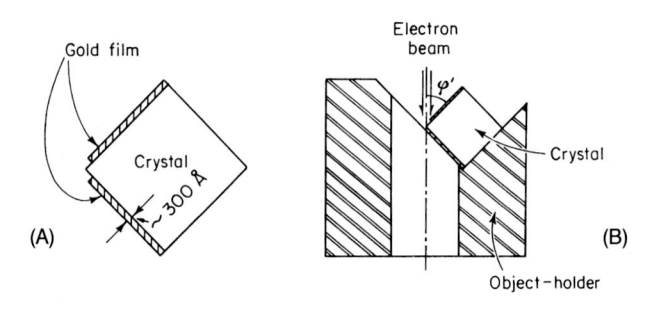

Figure 6 (A) Cross-section of the crystal, showing two surfaces coated with a gold film. (B) Specimen-holder for examining the crystal.

(*a*) *Experimental investigation.* We work with crystals obtained by cleavage, which are attached to a specimen-holder with a tilting-stage; this enables us to fix and maintain the orientation more easily. This procedure has been employed by Nonoyama and Uyeda (1965). Fig. 6A represents a cross-section of the crystal, ready for examination. To avoid charging effects which would otherwise occur when the electron beam falls on the object, the upper and lower surfaces of the crystal have been covered with a conducting metal film except in the neighborhood of the edge, which remains exposed. The crystal is attached to a special specimen-holder (Fig. 6B) in such a way that the incident beam falls on the entry surface at an angle φ' which is approximately equal to 45°, 37°, or 22°, in the various cases.

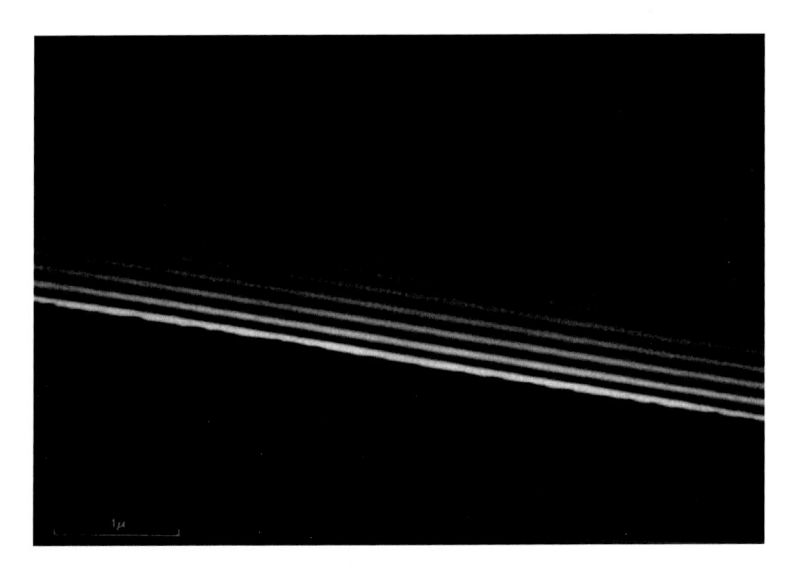

Figure 7 Magnesium oxide fringes (MgO) in dark field operation (R. Uyeda).

The photograph reproduced as Fig. 7, obtained by R. Uyeda, is an example of the type of image obtained in dark field operation. Fringes of equal thickness parallel to the edge are to be seen. The distribution of intensity over these fringes enables the absorption coefficient to be calculated (Ayroles et al., 1965).

(*b*) *The influence of simultaneous reflections.* With certain crystal orientations, intense reflections are produced simultaneously at planes other than those with indices (200). As Nonoyama and Uyeda (1965) have shown, the values of t_0 and $1/\mu_0$ are now "anomalous"; we have observed this phenomenon at the various voltages that we have used.

As the accelerating voltage is raised, spots of progressively higher orders appear on the diffraction patterns, and therefore simultaneous reflections become more frequent.

The corresponding photographs were not employed to determine t_0 and $1/\mu_0$, however.

(*c*) *Variation of the extinction distance with the accelerating voltage.* Our measurements have been made for accelerating voltages V lying between 100 kV and 1200 kV. The values of V, measured with an accuracy of better than 0.5%, were 101, 301, 503, 752, 1002, and 1204 kV. We have worked with various crystals, and for some of these, the measurements correspond to several different orientations.

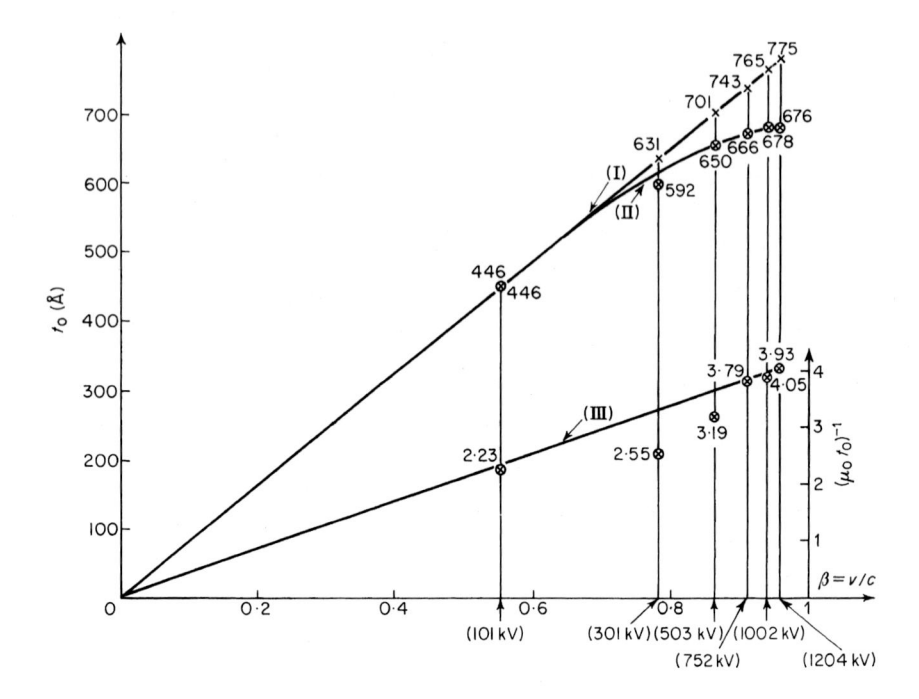

Figure 8 Variation in the extinction distance as a function of the voltage.

Fig. 8 represents t_0 as ordinate plotted against $\beta = v/c$ as abscissa. The theoretical expression (12) shows that t_0 varies linearly with v/c. The straight line (I) is obtained by using the value of V_h that is deduced from our measurements at 100 kV: $V_h = 7.66 \pm 0.27\ V$. This is in agreement with the value calculated by Honjo and Mihama (1954).

The experimental curve (II) is initially straight. At a voltage $V = 100$ kV, the mean of our results yields $t_0 = 445 \pm 7$ Å for the bright field set and $t_0 = 447 \pm 7$ Å for the dark field set. Beyond 100 kV, curve (II) bends away and separates appreciably from (I). *The result of this is that the extinction distance does not remain proportional to the electron velocity; it tends toward a limiting value at very high voltages.*

We have also plotted in Fig. 8 the experimental values of the quantity $(\mu_0 t_0)^{-1}$ curve (III). In theory, this quantity should be proportional to the ratio v/c; the straight line (III) represents this theoretical relationship. We notice that at 300 kV and 500 kV, however, the values of $(\mu_0 t_0)^{-1}$ differ noticeably from (III).

(d) *Variation of the mean absorption coefficient μ_0 with accelerating voltage.* All the usual precautions that are required in densitometry measurements were

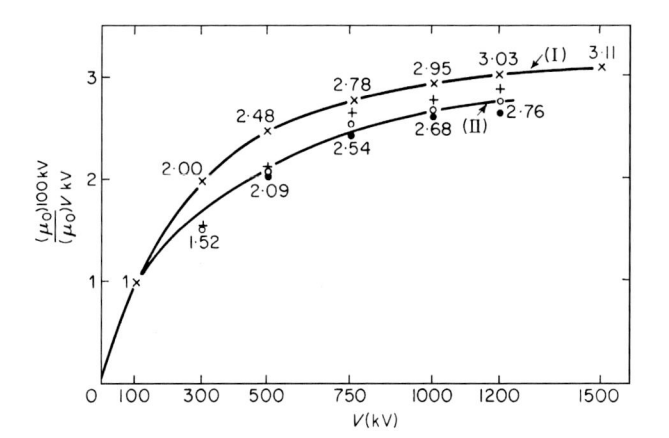

Figure 9 Variation of the mean absorption coefficient as a function of the voltage.

taken during the development of the photographic plates. We work on the linear portion of the characteristic curve of the latter. The difference D between the optical densities of two points on the photograph is therefore proportional to I_t (Eq. (15)).

We can determine $1/\mu_0$ with an accuracy of 5%. Fig. 9 shows how the transmission coefficient varies with the accelerating voltage, taking the transmission coefficient at 100 kV, $(1/\mu_0)_{100\ kV}$ as a reference value. At each voltage, the mean values of the bright field images and the dark field images are indicated differently.

Curves (I) and (II) represent the theoretical values and the general average of the experimental values respectively of the ratio $\frac{(\mu_0)_{100\ kV}}{(\mu_0)_{V\ kV}}$ at various operating voltages.

Examination of these curves shows that at 1200 kV, the transmission coefficient has been multiplied by a factor 2.76, whereas theory predicts 3.03. This difference is comparable with the discrepancies that we observe in the measured values: $1/\mu_0$ may vary not only from one crystal to another, but also from one region of a particular crystal to another. These variations are greater than the error in our measurements.

The results concerning the mean absorption coefficient can also be set out as in Fig. 10. Here, the abscissa is $(v/c)^2$ and the ordinate is $1/\mu_0$. The theoretical curve (I) of Fig. 9 now becomes a straight line and we have plotted the latter in such a way that it passes through the experimental point obtained at $V = 100$ kV. It is interesting to notice that within the range of voltages considered, the experimental curve (II) gets progressively

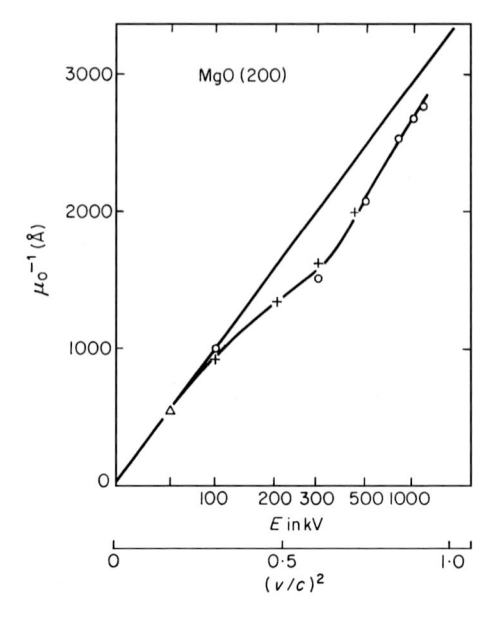

Figure 10 Variation of the transmission power $1/\mu_0$ as a function of the voltage. $\circ =$ Dupouy et al.; $\times =$ Uyeda et al.; $\triangle =$ Kamiya (1963).

closer to the theoretical straight line as the electron velocity increases. Some points obtained by R. Uyeda during recent experiments are also shown in Fig. 10.

In conclusion, we can say that the experimental results are only in partial agreement with theory; in particular, the extinction distance does not vary linearly as a function of the electron velocity: it tends toward a limit at very high voltages. So far as the penetration is concerned, however, our results are essentially the same as those predicted by the theory: *the penetration of electrons into a crystal is about three times higher at 1200 kV than at 100 kV.* This result is an additional argument in favor of using very high voltages in electron microscopy.

4. CHROMATIC ABERRATION

In studying the resolution, we have provisionally assumed that the chromatic aberration could be regarded as negligible; this is obviously not the case in general. We must therefore reconsider this point.

We naturally strive to produce incident electron beams that are as monochromatic as possible; however, there are three main factors that cause a spread of the electron velocities, along the trajectories between the electron source and the photographic plate.

4.1 Dispersion of Initial Velocities

With the heated cathodes that are in use today, the electrons leave the tungsten filament with initial velocities distributed according to a Maxwellian law about a most probable velocity V_m which depends upon the filament temperature (Boersch, 1954). Under the usual operating conditions V_m corresponds to an energy of only a few tenths of an electron-volt (Dietrich, 1958). As the accelerating voltage is raised, the chromatic aberration that can result from this effect, becomes progressively smaller. At 1000 kV, it is negligible.

4.2 Fluctuations in the Accelerating Voltage and in the Lens Currents

(*a*) The equation of motion for an electron in a magnetic electron lens with an axis of revolution $z'z$ shows that the focal length depends both on the accelerating voltage V and the magnetic induction B_z at each point along the axis between the pole pieces. If V and B_z vary slightly, the condition that must be satisfied if the trajectories and focal length are to remain invariant is $B_z^2/V = \text{const.}$, at each point on the axis and with time.

The induction B is a function of the current I passing through the magnetic ring coils. The fluctuations of the current I are thus in the same class as those of the accelerating voltage.

Hitherto, we have always striven to stabilize the voltage V and the current I independently, in order to keep the ratio B_z^2/V constant. If we wish to prevent the chromatic aberration that arises from independent variations of V and I from limiting the resolving power of the instrument, stabilization conditions must be satisfied which get progressively more exacting as the resolution becomes higher.

The radius r_c of the chromatic aberration disc is given by

$$r_c = \alpha \frac{\Delta V}{V} C_c \tag{16}$$

in which α denotes the angular aperture of the lens and C_c its chromatic aberration coefficient. We assume too that the energy of the electrons, eV, varies by an amount $e\Delta V$.

We set α equal to 5×10^{-3} rad, which corresponds closely to the optimum angular aperture at 1000 kV, and $C_c = 6$ mm. According to Eq. (16),

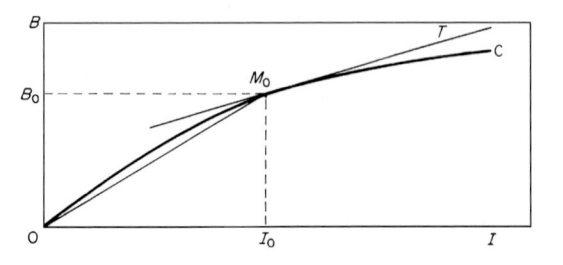

Figure 11 Magnetization curve of the iron of the magnetic lens; determination of the "working point".

the fluctuation of the operating voltage, $\Delta V/V$, must not be greater than 3.3×10^{-6} if we wish r_c not to exceed 1 Å.

In the Toulouse installation, we obtain stabilities of ± 3 V per million at present; the objective current I is stabilized to one or two parts in 10^6. The stability requirements that must be satisfied to obtain a resolving power close to the theoretical value have now been achieved. This result has been obtained only after many technical improvements.

(b) *A new method of correcting the effects arising from simultaneous variation of the accelerating voltage and the lens current.* It seems helpful to mention at this point a new method, developed by Trinquier (Trinquier, Dupouy, & Perrier, 1964), which brings in a contribution to the study of chromatic aberration. Let us suppose that B and V vary simultaneously and that their fluctuations remain small, of the order of one part per thousand for example. The achromatic condition, $B^2/V = $ constant, can be written

$$\frac{2\Delta B}{B} = \frac{\Delta V}{V}$$

When a fluctuation of the mains voltage produces a variation ΔV of the accelerating voltage, the magnetizing current I in a lens varies by ΔI and this produces a fluctuation ΔB of the magnetic induction B. The achromatic condition becomes

$$\frac{2\Delta B}{B} = \frac{\Delta I}{I}$$

or again

$$\frac{\Delta B}{B} = \frac{\Delta I}{2I}. \tag{17}$$

Consider now the magnetization curve (Fig. 11) for the soft iron of which the magnetic circuit of the lens is composed. The foregoing relation is satisfied, on curve C, at only one point, M_0, which we call the *working point*. At M_0, the slope of the tangent $M_0 T$ is equal to half that of the straight line OM_0.

Eq. (17) must be fulfilled at all times and the variations ΔV and ΔI must be exactly related. The requisite analysis for the study of this question is to he found in Trinquier's work. The solution proposed is a very simple and elegant one. It requires 110 complicated equipment and enables one to work with an A.C. voltage source, stabilized to only one part in 10^3. For this reason, we must bear it in mind in subsequent developments where it may well simplify the difficult problem that we have been discussing.

4.3 Electron Energy Losses in the Object

We recall that, generally speaking, if the incident beam is composed of monoenergetic electrons, the same is no longer true of the beam that emerges from the object. In traversing the latter, the electrons may lose varying amounts of energy, which will depend upon the nature of the specimen, its thickness, and the energy of the incident beam. This has the effect of creating chromatic aberration which affects the quality of the image.

We shall demonstrate that there is a considerable advantage to be gained in electron microscopy by using very high voltages, because the diameter of the chromatic aberration disc for a given specimen thickness decreases sharply as the electron energy is increased.

The mean energy losses have been calculated with the aid of Bethe's formula (1933) in which we have introduced the values of the correction term calculated by Sternheimer (1952, 1953). In Table 12, we list the mean energy losses of electrons passing through specimens of amorphous carbon, graphite, aluminum, silver, and gold. The curves plotted in Fig. 12 show how these energy losses vary as a function of the accelerating voltage. It can be seen that for each element considered, the energy loss decreases fairly rapidly as the voltage is raised from 50 kV to 1 MV, and passes through a minimum in the vicinity of 1 MV. This minimum is very broad for substances of a low atomic number, and becomes more marked in the case of gold, for example.

4.4 Variation of the Chromatic Aberration with Electron Energy

Using the figures given in Table 12, it is possible to deduce the mean energy ΔF (in electron-volts), for each voltage V, that would be lost by an electron

Table 12 Energy loss (in MeV/cm) as a function of the accelerating voltage V

V (kV)	C amorphous	C graphite	Al	Ag	Au
50	11.64	13.10	13.60	39.36	59.17
100	7.27	8.18	8.58	25.38	38.73
250	4.48	5.04	5.35	16.16	25.04
500	3.56	4.01	4.34	13.23	20.83
1000	3.22	3.62	4.01	12.31	19.77
1500	3.19	3.59	4.00	12.40	19.86
2000	3.19	3.59	4.03	12.58	20.32
3000	3.23	3.64	4.11	12.98	21.09

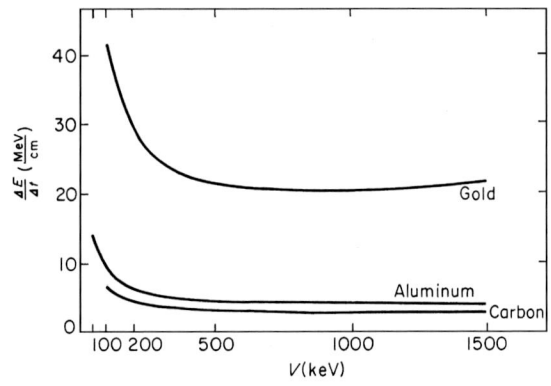

Figure 12 Mean energy loss $\Delta E/\Delta t$ in MeV/cm as a function of the accelerating voltage of the incident electrons, for carbon, aluminum, and gold.

on passing through an object whose thickness was equal to the mean free path t_m. Under these conditions, the electron is scattered only once, on the average. The radius of the chromatic aberration disc, r_c, is then deduced, for a lens for which the chromatic aberration coefficient C_c is equal to 2 mm and the angular aperture α is 5×10^{-3} rad. The results are collected in Table 13, for graphite, aluminum, and gold. Cosslett (1962) has published data calculated for aluminum. The values of energy losses agree with those of Table 13, but we have chosen a different basis for calculating the mean free path. The values adopted by Cosslett are based on the theory of Lenz; we have adopted those due to Burge and Smith. For a specimen of given thickness, however, our results agree with those of Cosslett.

So we now consider three films, of graphite, aluminum, and gold, of the same thickness $t = 1000$ Å. From Table 13, we can derive the values of r_c corresponding to 100 kV, 500 kV, and 1000 kV; see Table 14.

Table 13 Energy loss and radius of the chromatic aberration disc as a function of the voltage for graphite, aluminum, and gold. The specimen thickness is equal to the mean free path, as calculated from the theory of Burge and Smith (1962a)

V (kV)	t_m (Å)			ΔE (eV)			r_c (Å)		
	Graphite	Al	Au	Graphite	Al	Au	Graphite	Al	Au
50	700	420	31	92	56	19	184	112	64
100	1290	780	55	105	67	22	105	67	22
250	2710	1760	111	137	94	28	55	38	11
500	4770	3260	176	191	141	37	38	28	7.2
1000	9880	6230	283	357	250	56	36	25	5.6

Table 14 Values of the radius of the chromatic aberration disc at 100 kV, 500 kV, and 1000 kV, for graphite, aluminum, and gold foils of equal thickness ($t = 1000$ Å)

V (kV)	r_c (Å)		
	Graphite	Al	Au
100	82	86	390
500	8	8.3	41
1000	3.6	4	20

From these figures, we can make some important comments.

(a) In the majority of observations made at 100 kV, *the real resolution of the instrument is limited by chromatic aberration*. Thus a layer of graphite only 100 Å thick produces a chromatic aberration disc of radius 8.2 Å.

(b) For a given object, *the chromatic aberration is about 20 times less* at 1000 kV than at 100 kV. In fact it is under 20 times when we allow for the fact that C_c is larger at 1000 kV than at 100 kV.

We can confirm qualitatively that the chromatic aberration decreases as the operating voltage is raised by examining the quality of the images obtained at various voltages. We create an artificial dispersion of the electron velocities (Dupouy & Perrier, 1966) by placing a thin glass plate, the thickness of which is measured by an interference method, above or below the specimen. The object examined is a diatom (Fig. 13A) or a cluster of small magnesium oxide crystals (Fig. 13B). The images are blurred at 500 kV and become perceptibly better at 1000 kV or 1.2 MV.

We also restate a rule that has been established by several authors, and in particular by Cosslett (1956): at 100 kV, in the case of graphite, the real resolution is of the order of the thickness t of the specimen. If $t = 100$ Å,

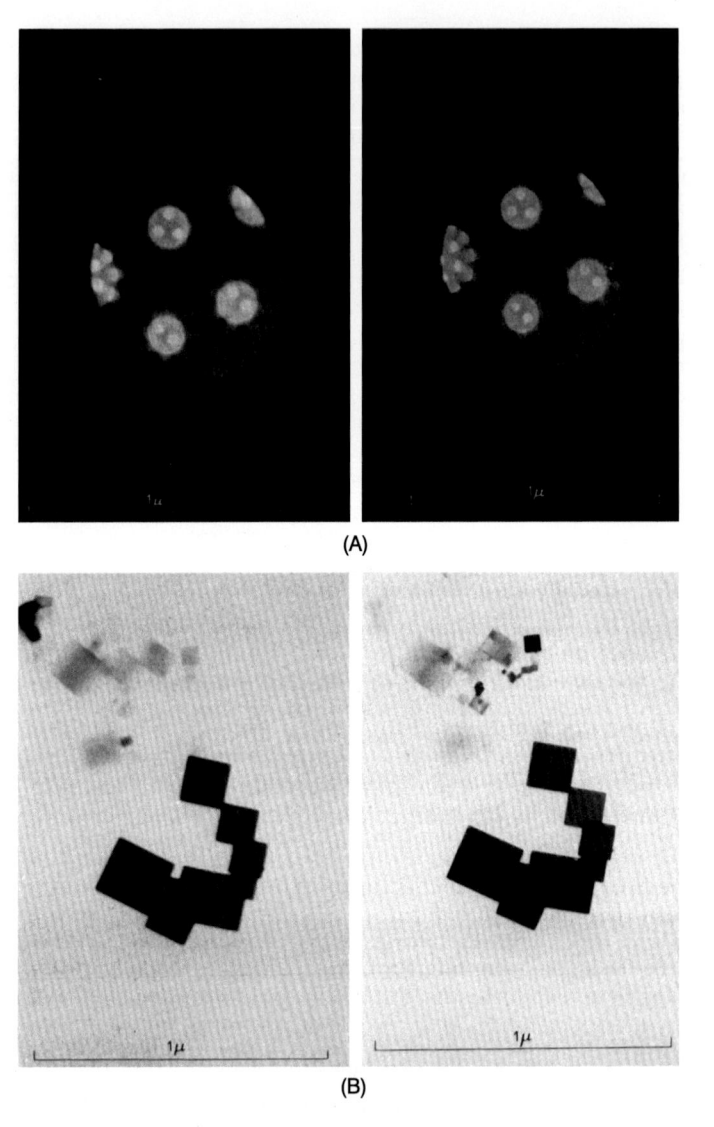

Figure 13 Modification of the quality of the image caused by the variation of the chromatic aberration as a function of the voltage V. (A) A diatom resting on a glass plate of thickness $t = 0.7\mu$; left: 500 kV; right: 1000 kV (Toulouse). (B) magnesium oxide crystals covered with a glass plate of thickness $t = 0.8\mu$; left: 500 kV; right: 1200 kV (Toulouse).

$r_c = 8.2$ Å. If, on the other hand, this same graphite film is examined at 1000 kV, we have $r_c = 0.36$ Å.

If we agree to work at resolution level 3, as discussed earlier, we deduce from Table 14 that a resolution of 100 Å will be obtained with an aluminum film of thickness $t = 2.5\mu$ for a working voltage $V = 1000$ kV. One of the basic advantages of working at very high voltages is precisely that the chromatic aberration decreases spectacularly as the electron energy is raised. It is for this reason that, at Toulouse, we can still obtain images when the object is several microns thick, especially when the latter is a crystalline material.

5. DETERIORATION OF THE OBJECT UNDER THE ELECTRON BEAM

The energy transferred to the object from the electron beam as the latter passes through the specimen may give rise to a variety of effects: increase in temperature and ionization, for example. In some cases, these effects may be disastrous, even to the point of destroying the specimen.

If the object being examined is living material, for example, the electrons will ionize atoms and molecules along their path as Cosslett (1951) has already pointed out; when this effect becomes large enough in some region of vital activity in the organism, the latter will be killed.

So far as the temperature rise is concerned, it has already proved possible to restrict the effect by using a modern microscope fitted with a double condenser. When the specimen thickness reaches several hundred Angstroms, however – and this is the case with bacteria – the electrons can still have quite a harmful effect.

It seems today that one of the major advantages of very high voltage electron microscopy is the appreciable reduction in the energy transferred to the object from the electron beam for a given specimen thickness. This is clearly visible in Table 12 and the curves plotted in Fig. 12. In the cases of graphite and carbon for example, the energy lost by the incident electrons is 3.6 times smaller at 1000 kV than at 50 kV; for aluminum and gold, the corresponding figures are 3.4 and 3 respectively.

Some conclusive experiments on this topic have been performed by Kobayashi and his collaborators (Hori et al., 1964). They check that the specimen still retains its crystalline structure by observing its diffraction pattern. For a certain "dose" of electrons, the crystal lattice is destroyed and the diffraction diagram vanishes. This dose is appreciably lower at 75 kV than at 250 kV. These experiments have been pursued in a slightly different form by Bahr and Zeitler (1964). They need to be extended to higher voltages.

Nevertheless, it is already clear that, by using very high voltages, the specimen is modified by the image-forming beam to a markedly lesser extent.

6. ELECTRON DIFFRACTION AND MICRODIFFRACTION AT VERY HIGH VOLTAGES

Electron diffraction at very high voltages has already been the subject of a range of papers (Finch, Lewis, & Webb, 1955; Möllenstedt, 1946; Papoular, 1955; Maruse et al., 1956; Kato, Nagakura, & Watanabe, 1956; Kasatotchine, Loukianovitch, Popov, & Tchmoutov, 1960; Kasatotchine, Loukianovitch, & Popov, 1960; Birkes & Demeter, 1960; Watanabe, 1960).

By using high energy electrons, one benefits from the following-major advantages:

a. Thicker objects and higher densities can be employed;
b. The contrast of the diffraction diagram against the continuous background becomes higher;
c. The broadening of the diffraction rings, which arises from interactions between the electrons and the object, is reduced;
d. Since the intensity of the diffracted beams is greater and the Bragg angles are smaller, the amount of information recorded on the plate is higher in given conditions of observation.

6.1 Diffraction Patterns

Fig. 14 shows a set of diffraction rings from gold, observed at voltages from 300 kV (A) to 1000 kV (B). At the latter voltage, more than 50 rings can be counted on the plate. Fig. 15A shows the diffraction pattern from a monocrystalline gold film, 1000 Å thick.

This was obtained by evaporation *in vacuo* and orientation on the cleavage surface of a sodium chloride crystal, gradually heated to 400°C during 30 min. Fig. 15B shows the diffraction pattern of a gold film that has been slightly bent.

6.2 Microdiffraction: The Effect of the Spherical Aberration of the Objective

When the specimen under examination possesses a crystalline structure, an electron microscope can provide us not only with an enlarged image of the object but also with the microdiffraction pattern corresponding to a

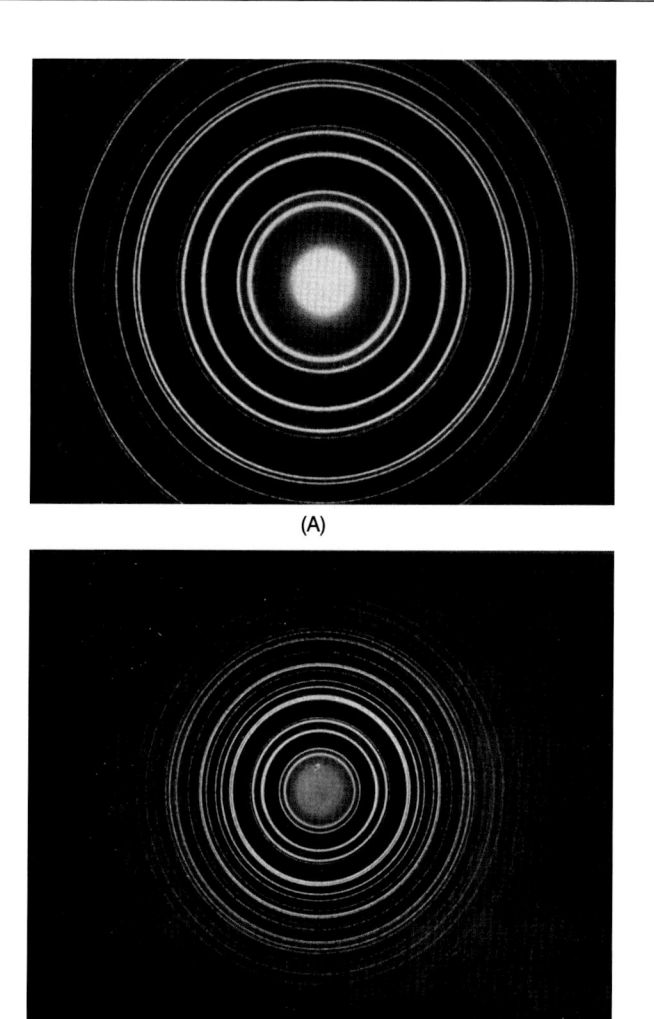

Figure 14 Diffraction rings from gold, photographed (A) at 300 kV; (B) at 1000 kV (Toulouse).

small region of the specimen. The arrangement that is used to observe these microdiffraction patterns, which come from "selected areas", was conceived by Boersch (1936) and applied in the electron microscope by Le Poole (1947). It is now in common use.

Let R denote the radius of the selected area aperture D; the conjugate region at the object is a small circle C_1 of radius $r = R/G$, where G is the magnification of the objective O (Fig. 16). Suppose now that C is a crystal

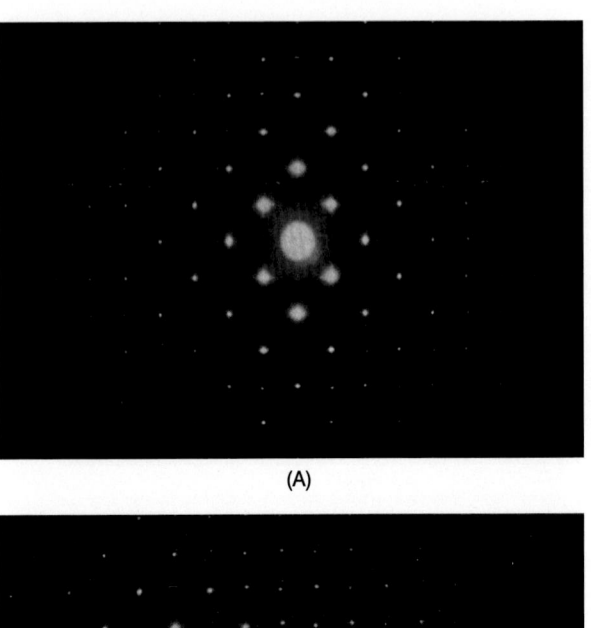

(A)

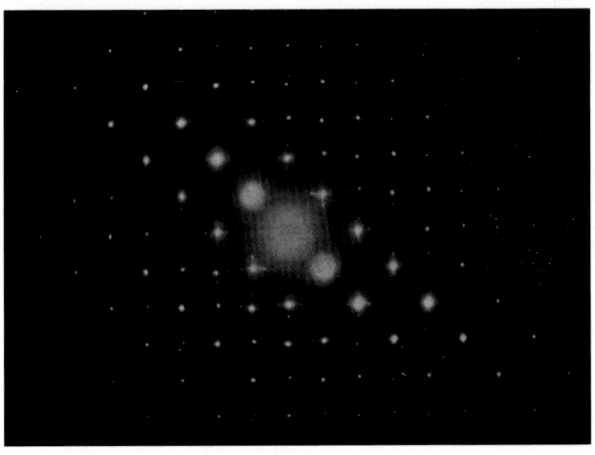

(B)

Figure 15 Diffraction pattern from a monocrystalline gold film ($t = 1000$ Å, $V = 1000$ kV; Toulouse). (A) Normal pattern; (B) the film displays two different orientations.

situated off the optic axis; the Gaussian image of its cross-section AB is formed at $A'B'$, and is intercepted by the diaphragm.

Consider now a family of diffracted beams traveling in directions making an angle $\alpha = 2\theta_{hkl}$ with the incident beam; θ_{hkl} denotes the Bragg angle corresponding to the lattice planes with indices (hkl). As several authors have shown (Agar, 1947; Riecke, 1961; Phillips, 1960), the beams refracted by the objective will be stopped or transmitted at the diaphragm D, because

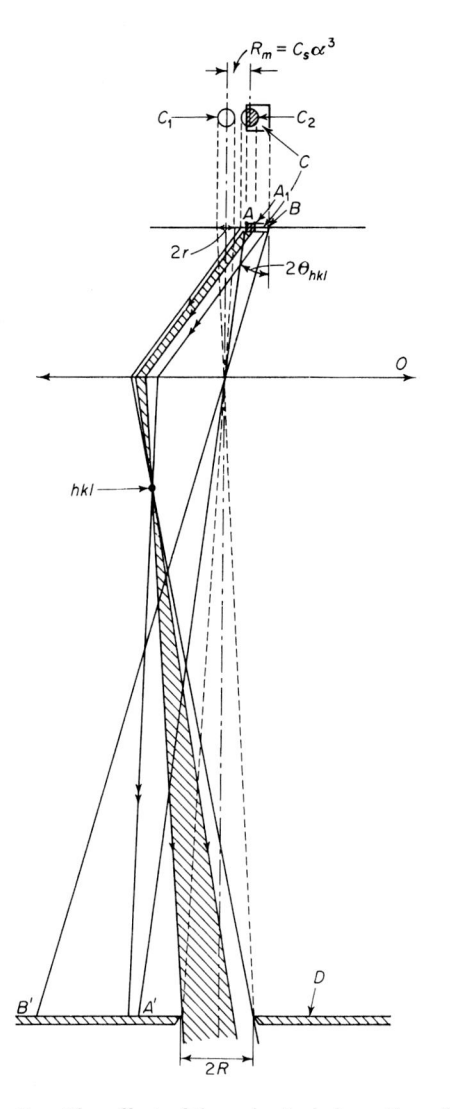

Figure 16 Microdiffraction. The effect of the spherical aberration of the objective lens. The upper part of the figure represents the projection of the object plane.

of the spherical aberration of the objective, according to the value of θ. In the case shown in Fig. 16, the transmitted beams come from the region AA_1 of the crystal. The area of this zone that is really responsible for the formation of the observed spot, with indices (hkl), is the shaded part of a circle C_2 of radius r. The center of C_2 is effectively at a distance $R_m = C_s\alpha^3$

Table 15 Reduction of the influence of spherical aberration as the working voltage V is raised

V (kV)	80	100	1000	1200
λ (Å)	0.04176	0.03701	0.008719	0.007592
R_m (Å)	2403	1673	44	28

from the center of C_1, where C_s is the spherical aberration coefficient of the objective (Ayroles & Mazel, 1965).

The example that we have chosen reveals that a microdiffraction pattern can be obtained from an object region lying outside the area selected by D in the Gaussian image observed in the microscope. This effect becomes particularly troublesome when structures are to be identified among a population of very small crystals very close together, or when changes of orientation of such crystals are to be determined. In these cases, confusion can arise because the diffraction pattern observed is the result of superimposing the patterns from several crystals.

6.3 The Advantage of Using Very High Voltages

It is here that the advantage of going to very high voltages appears (Dupouy & Perrier, 1964). We shall show that the distance R_m becomes very much smaller when the operating voltage is increased considerably.

We may write

$$\alpha = 2\theta = \lambda/d \quad \text{and} \quad R_m = C_s(\lambda/d)^3$$

in which λ denotes the associated wavelength and d the distance between the lattice planes (hkl). In the instruments that we have used,

$$C_s = 6.5 \text{ mm} \quad \text{for } V = 1000 \text{ kV}$$

and

$$G_s = 3.2 \text{ mm} \quad \text{for } V = 100 \text{ kV}$$

Since the wavelength λ decreases as the voltage is increased, the influence of spherical aberration decreases appreciably at higher voltages.

Table 15 gives the values of R_m corresponding to families of lattice planes for which $d = 1$ Å, which is almost the case for the (420) planes of magnesium oxide, MgO ($d_{(420)} = 0.94$ Å). We see that in going from voltages of 80 or 100 kV to those that are in everyday use in the Toulouse

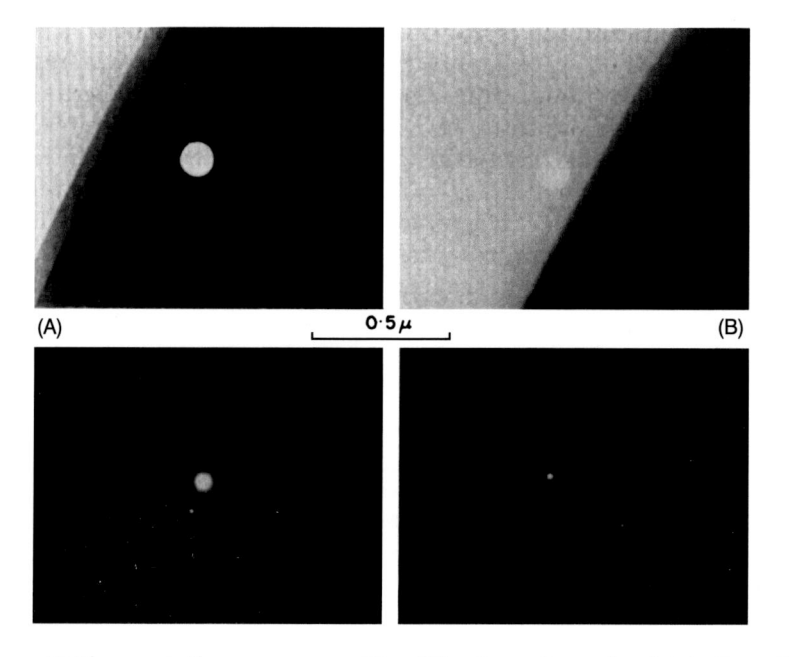

(A)   0·5 μ   (B)

Figure 17 Changes in the appearance of the diffraction pattern of molybdenite as the crystal is moved steadily away from the optic axis ($V = 100$ kV; Toulouse).

laboratory, R_m can be made 40 or 50 times smaller. With selected area apertures of very small diameter, the exposure time required to photograph the pattern does, of course, become longer; the electron energy is, however, such that the working conditions are tolerable.

6.4 Experimental Results

In all our experiments, we set the center of the selected area aperture D on the axis of the microscope (Fig. 16) and we gradually move the crystal away from the axis of the instrument. Following a method already described (Dupouy & Perrier, 1964), we photograph the object itself at each position of the crystal C. We then print on to this photograph the position of the selected area aperture.

Finally we record the corresponding diffraction pattern on a second plate.

Fig. 17 shows the modifications of the diffraction pattern corresponding to two positions (A) and (B) of a flake of molybdenite. These photographs were taken at 100 kV with an OPL microscope. In Fig. 18, photographs of molybdenite are shown, taken at 1000 kV. For selected areas of the same

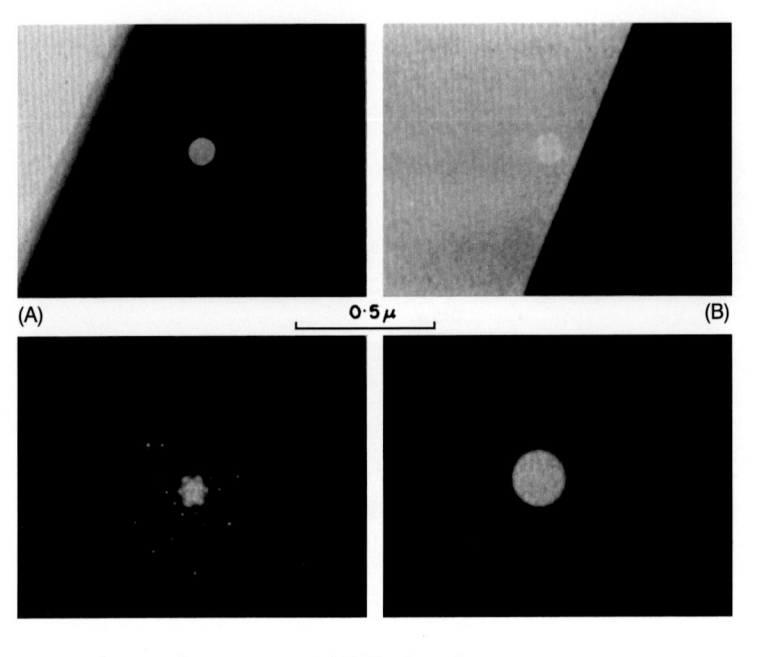

(A)   0·5 μ   (B)

Figure 18 Analogous changes at 1000 kV (Toulouse).

order of magnitude as those used at 100 kV, we obtain more detailed diffraction patterns. In photographs (B) of Figs. 17 and 18, the relative positions of the images of the object and of the selected area aperture are very close. We have indexed the spots closest to the central spot; their indices are (2020) at 100 kV and (61$\bar{7}$0) at 1000 kV.

This displays very clearly the reduction in the effect of spherical aberration at very high voltages.

7. CONTRAST IN ELECTRON MICROSCOPY

It is not enough that the instrumental qualities of the microscope permit us to distinguish two closely adjacent points on the object. In addition, it is vital that the contrast of these two points should be sufficiently high for them to be discernible on the viewing screen or on the photographic plate.

This is the reason why it seems essential to broach this topic here. We shall take this opportunity to discuss a method, developed in our laboratory, by which the image contrast can be appreciably enhanced at any operating

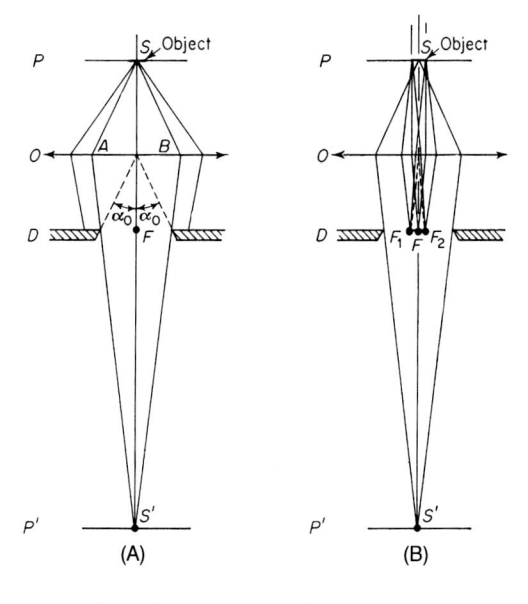

Figure 19 (A) The origin of amplitude contrast; (B) the method of the contrast stop.

voltage. We shall distinguish between two cases, amorphous substances and crystalline materials.

7.1 Amorphous Objects

In the light optical microscope, the origin of image contrast lies in the different amounts of light absorbed at different points, according to the nature and thickness of the region through which the incident light beam passes.

It is in the interactions of the electrons with the object that we must seek the origin of contrast in electron microscopy.

(*a*) *Amplitude contrast.* Consider a point S on the object (Fig. 19A). Of the electrons which travel in the immediate vicinity of S, some are scattered at angles such that after refraction through the objective O, they pass through the opening in the diaphragm D, placed in the back focal plane of the objective, and then converge to S' in the image plane P', conjugate to the object plane P with respect to O. All these electrons lie within the angle formed by the extreme rays SA and SB.

Other electrons are scattered at greater angles than the angular aperture α_0 of the objective, and are halted by the diaphragm D. They do not participate in the image formation, therefore. The regions of the object that scatter the electrons in this way appear dark in the image. The resulting contrast is known as *amplitude contrast*.

The contrast C between the images of two points may be defined as $C = \Delta I/\bar{I}$; $\bar{I}$ denotes the mean intensity of the beam that has passed through the two points in question and ΔI the difference between the intensities at these points.

Now consider two contiguous elements of the same object. We assume that only the thickness varies, and differs by Δt. The expression for the contrast becomes

$$C = 1 - e^{-\Delta t/t_m}.$$

For a given aperture α, the mean free path t_m increases with electron velocity, and so the contrast drops rapidly as the voltage is raised. This is a serious inconvenience when biological specimens, for example, are being examined. We have overcome this difficulty by a very simple method, the principle of which will be described below. First, however, we shall say a few words about *phase contrast*.

(b) *Phase contrast.* Just as in classical optics, it is possible to consider another kind of contrast, which is called phase contrast. This is caused by appropriate recombination in the image plane of waves that have passed through the aperture of the objective. A theoretical study of this has been made by Heidenreich (1964) and Hamming and Heidenreich (1965) in particular.

We shall set out only the essentials here. The basic idea is to transfer the arrangement suggested by Zernike (1942), which is now in regular use in optics, into the domain of electron microscopy. In the image focal plane of the objective is placed a narrow "phase plate" which just covers the region enclosing the image focus F; its thickness is so chosen that it introduces a phase difference of $\pi/2$ between the direct wave and a wave that has been scattered by a small detail in the object (Faget, Fagot, & Fert, 1960; Faget, Fagot, Ferré, & Fert, 1962). By recombining the direct and scattered waves, the contrast in the image can be enhanced to a certain extent. Unfortunately, the presence of the phase plate in the path of the electrons creates considerable chromatic aberration which affects the quality of the image.

Consider the case of a graphite quarter-wave plate, at an operating voltage of 100 kV. Its thickness is about $t = 200$ Å; the energy lost by the electrons emerging from the plate is $\Delta E = 16$ eV. If the optimum angular aperture of the objective is $\alpha_M = 5 \times 10^{-3}$, and the chromatic aberration coefficient $C_c = 2$ mm, the radius of the chromatic aberration spot will be given by Eq. (16). In the present case, we find $r_c = 16$ Å: it is the chromatic aberration that limits the real resolution of the instrument. We must therefore reject the possibility of using such a phase plate when observing objects whose dimensions correspond to the best resolution that modern electron microscopes can attain.

A solution must be sought along other lines.

(c) *The contrast stop method.* Consider a thin object (Fig. 19B) on which a beam of electrons impinges; the beam is fairly parallel, or at all events its angular aperture is small compared with that of the objective. This is equivalent to saying that the object is illuminated by coherent waves. The electrons that pass through the objective and the opening in the diaphragm D, and thus contribute to the image formation in the plane P', can themselves be divided into two groups.

i. The electrons which pass through the object *without interacting with the latter.* They converge at the image focus F before reaching the image plane P'; practically, their only effect is to fog the plate and diminish the contrast. In the case where $V = 1000$ kV and the graphite plate is 200 Å thick, while the objective angular aperture $\alpha_0 = 2 \times 10^{-3}$ rad, these electrons represent about 68% of the total number of the incident electrons.

ii. In a second group we place all the electrons that are scattered through angles α: $0 < \alpha \leq \alpha_0$. In the "contrast stop" method that we use, all the electrons that pass through the specimen without interaction, and some of those scattered through very small angles, are halted. All these electrons will converge onto the region $F_1 F_2$ surrounding F (Fig. 19B); we intercept them by means of a contrast stop which consists of an opaque metal disc of suitable diameter (6μ in our experiments).

Let α_1 be the angle at which the radius of the contrast stop is seen. The electrons which contribute to the image formation occupy annular aperture defined by the angles α_0 and α_1.

It is easy to show (Dupouy, Perrier, & Verdier, 1966) that for two object points at which the thickness is t and $t + \Delta t$, the contrast between the two corresponding image points reaches its greatest value when the object thickness is small. If the specimen thickness is steadily increased, the contrast vanishes and changes sign.

It is worth noting the difference between this method and the method known in classical optics as *strioscopy.* In the latter, the contrast stop is far

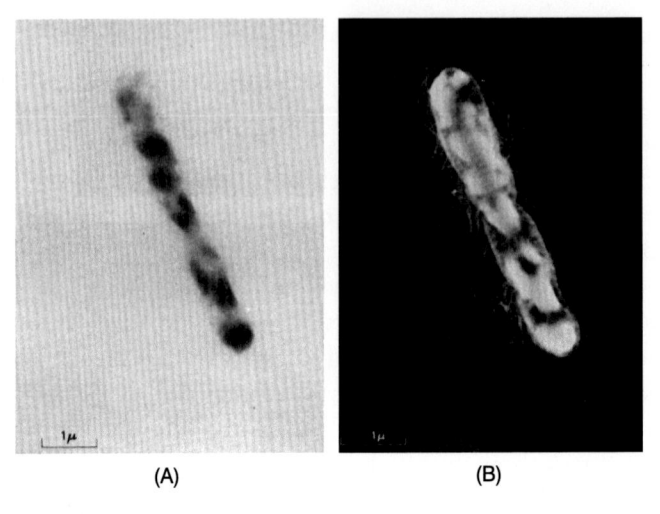

(A)   (B)

Figure 20 (A) Bright field photograph of a Proteus cell (1000 kV; Toulouse). The contrast is poor and no flagella are visible; (B) photograph of the same cell, obtained using the contrast stop method (1000 kV; Toulouse). The increase in contrast is striking. The flagella of the bacterium are clearly visible.

smaller; it must just cover the central disc of the Airy diffraction pattern in the focal plane of the objective in electron microscopy. Here, for easier setting, we use a wire of which the diameter is such that all the electrons of group (i) are halted by the contrast stop, together with a fraction of the electrons of group (ii).

Results. We reproduce here a few photographs obtained by this procedure, in the course of experiments at operating voltages V between 100 and 1200 kV. The results obtained are most striking.

Fig. 20A was taken in normal conditions at 1000 kV. The object is *B. Proteus*, which bears numerous flagella. The contrast is very poor and no flagella are discernible. In Fig. 20B, the diameter of the objective aperture is the same, but it is now fitted with a wire, the role of which has been described above. The appearance of this picture is arresting: the flagella and their points of attachment to the cell are now clearly visible. We should add that this bacterium has not been subjected to any special treatment: it has not been stained, nor shadowed with a metal. This underlines the effectiveness of the method.

Fig. 21A is the image of another *Proteus* cell, photographed as we have described. It did not seem necessary to reproduce the bright field photograph which is similar in contrast to Fig. 20A.

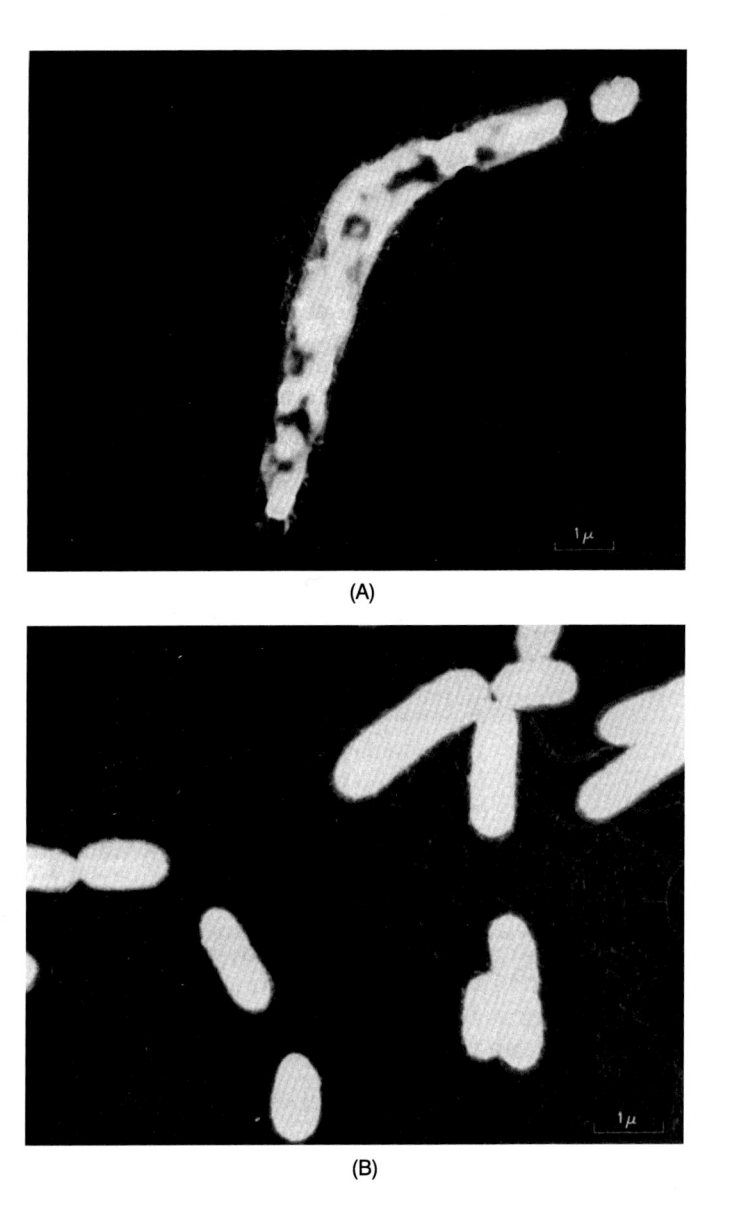

(A)

(B)

Figure 21 (A) Another photograph of a *Proteus* cell, using the contrast stop method (1000 kV; Toulouse); (B) a photograph taken under similar conditions at 100 kV (Toulouse).

We have also worked at 100 kV. Fig. 21B is again a *Proteus* cell, taken at $V = 100$ kV. Fig. 22 shows two pictures of the bacterium *Megatherium*, taken in bright field conditions (A) and with the contrast stop (B).

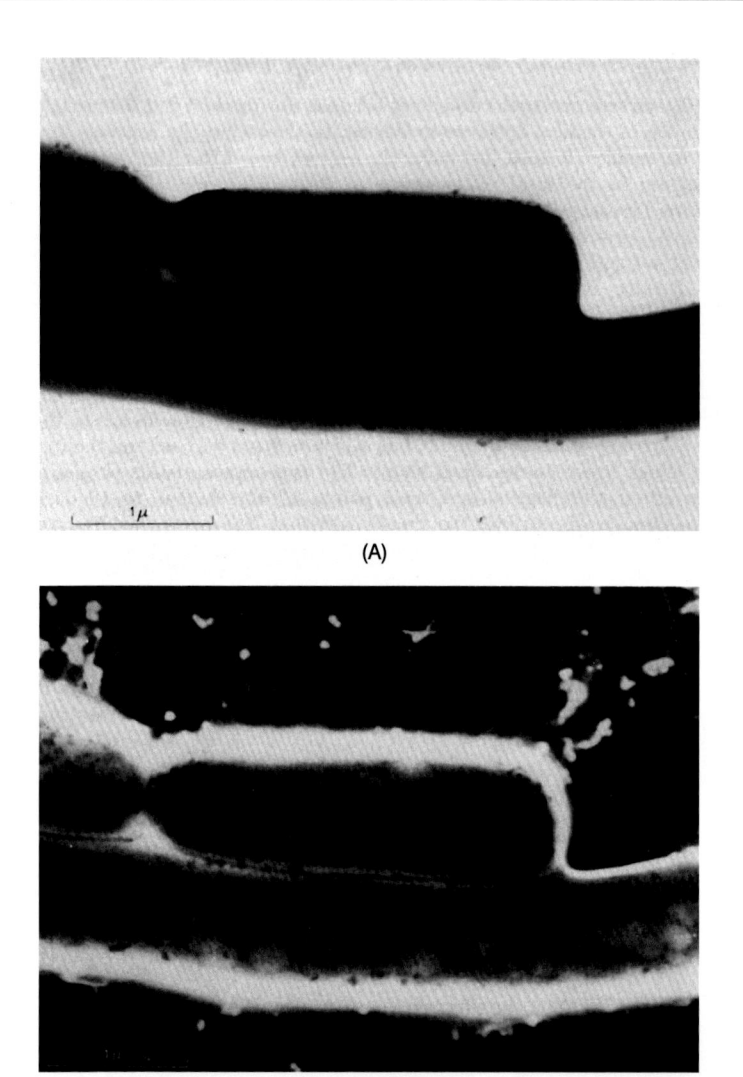

(A)

(B)

Figure 22 (A) Bright field photograph of a *Megatherium* bacterium (100 kV; Toulouse); (B) photograph of the same bacterium using the contrast stop method (100 kV; Toulouse).

(*d*) *The effect of plural scattering.* Hitherto, we have been assuming that the thickness t of the specimen is smaller than the mean free path t_m, so that an electron is scattered only *once* as it passes through the specimen. When $t > t_m$, we are involved with *plural scattering* and if $t \gg t_m$, the scattering is called *multiple*. We must ask what influence these types of scattering can have upon the image contrast. The problem has been discussed by several authors (Halliday & Quinn, 1960; Bahr & Zeitler, 1957; Burge & Smith, 1963; Cosslett, 1965).

We denote the mean number of collisions within the specimen by m. Then

$$\frac{I}{I_0} = e^{-t/t_m}$$

where I_0 and I are the intensities of the incident and emergent beams, and t_m is the mean free path between two collisions. We have $t/t_m = m$. Furthermore, the probability W_n that an electron will make n collisions is given by a Poisson distribution law:

$$W_n = \frac{m^n}{n!} e^{-m}. \tag{18}$$

A simple calculation shows that the maximum probability that a single scattering act will occur in the thickness t corresponds to $t/t_m = 1$ and in this case, $I/I_0 = 1/e$.

From Eq. (18) we can calculate the proportion of the number of incident electrons which pass through the object after interacting a given number of times. Let us consider the case $t = t_m$: we find that 37% of the electrons are not scattered at all; 37% are singly scattered, 18% are scattered twice; 6% are triply scattered and only 2% are scattered more than three times. Burge and Smith (1963) have shown that for given objective aperture and operating voltage V, the image contrast for increasing object thicknesses is slightly lower as a result of plural scattering than would be expected if only single scattering were taken into account. However, this effect is less noticeable at high (1000 kV) than at low (100 kV) voltages.

7.2 Crystalline Substances

Let us now examine the case in which the specimen, a metal or an alloy for example, forms a medium consisting of an assembly of crystals of varying degrees of perfection. The image contrast is basically due to local differences in the intensity of the beams which have undergone Bragg diffraction at suitably orientated families of lattice planes.

Consider a parallel beam of monochromatic incident electrons, falling on a crystal C which lies in a plane onto which the objective L is focused (Fig. 23). The orientation of the crystal C is such that the angle of incidence

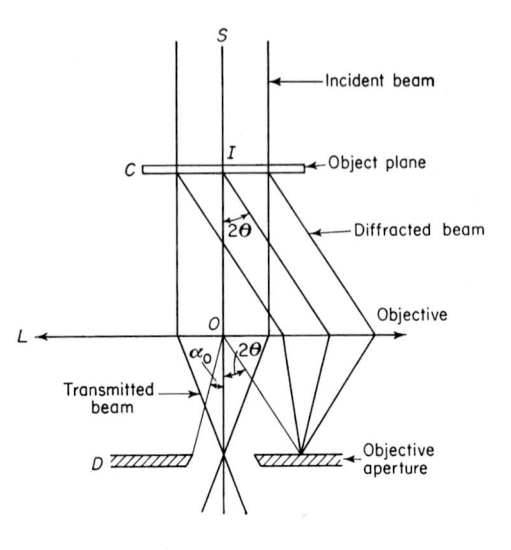

Figure 23 The origin of diffraction contrast.

at a given family of lattice planes (*hkl*) is equal to the Bragg angle, θ. In these conditions, the diffracted beam makes an angle 2θ with the optic axis *SIO*. If d_{hkl} is the distance between two successive lattice planes, we have $2d_{hkl} \times \sin\theta = n\lambda$, where λ denotes the associated electron wavelength and *n* is an integer (we shall put *n* equal to unity).

The objective aperture, lying in the image focal plane, enables us to define the angular aperture, α_0, of the objective. If $2\theta > \alpha_0$, the reflected electron beam is halted by the objective aperture. For electrons accelerated through $V = 1000$ kV, $\lambda = 0.00872$ Å. When the lattice planes are separated by a distance $d_{hkl} = 1$ Å, for example, $2\theta \simeq 9 \times 10^{-3}$ rad. The condition $2\theta > \alpha_0$ is therefore usually satisfied in practice. Generally speaking, we arrange that only the direct beam (bright field image) or the diffracted beam (dark field image) passes through the objective aperture. In the bright field image, those regions of the object at which a strong diffraction can occur appear dark.

This *diffraction contrast* enables us to observe all kinds of modification which occur in the crystal lattice without it being necessary to work at high enough resolutions for the component planes of the lattice to be separated.

We may mention two advantages which appear when we work at very high voltages. The contrast of the dark field images is particularly striking. We shall speak later about the fringes of equal thickness at the edge of a crystal; it is found that the contrast of these fringes is much higher in dark field than in bright field conditions (forthcoming publication).

Another advantage resides in the fact that the angle θ is now small because of the short wavelength. Therefore the effect of spherical aberration on the dark field image is considerably less at 1000 kV than at 100 kV. It is thus no longer necessary to operate with oblique incident illumination; we work with a beam normal to the surface of the object. This is a considerable simplification, since all the microscope settings remain the same. To transfer from bright field conditions to dark field observation, the objective aperture has merely to be shifted in its own plane; this requires only a few seconds.

We shall return to the question of diffraction contrast, and illustrate how it has been applied to the study of metals and alloys.

8. VERY HIGH VOLTAGE ELECTRON MICROSCOPES

We shall now say a few words about the questions arising in the construction of electron microscopes working at very high voltages – 500 kV or higher.

In this category, the microscopes at present in operation are as follows:

  i.   the microscope built by Popov in Moscow (550 kV);
 ii.   the microscopes designed in collaboration with professors from Japanese Universities by the commercial firms Shimadzu (500 kV), the Japan Electron Optics Laboratory Co. (JEOL) and Hitachi (1000 kV);
iii.   the Cavendish Laboratory microscope in Cambridge, constructed under the direction of Cosslett (750 kV) and
 iv.   the microscope at the Electronic Optics Laboratory in Toulouse, conceived by Dupouy and Perrier (1500 kV).

We first discuss the high voltage installation and then the microscope column. To avoid repetition, we shall not describe each of the instruments listed above in turn; we prefer to indicate, within the general framework, special features which are peculiar to them.

8.1 High Voltage Generators and Accelerators

In the region of 1 MV, one immediately thinks of the Van de Graaff type of generator, for reasons of simplicity. Unfortunately, no one has yet succeeded in obtaining the necessary stability, so far as we are aware. We have already seen that the relative variation $\Delta V / V$ of the high voltage must be less than 10^{-5}. Furthermore, the vibrations caused by the belt and the driving motor

Figure 24 The generator and the accelerating tube of the Cavendish Laboratory installation (Cambridge).

are sources of disturbance of the image quality and it seems difficult wholly
to eliminate them.

On the other hand, the new electrostatic machines invented by Félici
and developed in Grenoble by the SAMES company can provide voltages of
700 kV very simply, and can undoubtedly go beyond this figure. They offer
many advantages, but have not yet been used at these very high voltages.
They ought to be considered for future designs.

The generators employed up to the present date are of the Cockcroft–
Walton or Greinacher type.

(a) *High voltage generators in air.* The generators for the microscopes at
Toulouse and at Cambridge have been built by Haefely (Basel). The first
arrangement was described at the Delft Congress (Adler, Minkner, Reinhold, & Seitz, 1960). These are generators in air; they are very convenient
and well-adapted to receive all the modifications that, with use, one might
wish to make to perfect them. They have two disadvantages, however.

The metal walls of the room containing the generator form a Fara-
day cage at earth potential; the safe distance between the domes above the
columns of the high voltage generator and the walls of the room increases
rapidly with the voltage used. The dimensions of the room can hence become impressive, as they do with the Toulouse microscope (20 m across). In
the case of the Cavendish Laboratory installation in Cambridge, the room
containing the generator and the tube in which the electrons are accelerated through 750 kV is smaller in size: 8.7 × 7.7 × 5.4 m (high). Fig. 24
shows the general appearance of it.

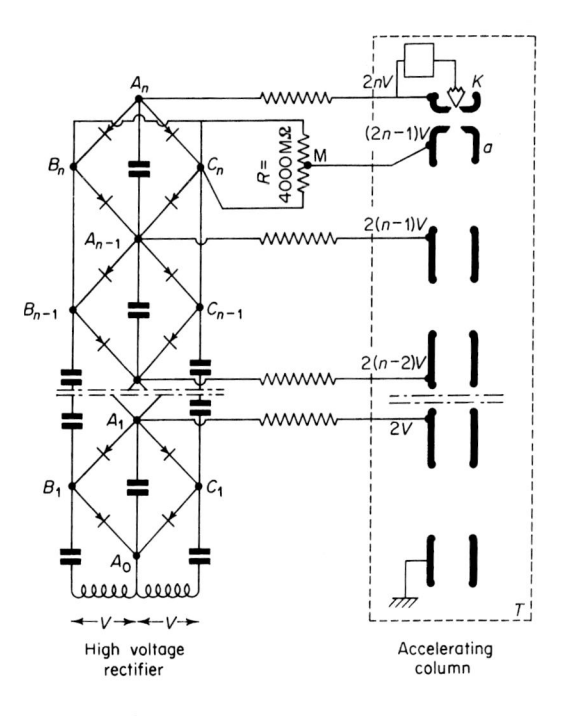

Figure 25 Schematic view of the high voltage generator and the accelerating tube.

A second inconvenience lies in the fact that the path of the electrons through the accelerating tube is several meters long. The beam has, therefore, to be protected from parasitic fields, and, in particular, from the terrestrial magnetic field. We shall later explain how we overcame this difficulty.

We now describe the generator at Toulouse; this account is also generally applicable to the Cambridge equipment, except that the latter contains only five accelerating stages instead of ten and has an input supply of much higher frequency.

A synchronous motor drives a single-phase alternator which provides A.C. at a frequency of 400 Hz and at different stabilized voltages (Fig. 25). Through two transformers, this A.C. output is applied to a ten stage cascade rectifier. The arrangement is of the Cockcroft–Walton type, but is symmetrical to reduce residual ripple; for the same reason, the plane of symmetry of the equipment has been set in a meridian plane of the sphere which comprises the laboratory. The voltages obtained at each of the ten stages are filtered and then used to supply the electrodes of the electron accelerating tube.

The electron source has already been described in several publications (Adler et al., 1960; Bas, 1960; Dupouy & Perrier, 1963; Dupouy, Perrier, & Seguela, 1966). It consists of a gun, which supplies electrons accelerated through a few tens of kV; a magnetic lens then focuses these electrons which subsequently enter an accelerating tube. Within this tube, they encounter ten successive accelerating stages, each consisting of coaxial cylindrical electrodes. The total path of the electrons, from the cathode to the first condenser lens of the microscope, is more than 6 m in length.

In our initial installation, the effect of the residual horizontal component of the terrestrial magnetic field was sufficiently large to cause an important deflection of the electron beam. At the condenser lens of the microscope, the discrepancy between the real point of impact of the electron beam and the theoretical point reached 8 mm at 1500 kV and more than 18 mm at 300 kV. The reason for this was that the original electrodes did not form an efficient enough magnetic screen for the electron beam because of their low magnetic permeability. We have replaced these electrodes by new ones, geometrically identical with the originals, and machined from solid "mumetal". When the accelerator had been fitted with these new electrodes, we were able to observe that the point of impact of the beam at the top of the microscope shifted less than 0.1 mm when the accelerating voltage changed from 250 to 1500 kV. Thanks to this good conservation of the alignment, examination at 1000 kV can follow observation at 300 kV after a few minutes only.

The original gun required an electrical power supply that was provided by an alternator mounted in the upper part of the generator and driven by a vertical insulating shaft nearly 7 m long. The rotation of this unit caused disturbing vibrations. This gun has been replaced by another gun of classical design equipped with a hairpin filament which requires less power; a battery of 6 V, 120 Ah accumulators is adequate to heat the filament. A second battery of 100 V and 10 Ah supplies a lens, situated between the gun and the accelerating tube (Bas, 1960).

Finally, the voltage of a few kV which is required by the gun is extracted from the main rectifier, as the diagram of Fig. 26 indicates. We have mounted a resistance R of 4000 MΩ between the tops B_n and C_n of the lateral columns (Fig. 25). Between the point A_n at the highest potential and the midpoint M of this resistance, a steady potential difference appears which varies from 12.5 to 75 kV when the voltage changes from 250 to

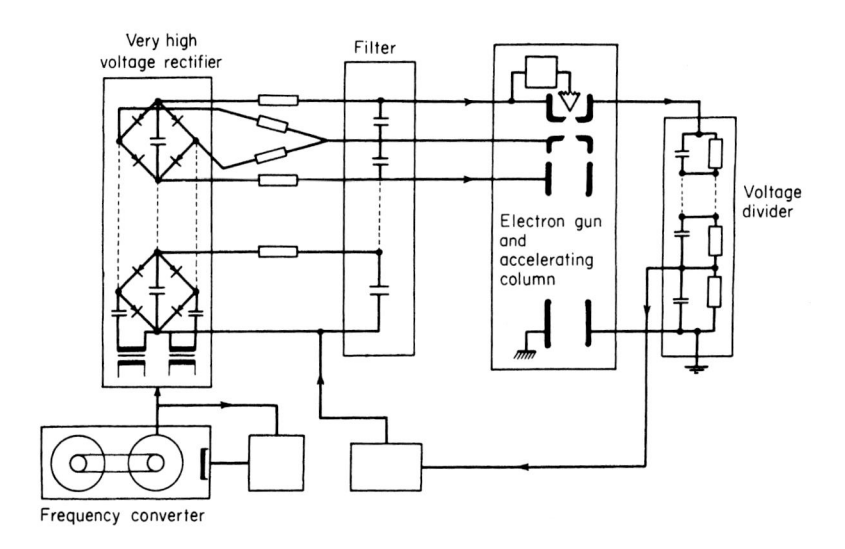

Figure 26 General layout of the very high voltage installation.

1500 kV. This places at our disposal a suitable source for supplying the anode of the gun.

By this arrangement, the alternator mentioned above is used only for preliminary adjustments, and never in the course of making observations with the microscope.

Stabilization of the very high voltage. A primary stabilization is carried out on the 400 Hz voltage; we have improved the performance of this first stabilizer by using high quality components (and in particular, by replacing carbon resistors by wire-wound resistances). In addition, the voltage reference valves have been suitably aged.

When this pre-stabilizer alone is working, the stability of the high voltage is $\pm 3 \times 10^{-4}$ during 1 min; the drift during 1 h is not greater than 2×10^{-3}.

In principle, therefore, only the small variations arising in particular from variations of beam intensity remain to be compensated. For this, the builders intended to modify the voltage applied to the tenth stage – the one nearest earth – in such a way that the total voltage had the prescribed value. The latter is determined by measuring the current through a 6000 MΩ resistance divided into ten parts; each part is shunted by a suitable condenser so that not only the very slow changes of voltage but also the faster variations are transmitted undistorted. We have retained the principle of this method of stabilization, and employed a stabilizer of laboratory construction.

Two arrangements have been employed in turn. The first uses a "Sefram" nanoammeter pen–recorder as direct current amplifier and an unsaturated standard Weston cell as reference. This arrangement gave the following results (Dupouy & Perrier, 1963, 1966): a stability of $\pm 1 \times 10^{-5}$ during 1 min; a drift of 1×10^{-5} during one hour; a residual ripple of 4×10^{-6} peak to peak value. The performance is limited by the slow response of the galvanometer. This arrangement is still sometimes employed temporarily.

The second stabilizer is wholly electronic; it contains a selected and aged neon tube as reference. The voltage supply to it is highly stabilized. The amplifier, which has a total gain of more than 5×10^6, is in two parts. The first, which operates at low voltage, is equipped with a mechanical chopper. The second works at 40 kV. The final tube, type 6BK 4, is mounted as a cathode follower to keep the output impedance low. Under these conditions, the open–circuit cut–off frequency lies above 2000 Hz. All the auxiliary supplies, including the 40 kV supply, are stabilized. With this device, the variations in the high tension are equal to or less than $\pm 3 \times 10^{-6}$ during 1 min; the drift is of the order of 1×10^{-5} in 1 h and the residual ripple, peak-to-peak, is 2.5×10^{-6}.

(*b*) *High voltage generators in a pressurized atmosphere.* The dimensions of the high voltage generator and of the accelerating tube can be very considerably reduced by using pressurized gas insulation. The generator is then placed in a kind of metal tank with fairly thick walls, filled with freon or a similar gaseous mixture, at a pressure of the order of 2 kg/cm^2. This tank is supported by suitable brackets directly above the microscope.

This is the system adopted by the Japanese designers, whom we have already mentioned. The generator is of the type described above. We should, however, mention that Shimadzu separates the generator from the accelerating tube properly speaking; this allows a bigger container to be used, in which the distances between the various electrical components of the generator are large enough for protective screens against parasitic effects to be inserted. The generator and the accelerating tube are connected by means of an insulated cable.

The accelerating tube itself consists of ten stages of electrodes, insulated by means of glass cylinders, each of which is 8 cm high. Another accelerating tube comprising twenty stages is in the course of construction, for use

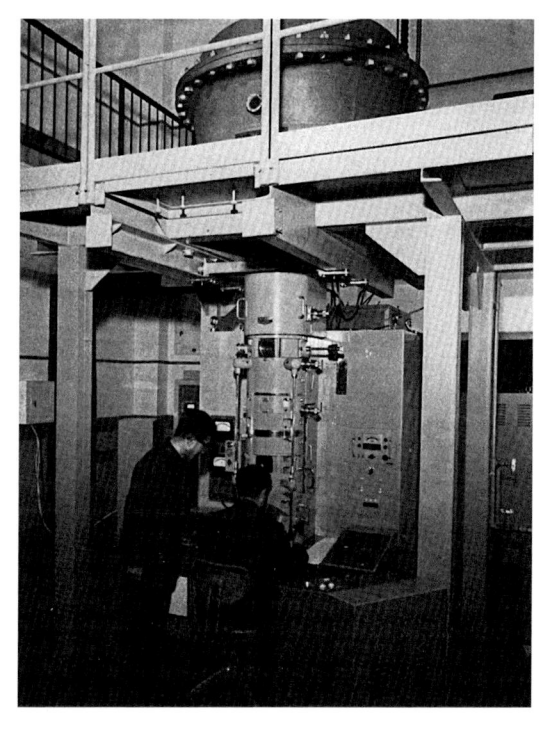

Figure 27 Photograph of the Hitachi microscope (500 kV), installed in the laboratory of Professor R. Uyeda (Nagoya).

with the microscope at higher voltages (Hori et al., 1964). The filament is changed by a simple arrangement, without the gas having to be drawn off.

In the Hitachi microscopes (500 and 1000 kV), the accelerating tube and the high tension generator are placed in the same tank; they stand on pillars which are placed round the microscope (Fig. 27). The accelerating tube also contains ten accelerating stages.

The JEOL model is intended to work at 1000 kV, and it contains twelve accelerating stages, each giving 87 kV except for the final one which operates at 43.5 kV. The diameter of the pressurized tank is 2.9 m, and its height is 3.3 m. This tank rests on a kind of mezzanine, built in the microscope room. The total height of the installation above ground level is 6.6 m (Fig. 28).

The various generators to which we have been referring are stabilized to better than 10^{-5} (Hori et al., 1964).

Figure 28 Photograph of the JEOL microscope (1000 kV), in the laboratories of the Japan Electron Optics Laboratory Co. (Tokyo).

8.2 Stabilization of the Lens Currents

In general, the stabilizer employed is of the conventional type: the current flowing in each lens also passes through a standard resistance. The potential difference across the terminals of this resistance is compared with the e.m.f. of a standard cell; after suitable amplification, the difference between these voltages drives a triode placed in series with the lens.

The tendency today is to employ transistorized equipment. To prevent the current from drifting, either the standard resistances are water-cooled (Cambridge) or the main circuit components are placed in an oil-bath (Shimadzu).

Generally speaking, the current supplies are stable to 5×10^{-6} and even better in some installations.

8.3 The Electron Microscope

If high energy electrons are to be focused without a considerable increase in the focal lengths of the magnetic lenses, the magnetic induction in the gap

Figure 29 Hydraulic device (fork lift trolley), with which the object chamber or other components can be extracted from the column of the Cambridge microscope.

and the number of ampère-turns in the lens-windings must naturally be increased. The result of this is that the lenses of very high voltage microscopes are much bigger than those of ordinary microscopes. It is therefore of interest to know the rules of similitude with which we can transform from conventional lenses to new ones of far greater dimensions, or conversely. We shall mention these rules later.

The structure of the super-microscopes that we are dealing with is essentially the same as that of standard commercial instruments. Each contains a double condenser, one objective, one intermediate lens, and one projector. Sometimes, as in the JEOL microscope, a sixth lens is available, which proves to be useful in electron diffraction.

The outer diameter of the lenses ranges from 30 cm (Cambridge) to 55 cm (Toulouse). The lenses are usually not cooled, although sometimes they are in certain instruments (Cambridge).

Various features have been incorporated which enable the components of the microscope column to be demounted and reassembled rapidly. The hydraulic fork lift trolley used at Cambridge is a case in point. In Fig. 29 the specimen chamber can be seen being removed from the column: a general view of the microscope is reproduced in Fig. 30.

The pole-pieces of the lenses can generally be changed, so as to modify certain characteristics of the instrument and to adapt it to various kinds of work.

In the Toulouse microscope (Fig. 31), the lenses are fixed (Dupouy & Perrier, 1962). They have been machined with the utmost care and they fit

Figure 30 The 750 kV electron microscope at the Cavendish Laboratory (Cambridge).

onto one another with the degree of precision corresponding to standard workshop practice. Each lens is provided with a "pole-unit", which contains the actual pole-pieces together with the various devices appropriate to the role of each lens, for example, the specimen-holder, aperture-holders, and stigmators. To align the microscope the pole-units of the various lenses are controlled by means of a micrometer mechanism.

All the control rods required for manipulation of the instrument in various ways are mounted on ball-bearings; their ends are connected via universal joints to handles which lie within reach of the operator. The same is true of the controls with which the lens currents are adjusted: they are grouped on a desk so that the observer need not leave his seat during his work.

Despite their size, these microscopes are at least as convenient to use as instruments of the conventional type. We shall restrict this account to a few

Figure 31 The Toulouse electron microscope.

details about the objective lens. The other components of the column are more or less analogous to those of ordinary microscopes.

(*a*) *The objective lens.* In the Cambridge microscope, the specimen is inserted at the top of the objective, and the specimen-holder moves over the plane upper surface of the lens. Many top plates containing holes of different sizes can be substituted for one another without seriously affecting the alignment of the microscope.

The specimen is inserted through an air-lock; it can be changed in less than 1 min. Various accessories such as goniometers are to be incorporated. Three different apertures can be used. The objective is equipped with a magnetic octopole stigmator. For selected area diffraction, a device fitted

with suitable apertures is available in a chamber machined into the base of
the objective.

The lens characteristics (at 750 kV) are as follows: focal length, 4.7 mm;
spherical aberration coefficient, $C_s = 6.6$ mm; chromatic aberration coefficient, 3.5 mm; number of ampère-turns, 12,600; resolution parameter,
$R = C_s^{\frac{1}{4}} \lambda^{\frac{3}{4}} = 3.1$ Å.

In the Toulouse microscope, the characteristics (at 1 MV) of the objective lens at present in use are as follows: focal length, $f = 7.2$ mm;
spherical aberration coefficient, 6.1 mm; chromatic aberration coefficient,
5.6 mm; number of ampère-turns, 17,000; resolution parameter $R = 2.5$ Å.
Through each of the pole-pieces is bored a hole 12 mm in diameter; the
gap is likewise 12 mm. With other pole-pieces and a 6 mm gap, the focal
length f can be made equal to 5.1 mm.

A stigmator is built into the objective. The object is introduced either
vertically, through the upper part of the objective, or horizontally, by means
of a suitable air-lock. The specimen is deposited into a specimen holder device and temporarily becomes an integral part of the objective pole-pieces.
Crystalline specimens can be orientated by means of a goniometer stage.

We can, furthermore, also operate the lens as a "condenser-objective",
at a voltage of about 700 kV, and with a focal length $f = 4.9$ mm. The
spherical aberration coefficient is, in its turn, reduced to $C_s = 2.6$ mm.
The chromatic aberration coefficient now takes the value $C_c = 3.5$ mm.
Indeed, we hope to work in the "condenser-objective" mode at 1 MV in
the near future.

(b) *Protection against X-rays.* Whenever the electron beam strikes a solid
diaphragm along its path, X-rays are emitted at the point of impact. At
voltages of several hundred kV the X-rays are very hard and penetrating.
The operators must therefore be protected, and a constant watch kept upon
the efficiency of this protection.

The principal source of X-rays is situated at the bottom of the accelerating tube, at a diaphragm which acts as entrance pupil. The other
sources are the second condenser aperture, the specimen-holder, the objective aperture-holder, and finally, the viewing screen. In the most dangerous
zones, lead walls are built consisting of 10 cm thick bricks (Toulouse). The
viewing ports in the microscope column and the viewing chamber are
lead glass discs, several centimeters thick. Special precautions are taken to
measure and record the X-ray dose emitted from those points on the microscope column at which the radiation is most intense. Should the acceptable
dose be exceeded, an automatic system warns observers.

8.4 Note on the Construction and Design of Magnetic Electron Lenses

(*a*) *Rules of lens construction.* We have stated that in building microscopes for very high voltages, much bigger electron lenses than those of conventional microscopes must be used. It will be useful to recall some rules of lens construction.

A magnetic electron lens is merely an electromagnet, rotationally symmetric about the optic axis of the lens. Through this electromagnet is bored an axial canal, along which the electrons travel. If we look at the situation in this way, the methods of construction adopted for electromagnets can clearly be applied to electron lenses. A large body of work has been published on this topic (Stefan, 1889; du Bois, 1911, 1913; Weiss, 1907; Perot, 1915; Vilard, 1924; Cotton, 1928a, 1928b, 1928c; Dreyfus, 1935a, 1935b; Bitter, 1936; Montgomery, 1962).

A very simple and most important rule was given for the first time by Lord Kelvin (Thompson, 1872). This *similitude rule* may be stated as follows: *when all the linear dimensions of an electromagnet (including those of the gap) and the number of ampère-turns are multiplied by the same factor* Λ, *the same magnetic field is obtained at corresponding points.*

The volume of the field in the gap will therefore increase as the cube of the linear dimensions.

The first formula giving the magnetic field in the middle of the gap is due to Stefan. If the electromagnet consists of a cylindrical magnetic circuit, terminated by truncated cones whose tips would coincide (Fig. 32), the field on the axis at the point O is given by

$$H = 4\pi M_0 \{ (1 - \cos\theta) + \sin^2\theta \cos\theta \log(R_2/R_1) \}$$

M_0 denotes the magnetization/unit volume of the iron, θ the semi-angle at the tip of the pole-piece, which is generally set equal to 60°; R_1 is the radius of the face of the pole-piece and R_2 that of the cylindrical section. It is possible to modify this formula to take the central hole through the pole-pieces into account.

Stefan makes the assumption that the magnetization inside the iron pole-pieces is constant and parallel to the axis of revolution. This hypothesis is unsatisfactory. Bitter (1936), Gorter and de Klerk (1960, 1962), and Dreyfus (1935a, 1935b) adopt a more correct approach. The formulae they finally obtain display a linear relationship between the field H and $\log(R_2/R_1)$.

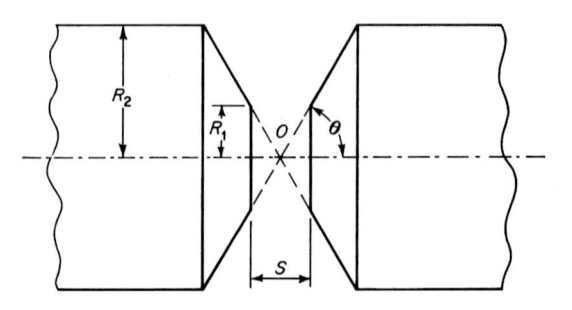

Figure 32 Cross-section of the magnetic circuit of an electromagnet in the neighborhood of the poles.

A remark due to Gorter and de Klerk (1962) is in place here. We may write

$$\log\left(\frac{R_2}{R_1}\right) = 1/3\log\left(\frac{R_2}{R_1}\right)^3$$

if the pole-pieces are arranged as shown in Fig. 32, R_1^3 is proportional to the volume v of the magnetic field region. For electromagnets with a center of similitude, R_2^3 is proportional to the total volume V of the magnet (iron and copper). The result of this is that, according to Kelvin's rule, the curve representing the field at O as a function of $\log(V/v)$ is a straight line. This is nearly what is observed, except when the gap is very small.

When the tips of the truncated conical portions of the pole-pieces do not coincide, the field distribution in the gap is a function of the parameter $2R_1/S$, where S denotes the distance between the poles. We again find that the curve giving the values of the field as a function of $\log(V/v)$ for similar magnets is a straight line, for a constant value of $2R_1/S$. This straight line is parallel to the line that is obtained when the apices of the cones coincide. In order to keep the magnetization as far from its saturation value as possible, one has to increase the area of cross-section of the magnetic circuit. For instance, the apex angle of the cones N_1, N_2, on which the pole-pieces are set, is made equal to 90° (Dreyfus, 1935a, 1935b).

Acting on the construction rules that we have just recapitulated briefly, we arrive at the simplified layout shown in Fig. 33, which represents a cross-section through a meridian plane of a magnetic electron lens. B_1, B_2 are the energizing coils, P_1, P_2 are the pole-pieces, $C \ldots C$ represents the remainder of the magnetic circuit. The whole assembly has rotational symmetry about the lens axis $Z'Z$, along which the electrons travel. The

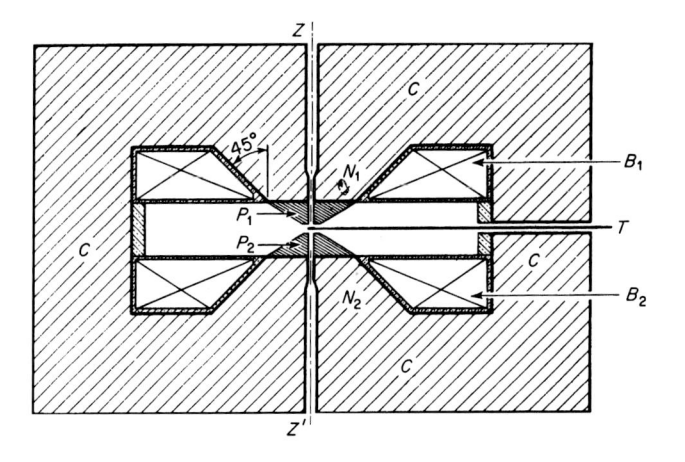

Figure 33 Cross-section through a meridian plane of a magnetic electron lens.

object to be examined is indicated schematically at the end of a rod T, which passes through the magnetic circuit.

With lenses constructed according to this design and suitable dimensions, it is possible to produce very strong fields at the center of the gap. Magnetic lenses of this type will therefore still be viable at voltages distinctly higher than 1 MV.

It is only when the accelerating voltage exceeds about 3 MV that it will doubtless become necessary to abandon this kind of lens, which will be excessively cumbersome. At that point, it is conceivable that superconducting metal windings, immersed in liquid helium, will take over.

(b) *Study of magnetic electron lens characteristics.* In any electron optics laboratory, it is very useful to be able to obtain rapidly the main characteristics of the lenses which are to be fitted to various instruments. We have developed an automatic device in the Toulouse laboratory with which the distribution of magnetic induction on the axis and its first and second derivatives can be plotted for magnetic electron lenses of various types. This device is coupled to an analogue computer which takes the data provided by the first unit and determines the electron trajectories, the focal lengths, and the principal planes of the lenses being studied.

The method involves the use of a vibrating probe (Gautier, 1954; Dupouy, Perrier, & Trinquier, 1963; Dupouy, Perrier, & Bousquet, 1963). We summarize the principle in a simple case. A cylindrical rod T, of small diameter, carries a closely-wound coil (Figs. 34 and 35). The axis of the solenoid S so formed coincides with that of the pole-pieces $Z'Z$ of the

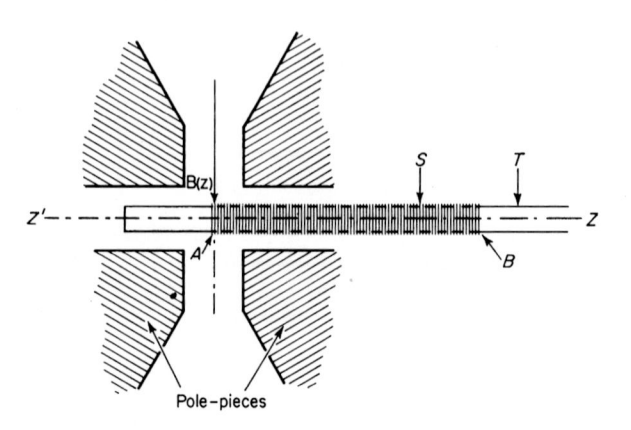

Figure 34 The vibrating probe method for measuring the magnetic induction within an electron lens.

lens. One end, A, of S is placed in the gap, the other end, B, is situated in a region where the induction is zero. The rod T is fixed to the membrane of an electrodynamic loudspeaker motor which transmits longitudinal sinusoidal oscillations to T along the axis $Z'Z$ at a frequency $N \simeq 350$ Hz. We assume that the amplitude a of the oscillations is small enough for the induction $B(z)$ to be considered uniform over a distance a (a is of the order of 1/100 mm). The amplitude of the sinusoidal e.m.f. induced in the solenoid S is proportional to the induction at the point with abscissa z. This e.m.f. is then amplified, rectified, and fed into a galvanometer pen-recorder.

By using windings with suitable characteristics, e.m.f.'s can be obtained whose amplitudes are proportional to the following quantities: $\Phi(z)$, the flux through a circle of radius equal to that of the torus, $B(z)$, $B'(z)$, and $B''(z)$. If such a device is to be made automatic, the motion of the probe-bearing mechanism must be connected to that of the pen-recorder trace in such a way that the abscissae of the curve plotted by the pen-recorder are proportional to the abscissae measured along the axis $Z'Z$ of the actual lens. The most practical way of achieving this is to drive the probe and the paper by means of two synchronous motors equipped with suitable scalers; the mains frequency now keeps the velocities of the two uniform motions proportional to one another.

The probes are glass or quartz rods, 2 or 3 mm in diameter, to within 1/100 mm; the coils are wound with wire of 0.3 to 0.5 mm in diameter. The e.m.f. from the probe, which is of the order of 1 mV, is magnified

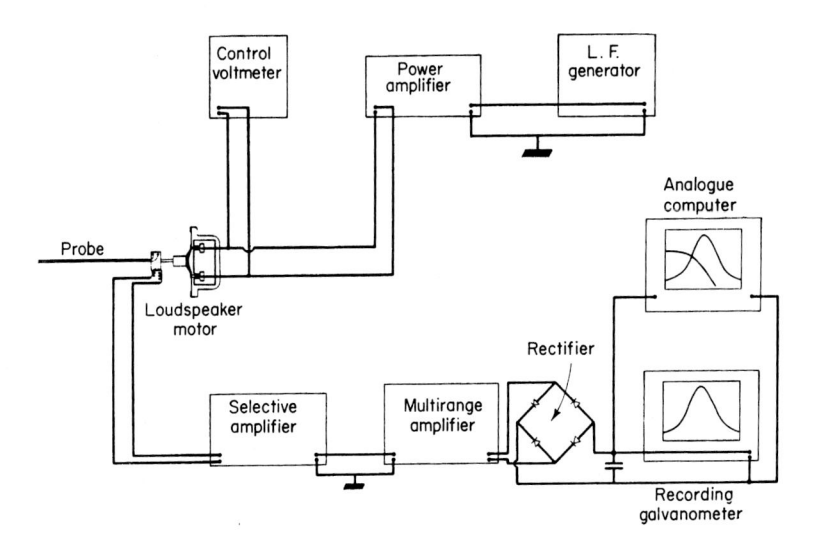

Figure 35 Diagram showing the electronic installation.

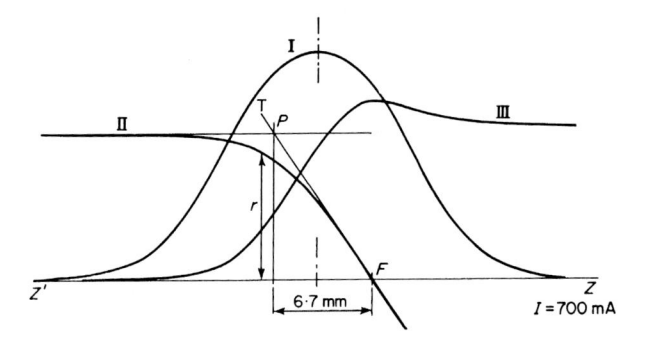

Figure 36 Curves from which the optical characteristics of a magnetic electron lens are calculated: (I) Distribution of magnetic induction. (II) An electron trajectory. (III) The derivative dr/dz at each point of the trajectory.

by a selective amplifier and then by a multirange millivoltmeter. The output voltage, of the order of 10 V, is then rectified by a germanium diode bridge and subsequently filtered and fed to the analogue computer or to the recording galvanometer. The electronic installation is illustrated diagrammatically in Fig. 35.

(c) *The characteristics of the objective of the Toulouse electron microscope.* As an example, we reproduce (Fig. 36) the curves from which the characteristics of the objective of the big Toulouse microscope were determined.

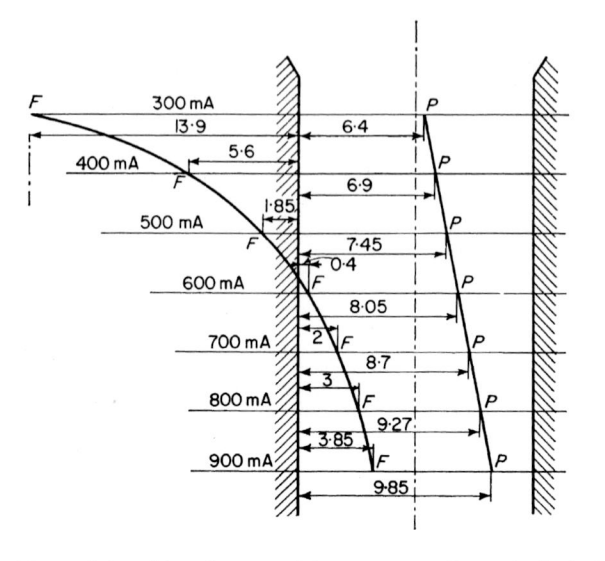

Figure 37 Positions of the object focus and the corresponding principal plane for various values of the objective lens excitation ($V = 1000$ kV). Lengths are given in mm.

Curve (I) shows the variation of the magnetic induction along the axis $Z'Z$ of the lens. Curve (II) represents the trajectory of an electron accelerated through one million volts; the intersection at F with $Z'Z$ fixes the position of the image focus. We draw the tangent FT to the trajectory at F, and the point P indicates the position of the corresponding principal plane. The focal length is then deduced. To remove the inaccuracy arising from the drawing of the tangent FT, the focal length is measured directly by plotting curve (III), representing the derivative dr/dz of the trajectory at each point; this derivative is given directly by the analogue computer. The abscissa of the maximum of the curve (III) is likewise that of the focus F.

In one of these experiments, the number of ampère-turns was 18,375 and the maximum induction $B_m = 15,200$ G.

These measurements have been repeated for various values of the objective excitation current. We thus know the focal length f, the positions of the focus F and of the corresponding principal plane for each value of the magnetization current (Fig. 37). We can therefore select the best working conditions for each kind of pole-piece used. This is just one example from among many of the services that such an installation can provide.

9. STUDY OF THE STRUCTURE OF METALS AND ALLOYS

The advantages of very high voltage microscopy may be exploited to examine objects a few microns thick. This is particularly valuable in the study of imperfections in the structures of metals and alloys: dislocations and stacking faults, for example. Observations of this kind have already made possible considerable progress in our knowledge of a large number of topics connected with metal physics and have provided solutions to various metallurgical problems.

Metals and alloys consist of an assembly of crystals of varying degrees of imperfection. The contrast in the image is here *diffraction contrast*, the mechanism of which we have already explained. From a more detailed analysis, in terms of the kinematic or the dynamical theory (Fukuhara et al., 1962; Hashimoto et al., 1960a; Hashimoto et al., 1962; Hirsch, 1962; Kainuma & Yoshioka, 1962; Kamiya, 1963; Kohra & Watanabe, 1961; Uyeda, 1962; Yoshioka, 1957), it has been shown that this diffraction contrast is sensitive to crystal thickness, movements of the atoms caused by external stresses applied to the crystal, and changes of orientation of the lattice planes, caused by bending a crystal for example.

Within a metal or alloy, therefore, it will be possible to observe the following features.

i. Grain junctions and grain boundaries; the latter will, in general, be indicated by fringes of equal thickness.

ii. Dislocations and stacking faults, in crystals that have been made imperfect by the action of various kinds of constraint.

iii. "Extinction contours" in best specimens.

iv. The precipitates occurring in specimens after suitable thermal treatment.

We reproduce photographs showing a few aspects of these various phenomena.

In Fig. 38 are to be seen the fringes of equal thickness near a grain boundary, in a specimen of stainless steel studied in bright field (A) and in dark field (B) operation.

Similar fringes are likewise visible in Fig. 39, which also contains dislocations. The enlarged portion (B) of one region of a contains fine moiré fringes.

Dislocations photographed at increasing voltages from 200 to 500 kV with a Hitachi microscope are shown in Fig. 40, which was obtained by

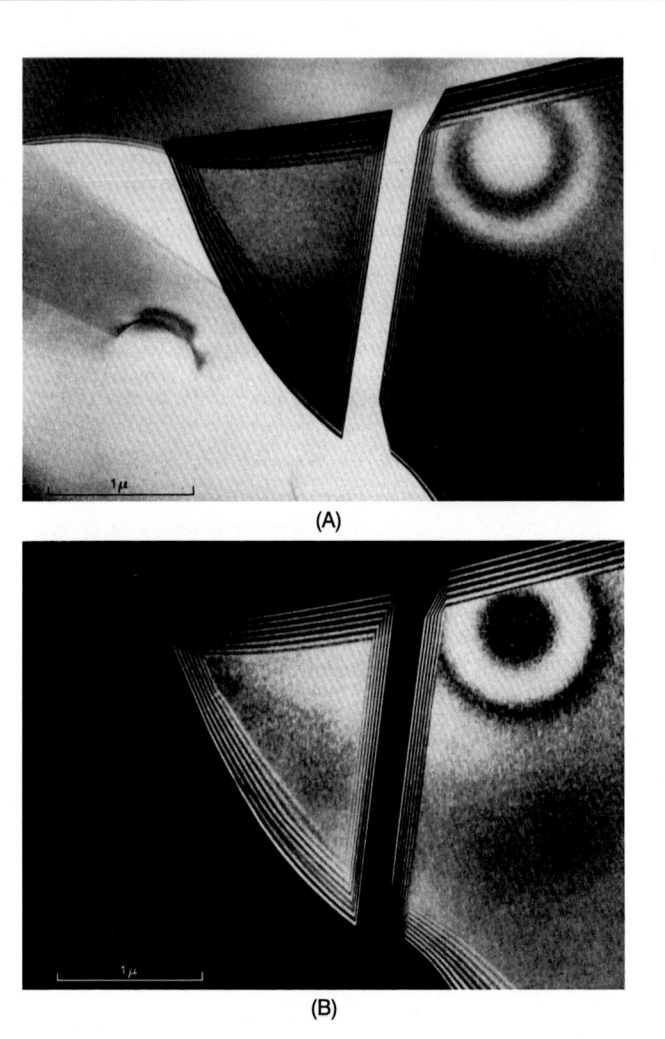

Figure 38 Fringes of equal thickness near the grain boundaries in stainless steel. (A) Bright field photograph; (B) dark field photograph (1000; kV Toulouse).

Imura and Hashimoto. The specimen is stainless steel, 1.05μ thick; the image quality can be seen to improve as the voltage is raised.

Fig. 41 shows one of the first images obtained with the Cambridge microscope at 500 kV. Dislocations in a nickel specimen are to be seen.

In Fig. 42, we see the effect of irradiating a copper specimen with neutrons. Small loops and a decorated twin boundary appear on the photograph, which was taken at 500 kV on the Cambridge microscope.

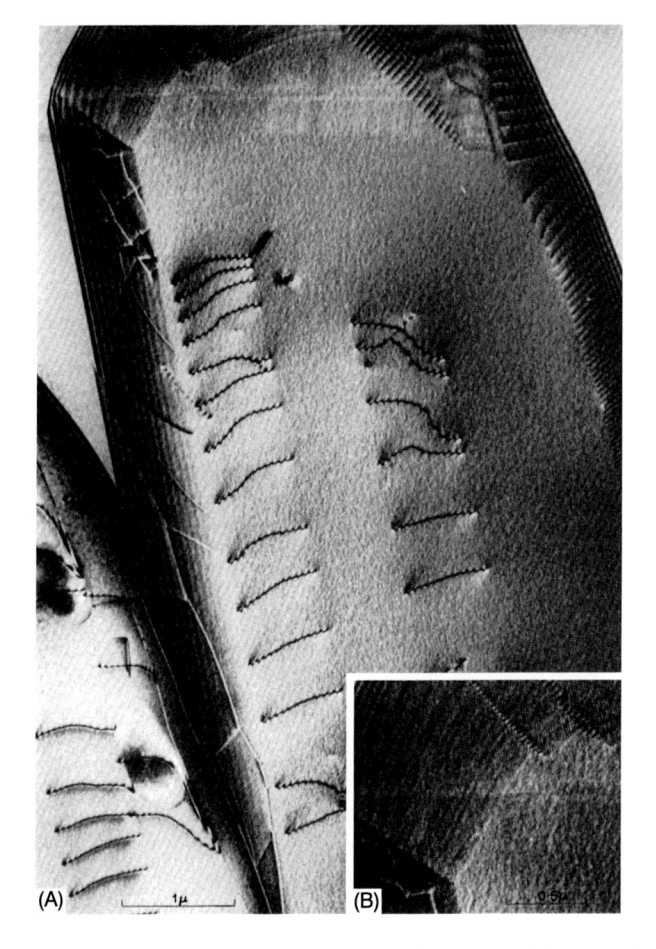

Figure 39 Fringes and dislocations in a stainless steel specimen. (B) is an enlarged view of part of (A); fine moiré fringes can be seen (1000 kV; Toulouse).

Fig. 43 shows dislocations in the form of loops and spirals in an Al–Cu alloy with 4% Cu (1000 kV, Toulouse).

The illustration of Fig. 44 is of a stacking-fault in a stainless steel specimen.

The photographs in Fig. 45 are reproductions of extinction contours in a thin aluminum film. These pictures were taken by R. Uyeda, who has used them to measure the extinction distance and the electron absorption coefficient at voltages of 100 kV (A) and 350 kV (B).

Fig. 46A displays numerous dislocations in a molybdenite specimen, thickness $t = 1.5\mu$; Fig. 46B also contains several dislocations and a stacking

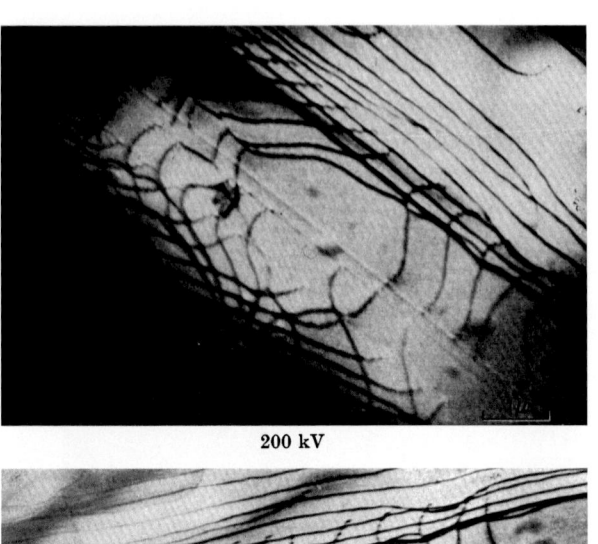

200 kV

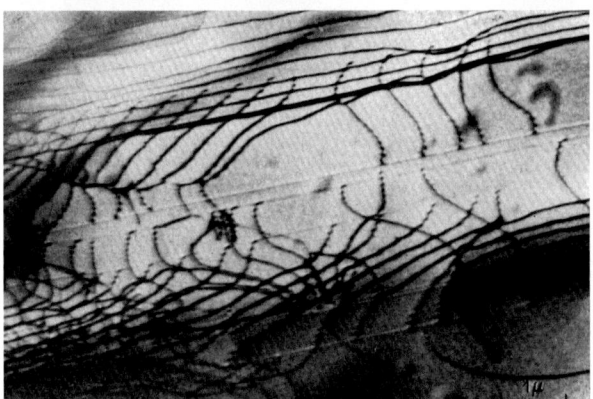

350 kV

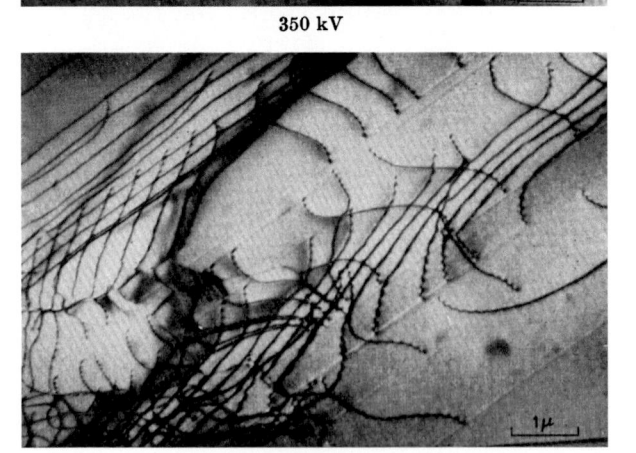

500 kV

Figure 40 Dislocations in a stainless steel specimen, thickness $t = 1.05\mu$. The image quality improves as the voltage is raised from 200 kV to 500 kV (Imura and Hashimoto).

Figure 41 Dislocations in a nickel specimen about 1.4μ thick. Photographed at 500 kV with the Cambridge microscope (specimen prepared by Dr. Hale of the National Physical Laboratory, Teddington).

fault in an alloy film, Cu–2% Al. These micrographs were taken at 300 kV by H. Hashimoto.

In Fig. 47 are to be seen numerous precipitates, formed as a result of suitable thermal treatment of an Al–Ag alloy, 25% silver.

Fig. 48 is a photograph of a region in an Al–Ag alloy, thickness $t = 4\mu$; grain junctions and numerous precipitates are to be seen.

Finally, Fig. 49 shows the diffraction pattern given by a monocrystalline aluminum specimen 6μ thick (1000 kV).

10. THE OBSERVATION OF LIVING MATERIAL

When living material (bacteria or cells, for example) is placed in the evacuated column of the electron microscope, the substance becomes dehydrated: only dead objects can be examined. Furthermore, structural modifications connected with the loss of water occur in the cells. If these

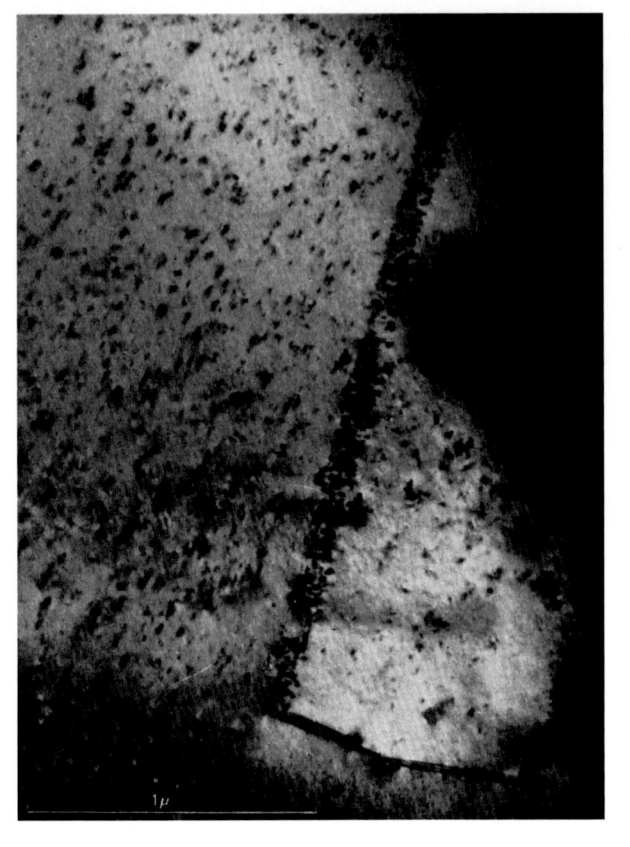

Figure 42 Copper foil, irradiated with neutrons and subsequently annealed for 250 min at 306°C, showing numerous small defects and a decorated twin boundary; thickness approx. 0.6μ. Taken at 500 kV at Cambridge; 45,000× on this print. (Specimen prepared by Dr. Eyre of the Atomic Energy Research Establishment, Harwell.)

difficulties are to be overcome, we must find a means of observing the specimen while it is in its normal living conditions. To achieve this, we replace the usual specimen–holder by a leak–proof box or a *specimen–holding microchamber*, within which is placed the biological material that is to be examined. Various authors have already recommended or built such microchambers. We mention, for example, Rüdenberg (1934), Krause (1937), Marton (1935), Ruska (1938, 1942), von Ardenne (1942), von Ardenne and Beischer (1940), Abrams and McBain (1944), Hiziya and Ito (1958), Nekrassova and Stojanova (1960), Nekrassova, Stojanova, and Zajdes (1960), and Heide (1960).

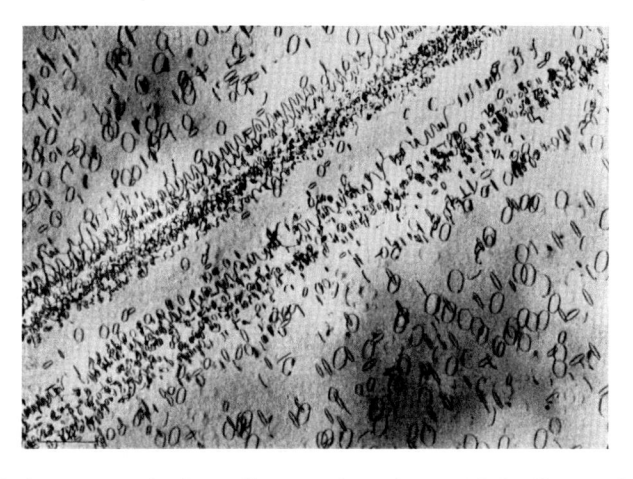

Figure 43 Dislocations in the form of loops and spirals in an Al–Cu alloy, 4% Cu (1000 kV) (Toulouse).

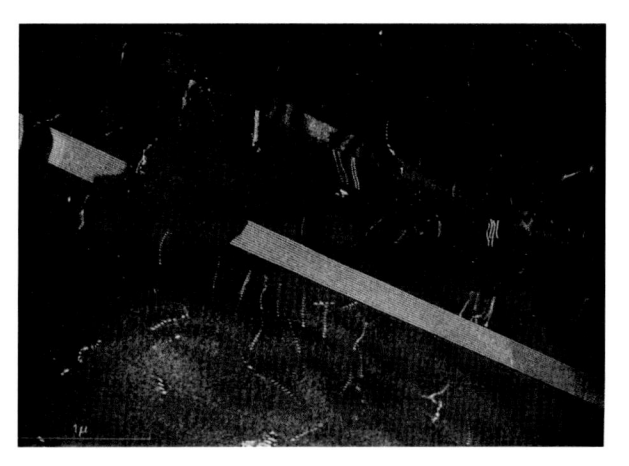

Figure 44 Stacking-fault in a stainless steel specimen (1000 kV). Specimen observed by Vingsbo (Toulouse and Uppsala).

We now give a brief description of the chamber that we have constructed for studying living material (Dupouy et al., 1960). Other chambers, with which objects can be observed in gaseous media, have been designed in this laboratory also (Dupouy, Durrieit, & Perrier, 1962; Dupouy & Perrier, 1962).

The two metal parts P_1 and P_2 (Fig. 50) have two very flat faces, parallel to each other and perpendicular to the electron beam. They are fitted into G, which guides them and ensures that they remain centered. Two

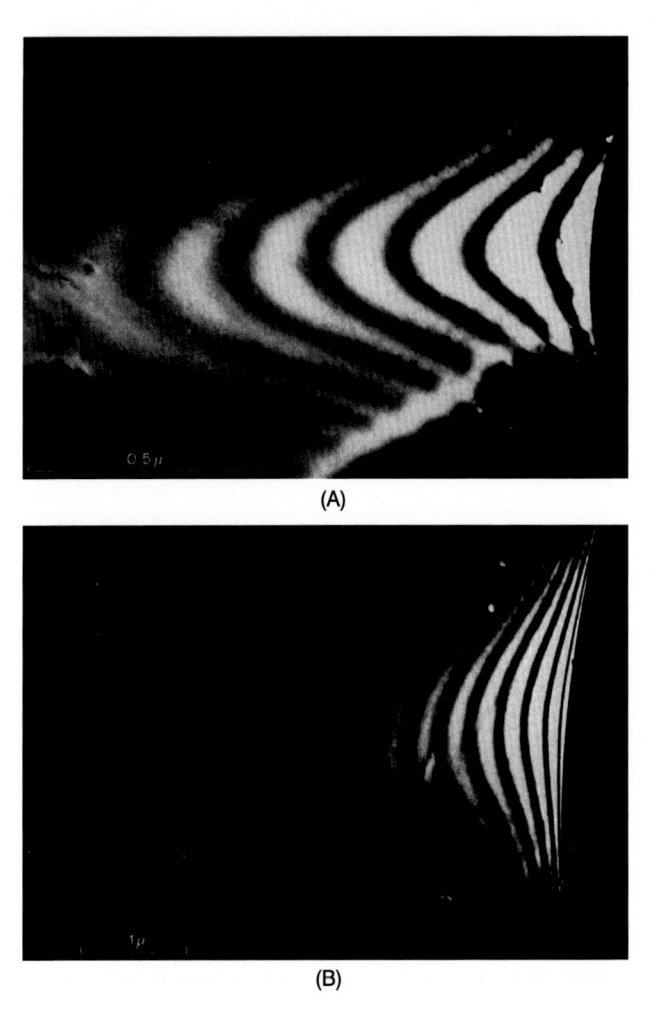

(A)

(B)

Figure 45 Extinction contours, observed in an aluminum foil: (A) 100 kV; (B) 350 kV. (Photographed by R. Uyeda, Nagoya.)

cylindrical openings O_1 and O_2 very small in diameter, provide a kind of narrow canal through which the electrons pass. Two circular joints R_1 and R_2, made of plastic, are punched out very accurately and fit into two collars, worked in G. Their purpose is to make a good seal; two caps C_1 and C_2 are screwed onto G and play a double role: they keep the whole assembly in position and they compress R_1 and R_2 to some extent. The cylindrical cavity surrounding P_1 and P_2 forms a chamber C filled with a reserve supply of air.

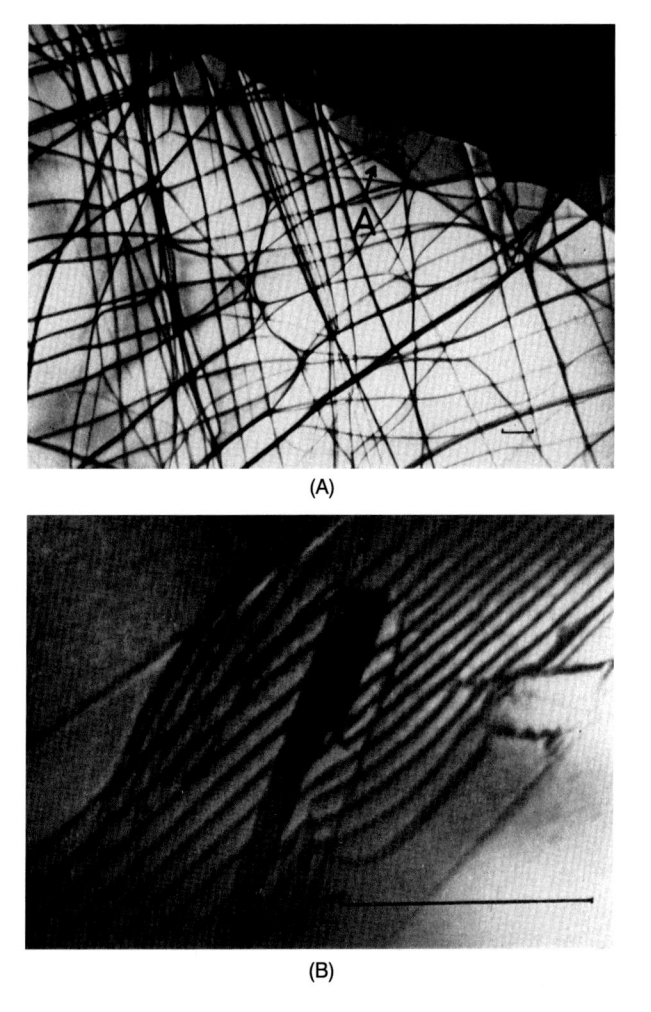

(A)

(B)

Figure 46 (A) Dislocations in a molybdenite specimen, thickness $t = 1.5\mu$ (300 kV); (B) dislocations and a stacking fault in a film of Cu–2% Al alloy. (Photographed by Hashimoto.)

The plane parallel faces of P_1 and P_2 are highly polished. Each is covered with a very thin film of parlodion-carbon, 600 Å thick. After drying, these films seem to be stretched like drum-heads over the openings O_1 and O_2 and they thus form two windows, F_1 and F_2, transparent to electrons. A permanent connection between the interior of the specimen-holder cell and a damp atmosphere is provided by means of a metal tube T of minute cross-section, the internal diameter of which is no bigger than 0.15 mm.

Figure 47 Precipitates formed by heat treatment in an Al–Ag alloy, 25% Ag (1000 kV; Toulouse).

Figure 48 Grain boundaries and precipitates in an Al–Ag alloy, 25% Ag, thickness $t = 4\mu$ (1000 kV; Toulouse).

Numerous difficulties arise in the construction of this microchamber; one of these is a consequence of the limited space available, both in the gap of the objective and in the axial channel through the pole-pieces. The total external volume of the microchamber is of the order of $1/10$ cm^3 only. Furthermore, although the windows of the cell have to be exceedingly thin, they must also be strong enough not to burst under the pressure when the assembly is placed in the vacuum. After various attempts, the diameter

Figure 49 Kikuchi line pattern obtained with an aluminum specimen of thickness $t =$ 6μ (1000 kV; Toulouse).

selected for the openings O_1 and O_2 is 0.1 mm. Under these conditions, the force exerted by atmospheric pressure on the surface of each window amounts to less than 8 cg. The components must be very accurately centered to ensure that the two windows in the microchamber are as perfectly coaxial as possible.

The examination of a few bacteria. We have worked with *Corynebacterium diphteriae* and *Bacillus subtilis*. The cross-section of the electron beam is slightly larger than the diameter of the holes in the diaphragms O_1 and O_2. The exposure time required for each photograph is not more than two or three seconds. The bacteria do of course experience the effects of the

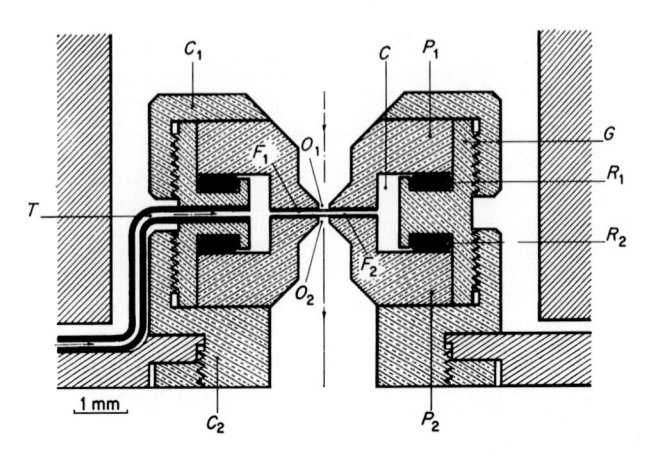

Figure 50 Cross-section of the microchamber, used in the photography of live bacteria.

electrons and of the X-rays produced at the points of impact on each diaphragm.

After observation, the specimen–cell is withdrawn from the microscope column and, using an appropriate procedure (Dupouy & Perrier, 1962), the irradiated bacteria which lie in the central part of the collodion window on which they had been placed are extracted. These bacteria are placed in a suitable culture medium. The result of this is as follows: *the bacteria remain alive after irradiation and they are capable of reproducing.* This work was performed in collaboration with Mme Enjalbert, Professor in the Medical Faculty of Toulouse.

Fig. 51 shows spores of *Bacillus subtilis* (A) and several pictures of *Corynebacterium diphteriae* (B, C, D).

This research is at present being continued in our laboratory. Perhaps it will be possible, one day, to observe the stages of the life of a bacterium or a cell, by taking photographs at regular time-intervals for several hours. This would lead to work, the future importance of which we have no need to stress.

11. CONCLUSION

In the course of this account, of necessity incomplete, we have tried to bring out the advantages which accrue from using very high voltages in electron microscope research.

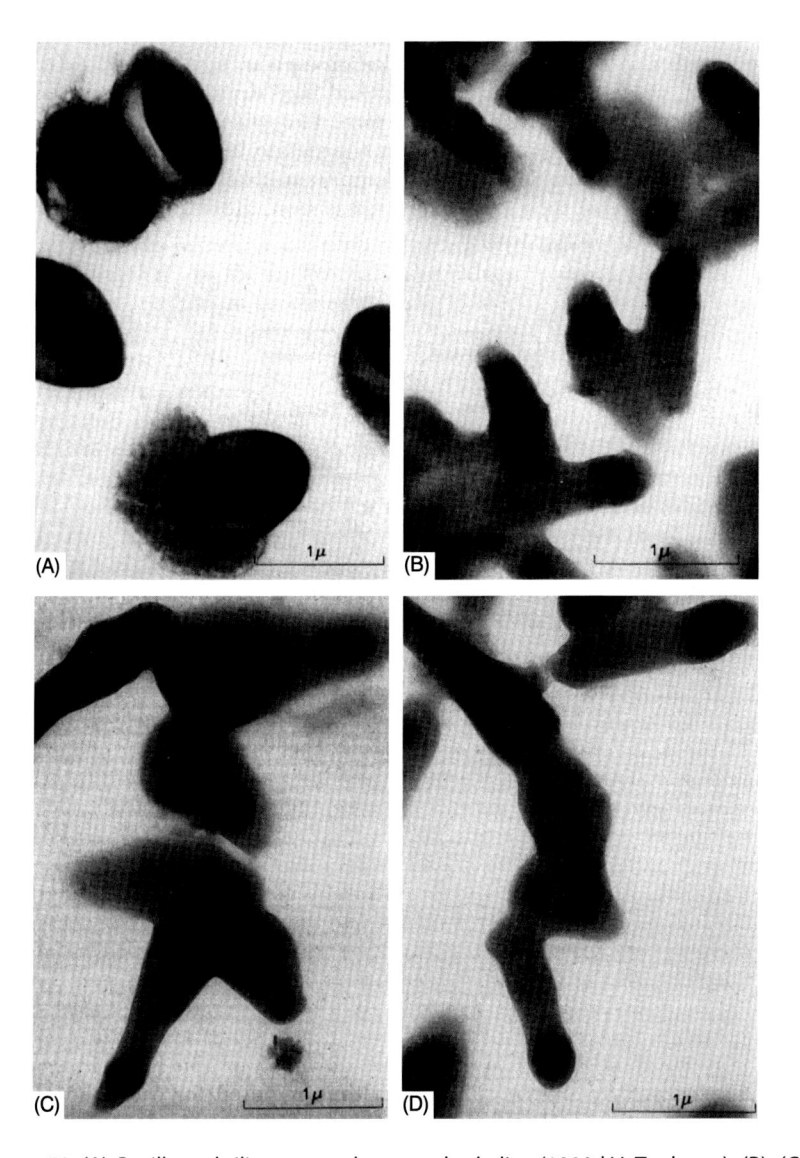

Figure 51 (A) *Bacillus subtilis* spores, photographed alive (1000 kV; Toulouse). (B), (C), and (D) *Corynebacterium diphteriae* photographed alive (1000 kV; Toulouse).

One of the most important, without any doubt, arises from the spectacular reduction of chromatic aberration. For a given specimen, it is about ten to twenty times smaller at 1000 kV than at 100 kV. This allows us to obtain good images of metal specimens several microns thick. At ordinary voltages, the practical resolving power for most specimens is limited by chromatic aberration whereas at very high voltages (1000 kV) we can hope to achieve a resolution very close to the theoretical value.

In addition, the object is distinctly less altered by the electron beam when the operating voltage is raised.

The disturbing effect of spherical aberration on selected area diffraction patterns from an assembly of closely packed microcrystals is 40 or 50 times smaller at very high voltage than at 100 kV.

In the case of biological objects, on the other hand, and of amorphous substances in general, the image contrast becomes very poor at 1000 kV. We have overcome this difficulty by the very simple contrast stop method.

The photographs of metals and alloys that illustrate this account reveal a great wealth of characteristic details: dislocations, stacking faults, precipitates. Further study of these will be very useful for the progress of solid state physics.

For the first time, by working at 1000 kV, we have been able to photograph bacteria alive: they could reproduce after irradiation. We are actively pursuing this work; it may well be the beginning of a new, thrilling adventure, in a domain hitherto inaccessible in electron microscopy: the study of living cells.

Today, only a small number of these microscopes are in operation. In various countries, however, the construction of similar instruments is contemplated. It is our conviction that, despite the noticeably higher cost of such installations, there is a great future for electron microscopy at very high voltages, by virtue of the advantages that are put at the disposal of research.

At the time when we built the microscope at Toulouse, the project seemed rather bold. At all events, research should admit of a taste for adventure. We are conscious that our efforts have been of some use in subsequent designs, and we hope that many investigators will follow the route that we have blazed.

ACKNOWLEDGMENTS

Many colleagues have been kind enough to provide documents, as well as photographs to illustrate the text: I would like to express my gratitude to them at this point. In particular, my thanks are due to Dr. V. E. Cosslett of the Cavendish Laboratory (Cambridge); Professor R. Uyeda of the University of Nagoya; Professor K. Kobayashi and Professor H. Hashimoto of the University of Kyoto, and to Mr. K. Ashinuma of the firm of Japan Electron Optics Laboratory (Tokyo). Professor R. E. Burge of the University of London has most kindly given much information which appears in various tables in this review. I would like to thank him most warmly.

His thorough knowledge of electron optics has enabled Dr. P. W. Hawkes to make an excellent translation of this article, and I am extremely grateful to him for this.

REFERENCES

Abrams, I. M., & McBain, J. W. (1944). *J. Appl. Phys.*, *15*, 607.
Adler, H., Minkner, R., Reinhold, G., & Seitz, J. (1960). *Proc. Eur. Reg. Conf. Electron Microsc. Delft*, *1*, 122.
Agar, A. W. (1947). *Br. J. Appl. Phys.*, *9*, 33.
Ayroles, R., Dupouy, G., Mazel, A., Perrier, F., & Uyeda, R. (1965). *J. Microsc.*, *4*, 429.
Ayroles, R., & Mazel, A. (1965). *J. Microsc.*, *4*(2), 193.
Bahr, G. F., & Zeitler, E. H. (1957). *Exp. Cell Res.*, *12*, 44.
Bahr, G. F., & Zeitler, E. H. (1964). In *Proceedings of the AMU–ANL high voltage electron microscope meeting* (p. 106). Argonne, IL: Argonne National Laboratory.
Bas, E. B. (1960). *Proc. Eur. Reg. Conf. Electron Microsc. Delft*, *1*, 126.
Beer, M. (1964). In *Proceedings of the AMU–ANL high voltage electron microscope meeting* (p. 44). Argonne, IL: Argonne National Laboratory.
Bethe, H. (1928). *Ann. Phys.*, *87*, 55.
Bethe, H. (1933). *Handb. Phys.*, *24*(1), 515.
Birkes, I., & Demeter, I. (1960). *Acta Phys. Hung.*, *12*, 99.
Bitter, F. (1936). *Rev. Sci. Instrum.*, 7, 479.
Black, G., & Linfoot, E. H. (1957). *Proc. R. Soc. A*, *239*, 522.
Boersch, H. (1936). *Ann. Phys.*, *27*, 78.
Boersch, H. (1954). *Z. Phys.*, *139*, 115.
Burge, R. E., & Smith, G. H. (1962a). *Proc. Phys. Soc.*, *79*, 673.
Burge, R. E., & Smith, G. H. (1962b). *Nature*, *195*, 140.
Burge, R. E., & Smith, G. H. (1963). *Proc. Phys. Soc.*, *81*, 612.
Cosslett, V. E. (1946). *Proc. Phys. Soc.*, *58*, 443.
Cosslett, V. E. (1951). *Practical electron microscopy*. London: Butterworth.
Cosslett, V. E. (1956). *Br. J. Appl. Phys.*, 7, 10.
Cosslett, V. E. (1962). *J. R. Microsc. Soc.*, *81*, 1–10.
Cosslett, V. E. (1965). *Lab. Invest.*, *14*, 772.
Cotton, A. (1928a). *C. R. Hebd. Séances Acad. Sci.*, *187*, 77.
Cotton, A. (1928b). *Rech. Invent.*, *9*, 453.
Cotton, A. (1928c). *Rev. Gén. Électr.*, *24*, 317.
Coupland, J. H. (1955). In *Proceedings of the international conference on electron microscopy, London* (p. 159). London: Royal Microscopical Society.
Dietrich, W. (1958). *Z. Phys.*, *152*, 306.
Dreyfus, L. (1935a). *ASEA-J.*, *12*, 8.
Dreyfus, L. (1935b). *Elektrotech. Maschinenbau*, *53*, 205, 213.
du Bois, H. (1911). *Z. Instrum.*, *31*, 362.
du Bois, H. (1913). *Ann. Phys.*, *42*, 953.
Dupouy, G., Durrieu, L., & Perrier, F. (1960). *C. R. Hebd. Séances Acad. Sci.*, *251*, 2836.
Dupouy, G., Durrieit, L., & Perrier, F. (1962). *C. R. Hebd. Séances Acad. Sci.*, *254*, 3786.
Dupouy, G., & Perrier, F. (1962). *J. Microsc.*, *1*, 167.

Dupouy, G., & Perrier, F. (1963). *Ann. Phys.*, *8*, 251.
Dupouy, G., & Perrier, F. (1964). *J. Microsc.*, *3*, 233.
Dupouy, G., & Perrier, F. (1966). In *Proceedings of the sixth international congress for electron microscopy, Kyoto: Vol. I* (p. 107). Tokyo: Maruzen.
Dupouy, G., Perrier, F., & Bousquet, L. (1963). *J. Microsc.*, *2*, 247.
Dupouy, G., Perrier, F., & Fabre, R. (1961). *C. R. Hebd. Séances Acad. Sci.*, *252*, 627.
Dupouy, G., Perrier, F., & Seguela, A. (1966). *C. R. Hebd. Séances Acad. Sci.*, *262*, 341.
Dupouy, G., Perrier, F., & Trinquier, J. (1963). *J. Microsc.*, *2*, 237.
Dupouy, G., Perrier, F., Trinquier, J., & Fayet, Y. (1967). *C. R. Hebd. Séances Acad. Sci.*, *265*, 675.
Dupouy, G., Perrier, F., Trinquier, J., & Murillo, R. (1967). *C. R. Hebd. Séances Acad. Sci.*, *265*, 1221.
Dupouy, G., Perrier, F., & Verdier, P. (1966). *J. Microsc.*, *5*, 655.
Engel, A., Koppen, G., & Wolff, O. (1962). In S. S. Breese (Ed.), *Proceedings of the fifth international congress for electron microscopy, Philadelphia: Vol. I*, E-13. New York: Academic Press.
Faget, J., Fagot, M., Ferré, J., & Fert, C. (1962). In S. S. Breese (Ed.), *Proceedings of the fifth international congress for electron microscopy, Philadelphia: Vol. I*, A-7. New York: Academic Press.
Faget, J., Fagot, M., & Fert, C. (1960). *Proc. Eur. Reg. Conf. Electron Microsc.*, *I*, 18.
Finch, G. I., Lewis, H. C., & Webb, D. P. D. (1955). *Proc. Phys. Soc. B*, *66*, 949.
Fujiwara, K. (1961). *J. Phys. Soc. Jpn.*, *16*, 2226.
Fukuhara, A., Kohra, K., & Watanabe, H. (1962). *J. Phys. Soc. Jpn.*, *17*(B-II), 195.
Gautier, P. (1954). *J. Phys. Radium*, *15*, 684.
Glaser, W. (1941). *Z. Phys.*, *117*, 285.
Glaser, W. (1949). *Acta Phys. Austriaca*, *4*, 38.
Gorter, C. J., & de Klerk, D. (1960). *Appl. Sci. Res.*, (B-8), 265.
Gorter, C. J., & de Klerk, D. (1962). *High magnetic fields*. New York and London: MIT Press and John Wiley.
Haine, M. E. (1961). *The electron microscope*. London: Spon.
Haine, M. E., & Mulvey, T. (1954). *J. Sci. Instrum.*, *31*, 236.
Halliday, J. S., & Quinn, T. F. J. (1960). *Br. J. Appl. Phys.*, *11*, 486.
Hamming, R. W., & Heidenreich, B. D. (1965). *Bell Syst. Tech. J.*, *44*, 207.
Hashimoto, H. (1964). *J. Appl. Phys.*, *35*, 277.
Hashimoto, H., Howie, A., & Whelan, M. J. (1960a). *Philos. Mag.*, *5*, 967.
Hashimoto, H., Howie, A., & Whelan, M. J. (1960b). *Proc. Eur. Reg. Conf. Electron Microsc. Delft*, *I*, 207.
Hashimoto, H., Iwanaga, M., Kobayashi, K., Shimadzu, S., Suito, E., & Tanaka, K. (1962). *J. Phys. Soc. Jpn.*, *17*(Suppl. B-II), 170.
Heide, H. G. (1960). *Naturwissenschaften*, *14*, 313.
Heidenreich, R. D. (1949). *J. Appl. Phys.*, *20*, 993.
Heidenreich, R. D. (1962). *J. Appl. Phys.*, *33*, 2321.
Heidenreich, R. D. (1964). *Fundamentals of transmission electron microscopy*. New York: John Wiley.
Hillier, J., Vance, A. W., & Zworykin, V. K. (1941). *J. Appl. Phys.*, *12*, 738.
Hirsch, P. B. (1962). *J. Phys. Soc. Jpn.*, *17*(Suppl. B-II), 143.
Hiziya, K., & Ito, T. (1958). *J. Electron Microsc.*, *6*, 4.
Honjo, G., & Mihama, K. (1954). *J. Phys. Soc. Jpn.*, *9*, 184.

Hori, T., Iwanaga, M., Kobayashi, K., Shimadzu, S., & Suito, E. (1964). *Bull. Instrum. Chem. Res.*, *42*, 439.

Howie, A., & Whelan, M. J. (1960). *Proc. Eur. Reg. Conf. Electron Microsc. Delft*, *1*, 181, 194.

Howie, A., & Whelan, M. J. (1961). *Proc. R. Soc. A*, *263*, 217.

Howie, A., & Whelan, M. J. (1962). *Proc. R. Soc. A*, *267*, 206.

Kainuma, Y., & Yoshioka, H. (1962). *J. Phys. Soc. Jpn.*, *17*(Suppl. B-II), 134.

Kamiya, Y. (1963). *J. Appl. Phys. Japan*, *2*, 386.

Kamiya, Y., & Uyeda, R. (1962). In S. S. Breese (Ed.), *Proceedings of the fifth international congress for electron microscopy, Philadelphia: Vol. I*, AA-11. New York: Academic Press.

Kasatotchine, V. J., Loukianovitch, V. M., & Popov, N. M. (1960). *Dokl. Akad. Nauk SSSR*, *131*(3), 609.

Kasatotchine, V. J., Loukianovitch, V. M., Popov, N. M., & Tchmoutov, K. V. (1960). *J. Chem. Phys.*, *57*(10), 822.

Kato, N., Nagakura, S., & Watanabe, H. (1956). In *Proc. first reg. conf. Asia and Oceania, Tokyo* (p. 125).

Kobayashi, K., & Sakaoku, K. (1964). *Bull. Instrum. Chem. Res.*, 473.

Kohra, K., & Watanabe, H. (1959). *J. Phys. Soc. Jpn.*, *14*, 1119.

Kohra, K., & Watanabe, H. (1961). *J. Phys. Soc. Jpn.*, *16*, 580.

Komoda, T., & Otsuki, M. (1964). *Jpn. J. Appl. Phys.*, *3*, 667.

Krause, F. (1937). *Naturwissenschaften*, *25*, 817.

Lenz, F. (1954). *Z. Naturforsch.*, *9a*, 185.

Le Poole, J. B. (1947). *Philips Tech. Rev.*, *9*, 33.

Le Poole, J. B., Oosterkamp, W. J., & Van Dorsten, A. C. (1947). *Philips Tech. Rev.*, *9*, 1.

Marton, L. (1935). *Bull. Cl. Sci., Acad. R. Belg.*, *21*, 553.

Marton, L. (1946). *Phys. Soc. Prog. Rep.*, *10*, 204.

Maruse, S., Morito, N. J., Sakaki, Y., & Tadano, B. (1956). *J. Electron Microsc.*, *4*, 5.

McGillavry, C. H. (1940). *Physica Eindhoven*, *7*, 329.

Miyake, S. (1959). *J. Phys. Soc. Jpn.*, *14*(10), 1347.

Möllenstedt, G. (1946). *Nachr. Akad. Wiss. Gött. Math.-Phys. Kl.*, *1*, 83.

Montgomery, D. B. (1962). *High magnetic fields*. New York and London: MIT Press and John Wiley.

Mott, N. F., & Massey, H. S. W. (1950). *The theory of atomic collisions*. Oxford: Clarendon Press.

Müller, H. O., & Ruska, E. (1941). *Kolloid Z.*, *95*, 21.

Nekrassova, T. A., & Stojanova, I. G. (1960). *Dokl. Akad. Nauk SSSR*, *134*, 467.

Nekrassova, T. A., Stojanova, I. G., & Zajdes, A. L. (1960). *Dokl. Akad. Nauk SSSR*, *130*, 6.

Nonoyama, M., & Uyeda, R. (1965). *Jpn. J. Appl. Phys.*, *4*, 498.

Papoular, M. (1955). Thesis, No. 3717. Paris.

Perot, A. (1915). *Mém. Acad. Sci. Inst. Fr.*, *153*, 35.

Phillips, R. (1960). *Br. J. Appl. Phys.*, *11*, 504.

Popov, N. M. (1959). *Izv. Akad. Nauk SSSR*, *23*, 436, 494.

Popov, N. M., & Zviagin, B. B. (1959). *Izv. Akad. Nauk SSSR*, *23*, 270.

Riecke, W. D. (1961). *Optik*, *18*, 278.

Riecke, W. D., & Ruska, E. (1966). In *Proceedings of the sixth international congress for electron microscopy, Kyoto: Vol. I* (p. 19). Tokyo: Maruzen.

Rüdenberg, R. (1934). Austrian patent, No. 137611.

Ruska, E. (1938). Swedish patent, No. 101576.

Ruska, E. (1942). *Kolloid Z.*, *100*, 212.

Ruska, E. (1964). *J. Microsc.*, *3*, 357.

Ruska, E. (1965). *Optik*, *22*, 319.

Smith, K. C. A., Considine, K., & Cosslett, V. E. (1966). In *Proceedings of the sixth international congress for electron microscopy, Kyoto: Vol. I* (p. 99). Tokyo: Maruzen.

Stefan, J. (1889). *Ann. Phys.*, *38*, 440.

Sternheimer, R. M. (1952). *Phys. Rev.*, *88*, 855.

Sternheimer, R. M. (1953). *Phys. Rev.*, *91*, 256.

Thomson, W. (Lord Kelvin) (1872). *Papers on electrostatics and magnetism* (p. 564).

Trinquier, J., Dupouy, G., & Perrier, F. (1964). *J. Microsc.*, *3*, 115.

Uyeda, R. (1962). *J. Phys. Soc. Jpn.*, *17*(Suppl. B-II), 155.

Vilard, P. (1924). *C. R. Hebd. Séances Acad. Sci.*, *179*, 1365.

von Ardenne, M. (1941). *Z. Phys.*, *117*, 657.

von Ardenne, M. (1942). *Z. Phys. Chem.*, *3*(51), 61.

von Ardenne, M., & Beischer, D. (1940). *Z. Elektrochem.*, *46*, 270.

Watanabe, H. (1960). *J. Phys. Soc. Jpn.*, *15*(12), 2368.

Weiss, P. (1907). *J. Phys. Radium*, *6*, 353.

Yoshioka, H. (1957). *J. Phys. Soc. Jpn.*, *12*, 618.

Zernike, F. (1942). *Physica*, *9*, 974.

INDEX

Printed in the United States
By Bookmasters